T0202816

Graduate Texts in Mathematics

Volume 278

Graduate Texts in Mathematics

Graduate Texts in Mathematics bridge the gap between passive study and creative understanding, offering graduate-level introductions to advanced topics in mathematics. The volumes are carefully written as teaching aids and highlight characteristic features of the theory. Although these books are frequently used as textbooks in graduate courses, they are also suitable for individual study.

More information about this series at http://www.springer.com/series/136

William P. Ziemer

Modern Real Analysis

Second Edition

With contributions by Monica Torres

Author
William P. Ziemer
Department of Mathematics
Indiana University
Bloomington, IN
USA

With contributions by
Monica Torres
Department of Mathematics
Purdue University
West Lafayette, IN
USA

ISSN 0072-5285 ISSN 2197-5612 (electronic)
Graduate Texts in Mathematics
ISBN 978-3-319-87840-9 ISBN 978-3-319-64629-9 (eBook)
https://doi.org/10.1007/978-3-319-64629-9

Mathematics Subject Classification (2010): 4601, 46EXX, 46GXX

Printed on acid-free paper

This Springer imprint is published by Springer Nature
The registered company is Springer International Publishing AG
The registered company address is: Gewerbestrasse 11, 6330 Cham, Switzerland

William Paul Ziemer
1934–2017

Bill Ziemer had a distinguished career as a mathematician across a broad spectrum of fields, with an international research reputation.

He was a constant student of mathematics, from undergraduate years at the University of Wisconsin–Madison, through his Ph.D. from Brown University in 1961 and nearly 50 years at Indiana University–Bloomington. He held many teaching and research honors including the Indiana University Distinguished Faculty Award in 1996 and the Outstanding Paper Prize in 2016 from the Mathematical Society of Japan.

Bill had a zest for life. Daughter Laura is an environmental lawyer and daughter Sarah, a mechanical engineer. Son, William Karl Ziemer, is a mathematics professor at California State University–Long Beach. Bill's wife, Suzanne, was his loving companion for over 60 years. He is survived by six grandchildren.

In the late afternoons, he could be found on tennis or squash courts and weekends enjoying competitive sailing.

Preface

This text is an essentially self-contained treatment of material that is normally found in a first-year graduate course in real analysis. Although the presentation is based on a modern treatment of measure and integration, it has not lost sight of the fact that the theory of functions of one real variable is the core of the subject. It is assumed that the student has had a solid course in Advanced Calculus. Although the book's primary purpose is to serve as a graduate text, we hope that it will also serve as useful reference for the more experienced mathematician.

The book begins with a chapter on preliminaries and then proceeds with a chapter on the development of the real number system. This also includes an informal presentation of cardinal and ordinal numbers. The next chapter provides the basics of general topological and metric spaces. Thus, by the time the first three chapters have been concluded, the students will be ready to pursue the main thrust of the book.

The text then proceeds to develop measure and integration theory in the next three chapters. Measure theory is introduced by first considering outer measures on an abstract space. The treatment here is abstract, yet short, simple, and basic. By focusing first on outer measures, the development underscores in a natural way the fundamental importance and significance of σ-algebras. Lebesgue measure, Lebesgue-Stieltjes measure, and Hausdorff measure are immediately developed as important, concrete examples of outer measures. Integration theory is presented by using countably simple functions, that is, functions that assume only a countable number of values. Conceptually they are no more difficult than simple functions, but their use leads to a more direct development. Important results such as the Radon-Nikodym theorem and Fubini's theorem have received treatments that avoid some of the usual technical difficulties.

A chapter on elementary functional analysis is followed by one on the Daniell integral and the Riesz Representation theorem. This introduces the student to a completely different approach to measure and integration theory. In order for the student to become more comfortable with this new framework, the linear functional approach is further developed by including a short chapter on Schwartz Distributions. Along with introducing new ideas, this reinforces the student's previous encounter with measures as linear functionals. It also maintains connection with previous material by casting some old ideas in a new light. For example, BV functions and absolutely continuous functions are characterized as functions whose distributional derivatives are measures and functions, respectively.

The introduction of Schwartz distributions invites a treatment of functions of several variables. Since absolutely continuous functions are so important in real analysis, it is natural to ask whether they have a counterpart among functions of several variables. In the last chapter, it is shown that this is the case by developing the class of functions whose partial derivatives (in the sense of distributions) are functions, thus providing a natural analog of absolutely continuous functions of a single variable. The analogy is strengthened by proving that these functions are absolutely continuous in each variable separately. These functions, called Sobolev functions, are of fundamental importance to many areas of research today. The chapter is concluded with a glimpse of both the power and the beauty of Distribution theory by providing a treatment of the Dirichlet Problem for Laplace's equation. This presentation is not difficult, but it does call upon many of the topics the student has learned throughout the text, thus providing a fitting end to the book.

We will use the following notation throughout. The symbol □ denotes the end of a proof and $a := b$ means $a = b$ by definition. All theorems, lemmas, corollaries, definitions, and remarks are numbered as $a.b$ where a denotes the chapter number. Equation numbers are numbered in a similar way and appear as $(a.b)$. Sections marked with * are not essential to the main development of the material and may be omitted.

The authors would like to thank Patricia Huesca for invaluable assistance in typesetting of the manuscript.

Bloomington, USA William P. Ziemer

Contents

CHAPTER 1

Preliminaries

1.1. Sets

This is the first of three sections devoted to basic definitions, notation, and terminology used throughout this book. We begin with an elementary and intuitive discussion of sets and deliberately avoid a rigorous treatment of "set theory" that would take us too far from our main purpose.

We shall assume that the notion of set is already known to the reader, at least in the intuitive sense. Roughly speaking, a **set** is any identifiable collection of objects, called the elements or members of the set. Sets will usually be denoted by capital roman letters such as A, B, C, U, $V, \ldots$, and if an object x is an element of A, we will write $x \in A$. When x is not an element of A we write $x \notin A$. There are many ways in which the objects of a set may be identified. One way is to display all objects in the set. For example, $\{x_1, x_2, \ldots, x_k\}$ is the set consisting of the elements $x_1, x_2, \ldots, x_k$. In particular, $\{a, b\}$ is the set consisting of the elements a and b. Note that $\{a, b\}$ and $\{b, a\}$ are the same set. A set consisting of a single element x is denoted by $\{x\}$ and is called a **singleton**. Often it is possible to identify a set by describing properties that are possessed by its elements. That is, if $P(x)$ is a property possessed by an element x, then we write $\{x : P(x)\}$ to describe the set that consists of all objects x for which the property $P(x)$ is true. Obviously, we have $A = \{x : x \in A\}$ and $\{x : x \neq x\} = \emptyset$, the **empty set** or **null set**.

The **union** of sets A and B is the set $\{x : x \in A \text{ or } x \in B\}$, and this is written as $A \cup B$. Similarly, if $\mathcal{A}$ is an arbitrary family of sets, the union of all sets in this family is

$$\{x : x \in A \text{ for some } A \in \mathcal{A}\} \tag{1.1}$$

and is denoted by

$$\bigcup_{A \in \mathcal{A}} A \quad \text{or by} \quad \bigcup \{A : A \in \mathcal{A}\}. \tag{1.2}$$

Sometimes a family of sets will be defined in terms of an indexing set I, and then we write

$$\{x : x \in A_\alpha \text{ for some } \alpha \in I\} = \bigcup_{\alpha \in I} A_\alpha. \tag{1.3}$$

W.P. Ziemer, *Modern Real Analysis*, Graduate Texts in Mathematics 278,
https://doi.org/10.1007/978-3-319-64629-9_1

If the index set I is the set of positive integers, then we write (1.3) as

$$\bigcup_{i=1}^{\infty} A_i. \tag{1.4}$$

The **intersection** of sets A and B is defined by $\{x : x \in A \,\text{and}\, x \in B\}$ and is written $A \cap B$. Similar to (1.1) and (1.2) we have

$$\{x : x \in A \text{ for all } A \in \mathcal{A}\} = \bigcap_{A \in \mathcal{A}} A = \bigcap\{A : A \in \mathcal{A}\}.$$

A family $\mathcal{A}$ of sets is said to be **disjoint** if $A_1 \cap A_2 = \emptyset$ for every pair A_1 and A_2 of distinct members of $\mathcal{A}$.

If every element of the set A is also an element of B, then A is called a **subset** of B, and this is written as $A \subset B$ or $B \supset A$. With this terminology, the possibility that $A = B$ is allowed. The set A is called a **proper subset** of B if $A \subset B$ and $A \neq B$.

The **difference** of two sets is

$$A \setminus B = \{x : x \in A \text{ and } x \notin B\},$$

while the **symmetric difference** is

$$A \Delta B = (A \setminus B) \cup (B \setminus A).$$

In most discussions, a set A will be a subset of some underlying set X, and in this context, we will speak of the **complement** of A (relative to X) as the set $\{x : x \in X \text{ and } x \notin A\}$. This set is denoted by $\tilde{A}$, and this notation will be used if there is no doubt that complementation is taken with respect to X. In case of possible ambiguity, we write $X \setminus A$ instead of $\tilde{A}$. The following identities, known as **de Morgan's laws**, are very useful and easily verified:

$$\begin{aligned} \Big(\bigcup_{\alpha \in I} A_\alpha\Big)^{\sim} &= \bigcap_{\alpha \in I} \widetilde{A_\alpha}, \\ \Big(\bigcap_{\alpha \in I} A_\alpha\Big)^{\sim} &= \bigcup_{\alpha \in I} \widetilde{A_\alpha}. \end{aligned} \tag{1.5}$$

We shall denote the set of all subsets of X, called the **power set of** X, by $\mathcal{P}(X)$. Thus,

$$\mathcal{P}(X) = \{A : A \subset X\}. \tag{1.6}$$

The notions of **limit superior (lim sup)** and **lim inferior (lim inf)** are defined for sets as well as for sequences:

$$\begin{aligned} \limsup_{i \to \infty} E_i &= \bigcap_{k=1}^{\infty} \bigcup_{i=k}^{\infty} E_i, \\ \liminf_{i \to \infty} E_i &= \bigcup_{k=1}^{\infty} \bigcap_{i=k}^{\infty} E_i. \end{aligned} \tag{1.7}$$

It is easily seen that

$$\begin{aligned}&\limsup_{i\to\infty} E_i = \{x : x \in E_i \text{ for infinitely many } i\,\},\\ &\liminf_{i\to\infty} E_i = \{x : x \in E_i \text{ for all but finitely many } i\,\}.\end{aligned} \tag{1.8}$$

We use the following notation throughout:

$\emptyset =$ the empty set,
$\mathbb{N} =$ the set of positive integers (not including zero),
$\mathbb{Z} =$ the set of integers,
$\mathbb{Q} =$ the set of rational numbers,
$\mathbb{R} =$ the set of real numbers.

We assume that the reader has knowledge of the sets $\mathbb{N}$, $\mathbb{Z}$, and $\mathbb{Q}$, while $\mathbb{R}$ will be carefully constructed in Section 2.1.

Exercises for Section 1.1

1. Two sets A and B are said to be equal if all the elements of set A are in set B and vice versa. Prove that $A = B$ if and only if $A \subset B$ and $B \subset A$.
2. Prove that $A \subset B$ if and only if $A = A \cup B$.
3. Prove de Morgan's laws, (1.5).
4. Let E_i, $i = 1, 2, \ldots$, be a family of sets. Use definitions (1.7) to prove

$$\liminf_{i\to\infty} E_i \subset \limsup_{i\to\infty} E_i.$$

1.2. Functions

In this section an informal discussion of relations and functions is given, a subject that is encountered in several forms in elementary analysis. In this development, we adopt the notion that a relation or function is indistinguishable from its graph.

If X and Y are sets, the **Cartesian product** of X and Y is

$$X \times Y = \{\text{ all ordered pairs } (x, y) : x \in X, y \in Y\}. \tag{1.9}$$

The ordered pair (x, y) is thus to be distinguished from (y, x). We will discuss the Cartesian product of an arbitrary family of sets later in this section.

A **relation** from X to Y is a subset of $X \times Y$. If f is a relation, then the **domain** and **range** of f are

$$\begin{aligned}\operatorname{dom} f &= X \cap \{x : (x, y) \in f \text{ for some } y \in Y\,\},\\ \operatorname{rng} f &= Y \cap \{y : (x, y) \in f \text{ for some } x \in X\,\}.\end{aligned}$$

Frequently symbols such as $\sim$ and $\leq$ are used to designate a relation. In these cases the notation $x \sim y$ or $x \leq y$ will mean that the element (x, y) is a member of the relation $\sim$ or $\leq$, respectively.

A relation f is said to be **single-valued** if $y = z$ whenever (x, y) and $(x, z) \in f$. A single-valued relation is called a **function**. The terms **mapping, map, transformation** are frequently used interchangeably with function, although the term function is usually reserved for the case in which the range of f is a subset of $\mathbb{R}$. If f is a mapping and $(x, y) \in f$, we let $f(x)$ denote y. We call $f(x)$ the image of x under f. We will also use the notation $x \mapsto f(x)$, which indicates that x is mapped to $f(x)$ by f. If $A \subset X$, then the **image** of A under f is

$$f(A) = \{y : y = f(x), \text{ for some } x \in \operatorname{dom} f \cap A\}. \tag{1.10}$$

Also, the **inverse image** of B under f is

$$f^{-1}(B) = \{x : x \in \operatorname{dom} f, f(x) \in B\}. \tag{1.11}$$

If the set B consists of a single point y, or in other words $B = \{y\}$, we will simply write $f^{-1}\{y\}$ instead of the full notation $f^{-1}(\{y\})$. If $A \subset X$ and f is a mapping with $\operatorname{dom} f \subset X$, then the **restriction** of f to A, denoted by $f \llcorner A$, is defined by $f \llcorner A(x) = f(x)$ for all $x \in A \cap \operatorname{dom} f$.

If f is a mapping, then we have $\operatorname{dom} f = D \subseteq X$ and $\operatorname{rng} f \subseteq Y$. We use the notation $f\colon D \to Y$ to denote a mapping. The mapping f is called an **injection** or is said to be **univalent** if $f(x) \neq f(x')$ whenever $x, x' \in \operatorname{dom} f$ with $x \neq x'$. The mapping f is called a **surjection** or **onto** Y if for each $y \in Y$, there exists $x \in X$ such that $f(x) = y$. In other words, f is a surjection if $f(D) = Y$ (i.e., $\operatorname{rng} f = Y$), while it is not a surjection if $\operatorname{rng} f \neq Y$. Finally, we say that f is a **bijection** if f is both an injection and a surjection. A bijection $f\colon D \to Y$ is also called a **one-to-one correspondence** between D and Y.

If f is a mapping from X to Y and g a mapping from Y to Z, then the **composition** of g with f is a mapping from X to Z defined by

$$g \circ f = \{(x, z) : (x, y) \in f \text{ and } (y, z) \in g \text{ for some } y \in Y\}. \tag{1.12}$$

There is one type of relation that is particularly important and is so often encountered that it requires a separate definition.

1.1. Definition. If X is a set, an **equivalence relation** on X (often denoted by $\sim$) is a relation characterized by the following conditions:

(i) $x \sim x$ for every $x \in X$, (reflexive)

(ii) if $x \sim y$, then $y \sim x$, (symmetric)

(iii) if $x \sim y$ and $y \sim z$, then $x \sim z$. (transitive)

Given an equivalence relation $\sim$ on X, a subset A of X is called an **equivalence class** if there is an element $x \in A$ such that A consists precisely of those elements y such that $x \sim y$. One can easily verify that distinct equivalence classes are disjoint and that X can be expressed as the union of equivalence classes.

A **sequence** in a space X is a mapping $f: \mathbb{N} \to X$. It is convenient to represent a sequence f as a list. Thus, if $f(k) = x_k$, we speak of the sequence $\{x_k\}_{k=1}^{\infty}$ or simply $\{x_k\}$. A subsequence is obtained by discarding some elements of the original sequence and ordering the elements that remain in the usual way. Formally, we say that $x_{k_1}, x_{k_2}, x_{k_3}, \ldots$ is a **subsequence** of $x_1, x_2, x_3, \ldots$ if there is a mapping $g: \mathbb{N} \to \mathbb{N}$ such that for each $i \in \mathbb{N}$, $x_{k_i} = x_{g(i)}$ and $g(i) < g(j)$ whenever $i < j$.

Our final topic in this section is the Cartesian product of a family of sets. Let $\mathcal{X}$ be a family of sets X_α indexed by a set I. The **Cartesian product** of $\mathcal{X}$ is denoted by

$$\prod_{\alpha \in I} X_\alpha$$

and is defined as the set of all mappings

$$x: I \to \bigcup X_\alpha$$

with the property that

$$x(\alpha) \in X_\alpha \tag{1.13}$$

for each $\alpha \in I$. Each mapping x is called a **choice mapping** for the family $\mathcal{X}$. Also, we call $x(\alpha)$ the αth **coordinate** of x. This terminology is perhaps easier to understand if we consider the case $I = \{1, 2, \ldots, n\}$. As in the preceding paragraph, it is useful to represent the choice mapping x as a list $\{x(1), x(2), \ldots, x(n)\}$, and even more useful if we write $x(i) = x_i$. The mapping x is thus identified with the ordered n-tuple $(x_1, x_2, \ldots, x_n)$. Here, the word "ordered" is crucial, because an n-tuple consisting of the same elements but in a different order produces a different mapping x. Consequently, the Cartesian product becomes the set of all ordered n-tuples:

$$\prod_{i=1}^{n} X_i = \{(x_1, x_2, \ldots, x_n) : x_i \in X_i, i = 1, 2, \ldots, n\}. \tag{1.14}$$

In the special case $X_i = \mathbb{R}, i = 1, 2, \ldots, n$, an element of the Cartesian product is a mapping that can be identified with an ordered n-tuple of real numbers. We denote the set of all ordered n-tuples (also referred to as vectors) by

$$\mathbb{R}^n = \{(x_1, x_2, \ldots, x_n) : x_i \in \mathbb{R}, i = 1, 2, \ldots, n\}.$$

The set $\mathbb{R}^n$ is called **Euclidean n-space**. The **norm** of a vector x is defined as

$$|x| = \sqrt{x_1^2 + x_2^2 + \cdots + x_n^2}; \tag{1.15}$$

the **distance** between two vectors x and y is $|x - y|$. As we mentioned earlier in this section, the Cartesian product of two sets X_1 and X_2 is denoted by $X_1 \times X_2$.

1.2. REMARK. A fundamental issue that we have not addressed is whether the Cartesian product of an arbitrary family of sets is nonempty. This involves concepts from set theory and is the subject of the next section.

Exercises for Section 1.2

1. Prove that $f \circ (g \circ h) = (f \circ g) \circ h$ for mappings f, g, and h.
2. Prove that $(f \circ g)^{-1}(A) = g^{-1}[f^{-1}(A)]$ for mappings f and g and an arbitrary set A.
3. Prove: If $f\colon X \to Y$ is a mapping and $A \subset B \subset X$, then $f(A) \subset f(B) \subset Y$. Also, prove that if $E \subset F \subset Y$, then $f^{-1}(E) \subset f^{-1}(F) \subset X$.
4. Prove: If $\mathcal{A} \subset \mathcal{P}(X)$, then
$$f(\bigcup_{A\in\mathcal{A}} A) = \bigcup_{A\in\mathcal{A}} f(A) \text{ and } f(\bigcap_{A\in\mathcal{A}} A) \subset \bigcap_{A\in\mathcal{A}} f(A)$$
and
$$f^{-1}(\bigcup_{A\in\mathcal{A}} A) = \bigcup_{A\in\mathcal{A}} f^{-1}(A) \text{ and } f^{-1}(\bigcap_{A\in\mathcal{A}} A) = \bigcap_{A\in\mathcal{A}} f^{-1}(A).$$
Give an example that shows that the above inclusion cannot be replaced by equality.
5. Consider a nonempty set X and its power set $\mathcal{P}(X)$. For each $x \in X$, let $B_x = \{0,1\}$ and consider the Cartesian product $\prod_{x\in X} B_x$. Exhibit a natural one-to-one correspondence between $\mathcal{P}(X)$ and $\prod_{x\in X} B_x$.
6. Let $X \xrightarrow{f} Y$ be an arbitrary mapping and suppose there is a mapping $Y \xrightarrow{g} X$ such that $f \circ g(y) = y$ for all $y \in Y$ and that $g \circ f(x) = x$ for all $x \in X$. Prove that f is one-to-one from X onto Y and that $g = f^{-1}$.
7. Show that $A \times (B \cup C) = (A \times B) \cup (A \times C)$. Also, show that in general, $A \cup (B \times C) \neq (A \cup B) \times (A \cup C)$.

1.3. Set Theory

The material discussed in the previous two sections is based on tools found in elementary set theory. However, in more advanced areas of mathematics this material is not sufficient to discuss or even formulate some of the concepts that are needed. An example of this occurred in the previous section during the discussion of the Cartesian product of an arbitrary family of sets. Indeed, the Cartesian product of families of sets requires the notion of a choice mapping whose existence is not obvious. Here, we give a brief review of the axiom of choice and some of its logical equivalences.

A fundamental question that arises in the definition of the Cartesian product of an arbitrary family of sets is the existence of choice mappings. This is an example of a question that cannot be answered within the context of elementary set theory. At the beginning of the twentieth century, Ernst Zermelo formulated an axiom of set theory called the axiom of choice, which asserts that the Cartesian product of an arbitrary family of nonempty sets exists and is nonempty. The formal statement is as follows.

1.3. THE AXIOM OF CHOICE. *If X_α is a nonempty set for each element α of an index set I, then*

$$\prod_{\alpha\in I} X_\alpha$$

is nonempty.

1.4. PROPOSITION. *The following statement is equivalent to the axiom of choice: If $\{X_\alpha\}_{\alpha\in A}$ is a disjoint family of nonempty sets, then there is a set $S \subset \cup_{\alpha\in A} X_\alpha$ such that $S \cap X_\alpha$ consists of precisely one element for every $\alpha \in A$.*

PROOF. The axiom of choice states that there exists $f \colon A \to \cup_{\alpha\in A} X_\alpha$ such that $f(\alpha) \in X_\alpha$ for each $\alpha \in A$. The set $S := f(A)$ satisfies the conclusion of the statement. Conversely, if such a set S exists, then the mapping $A \xrightarrow{f} \cup_{\alpha\in A} X_\alpha$ defined by assigning the point $S \cap X_\alpha$ the value of $f(\alpha)$ implies the validity of the axiom of choice. □

1.5. DEFINITION. Given a set S and a relation $\leq$ on S, we say that $\leq$ is a **partial ordering** if the following three conditions are satisfied:

(i) $x \leq x$ for every $x \in S$, (reflexive)
(ii) if $x \leq y$ and $y \leq x$, then $x = y$, (antisymmetric)
(iii) if $x \leq y$ and $y \leq z$, then $x \leq z$. (transitive)
If, in addition,
(iv) either $x \leq y$ or $y \leq x$, for all $x, y \in S$, (trichotomy)
then $\leq$ is called a **linear** or **total** ordering.

For example, $\mathbb{Z}$ is linearly ordered with its usual ordering, whereas the family of all subsets of a given set X is partially ordered (but not linearly ordered) by $\subset$. If a set X is endowed with a linear ordering, then each subset A of X inherits the ordering of X. That is, the restriction to A of the linear ordering on X induces a linear ordering on A. The following two statements are known to be equivalent to the axiom of choice.

1.6. HAUSDORFF MAXIMAL PRINCIPLE. *Every partially ordered set has a maximal linearly ordered subset.*

1.7. ZORN'S LEMMA. *If X is a partially ordered set with the property that each linearly ordered subset has an upper bound, then X has a maximal element. In particular, this implies that if $\mathcal{E}$ is a family of sets (or a collection of families of sets) and if $\{\cup F : F \in \mathcal{F}\} \in \mathcal{E}$ for every subfamily $\mathcal{F}$ of $\mathcal{E}$ with the property that*

$$F \subset G \quad \text{or} \quad G \subset F \qquad \text{whenever} \qquad F, G \in \mathcal{F},$$

then there exists $E \in \mathcal{E}$ that is maximal in the sense that it is not a subset of any other member of $\mathcal{E}$.

In the following, we will consider other formulations of the axiom of choice. This will require the notion of a linear ordering on a set.

A nonempty set X endowed with a linear order is said to be **well ordered** if each subset of X has a first element with respect to its induced linear order. Thus, the integers, $\mathbb{Z}$, with the usual ordering is not a well-ordered set, whereas the set $\mathbb{N}$ is well ordered. However, it is possible to define a linear ordering on $\mathbb{Z}$ that produces a well ordering. In fact, it is possible to do this for an arbitrary set if we assume the validity of the axiom of choice. This is stated formally in the well-ordering theorem.

1.8. THEOREM (The Well-Ordering Theorem). *Every set can be well ordered. That is, if A is an arbitrary set, then there exists a linear ordering of A with the property that each nonempty subset of A has a first element.*

Cantor put forward the continuum hypothesis in 1878, conjecturing that every infinite subset of the continuum (i.e., the set of real numbers) is either countable (i.e., can be put in one-to-one correspondence with the natural numbers) or has the cardinality of the continuum (i.e., can be put in one-to-one correspondence with the real numbers). The importance of this was seen by Hilbert, who made the continuum hypothesis the first in the list of problems that he proposed in his Paris lecture of 1900. Hilbert saw this as one of the most fundamental questions that mathematicians should attack in the twentieth century, and he went further in proposing a method to attack the conjecture. He suggested that first one should try to prove another of Cantor's conjectures, namely that every set can be well ordered.

Zermelo began to work on the problems of set theory by pursuing, in particular, Hilbert's idea of resolving the problem of the continuum hypothesis. In 1902 Zermelo published his first work on set theory, which was on the addition of transfinite cardinals. Two years later, in 1904, he succeeded in taking the first step suggested by Hilbert toward the continuum hypothesis when he proved that every set can be well ordered. This result brought fame to Zermelo and also earned him a quick promotion; in December 1905, he was appointed to a professorship in Göttingen.

The axiom of choice is the basis for Zermelo's proof that every set can be well ordered; in fact, the axiom of choice is equivalent to the well-ordering property, so we now know that this axiom must be used. His proof of the well-ordering property used the axiom of choice to construct sets by transfinite induction. Although Zermelo certainly gained fame for his proof of the well-ordering property, set theory at that time was in the rather unusual position that many mathematicians rejected the type of proofs that Zermelo had discovered. There were strong feelings as to whether such nonconstructive parts of mathematics were legitimate areas for study, and Zermelo's ideas were certainly not accepted by quite a number of mathematicians.

The fundamental discoveries of K. Gödel [32] and P.J. Cohen [15], [17] shook the foundations of mathematics with results that placed the axiom of choice in a very interesting position. Their work shows that the axiom of choice, in fact, is a new principle in set theory because it can neither be proved nor disproved from the usual Zermelo–Fraenkel axioms of set theory. Indeed, Gödel showed, in 1940, that the axiom of choice cannot be disproved using the other axioms of set theory, and then in 1963, Paul Cohen proved that the axiom of choice is independent of the other axioms of set theory. The importance of the axiom of choice will readily be seen throughout the following development, as we appeal to it in a variety of contexts.

Exercises for Section 1.3

1. Use a one-to-one correspondence between $\mathbb{Z}$ and $\mathbb{N}$ to exhibit a linear ordering of $\mathbb{N}$ that is not a well ordering.
2. Use the natural partial ordering of $\mathcal{P}(\{1,2,3\})$ to exhibit a partial ordering of $\mathbb{N}$ that is not a linear ordering.
3. For $(a,b),(c,d) \in \mathbb{N}\times\mathbb{N}$, define $(a,b) \leq (c,d)$ if either $a < c$ or $a = c$ and $b \leq d$. With this relation, prove that $\mathbb{N}\times\mathbb{N}$ is a well-ordered set.
4. Let P denote the space of all polynomials defined on $\mathbb{R}$. For $p_1, p_2 \in P$, define $p_1 \leq p_2$ if there exists x_0 such that $p_1(x) \leq p_2(x)$ for all $x \geq x_0$. Is $\leq$ a linear ordering? Is P well ordered?
5. Let C denote the space of all continuous functions on $[0,1]$. For $f_1, f_2 \in C$, define $f_1 \leq f_2$ if $f_1(x) \leq f_2(x)$ for all $x \in [0,1]$. Is $\leq$ a linear ordering? Is C well ordered?
6. Prove that the following assertion is equivalent to the axiom of choice: If A and B are nonempty sets and $f\colon A \to B$ is a surjection (that is, $f(A) = B$), then there exists a function $g\colon B \to A$ such that $g(y) \in f^{-1}(y)$ for each $y \in B$.
7. Use the following outline to prove that for every pair of sets A and B, either $\operatorname{card} A \leq \operatorname{card} B$ or $\operatorname{card} B \leq \operatorname{card} A$: Let $\mathcal{F}$ denote the family of all injections from subsets of A into B. Since $\mathcal{F}$ can be considered, a family of subsets of $A \times B$, it can be partially ordered by inclusion. Thus, we can apply Zorn's lemma to conclude that $\mathcal{F}$ has a maximal element, say f. If $a \in A \setminus \operatorname{dom} f$ and $b \in B \setminus f(A)$, then extend f to $A \cup \{a\}$ by defining $f(a) = b$. Then f remains an injection and thus contradicts maximality. Hence, either $\operatorname{dom} f = A$ in which case $\operatorname{card} A \leq \operatorname{card} B$, or $B = \operatorname{rng} f$, in which case f^{-1} is an injection from B into A, which would imply $\operatorname{card} B \leq \operatorname{card} A$.
8. Complete the details of the following proposition: If $\operatorname{card} A \leq \operatorname{card} B$ and $\operatorname{card} B \leq \operatorname{card} A$, then $\operatorname{card} A = \operatorname{card} B$.

   Let $f\colon A \to B$ and $g\colon B \to A$ be injections. If $a \in A \cap \operatorname{rng} g$, we have $g^{-1}(a) \in B$. If $g^{-1}(a) \in \operatorname{rng} f$, we have $f^{-1}(g^{-1}(a)) \in A$. Continue this process as far as possible. There are three possibilities: the process continues indefinitely, it terminates with an element of $A \setminus \operatorname{rng} g$ (possibly with a itself), or it terminates with an element of $B \setminus \operatorname{rng} f$. These three cases determine disjoint sets A_∞, A_A, and A_B whose union is A. In a similar manner, B can be decomposed into B_∞, B_B, and B_A. Now f maps A_∞ onto B_∞ and A_A onto B_A, and g maps B_B onto A_B. If we define $h\colon A \to B$ by $h(a) = f(a)$ if $a \in A_\infty \cup A_A$ and $h(a) = g^{-1}(a)$ if $a \in A_B$, we find that h is injective.

CHAPTER 2

Real, Cardinal, and Ordinal Numbers

2.1. The Real Numbers

A brief development of the construction of the real numbers is given in terms of equivalence classes of Cauchy sequences of rational numbers. This construction is based on the assumption that properties of the rational numbers, including the integers, are known.

In our development of the real number system, we shall assume that properties of the natural numbers, integers, and rational numbers are known. In order to agree on what those properties are, we summarize some of the more basic ones. Recall that the natural numbers are defined as

$$\mathbb{N}: = \{1, 2, \ldots, k, \ldots\}.$$

They form a **well-ordered set** when endowed with the usual ordering. The ordering on $\mathbb{N}$ satisfies the following properties:

(i) $x \le x$ for every $x \in S$.
(ii) If $x \le y$ and $y \le x$, then $x = y$.
(iii) If $x \le y$ and $y \le z$, then $x \le z$.
(iv) for all $x, y \in S$, either $x \le y$ or $y \le x$.

The four conditions above define a linear ordering on S, a topic that was introduced in Section 1.3 and will be discussed in greater detail in Section. 2.3. The linear order $\le$ of $\mathbb{N}$ is compatible with the addition and multiplication operations in $\mathbb{N}$. Furthermore, the following three conditions are satisfied:

(i) Every nonempty subset of $\mathbb{N}$ has a first element; i.e., if $\emptyset \ne S \subset \mathbb{N}$, there is an element $x \in S$ such that $x \le y$ for every element $y \in S$. In particular, the set $\mathbb{N}$ itself has a first element that is unique, in view of (ii) above, and is denoted by the symbol 1.
(ii) Every element of $\mathbb{N}$, except the first, has an immediate predecessor. That is, if $x \in \mathbb{N}$ and $x \ne 1$, then there exists $y \in \mathbb{N}$ with the property that $y \le x$ and $z \le y$ whenever $z \le x$.
(iii) $\mathbb{N}$ has no greatest element; i.e., for every $x \in \mathbb{N}$, there exists $y \in \mathbb{N}$ such that $x \ne y$ and $x \le y$.

The reader can easily show that (i) and (iii) imply that each element of $\mathbb{N}$ has an **immediate successor**, i.e., that for each $x \in \mathbb{N}$, there exists $y \in \mathbb{N}$ such that $x < y$ and that if $x < z$ for some $z \in \mathbb{N}$ where $y \ne z$, then $y < z$.

W.P. Ziemer, *Modern Real Analysis*, Graduate Texts in Mathematics 278,
https://doi.org/10.1007/978-3-319-64629-9_2

The immediate successor y of x will be denoted by x'. A nonempty set $S \subset \mathbb{N}$ is said to be **finite** if S has a greatest element.

From the structure established above follows an extremely important result, the so-called **principle of mathematical induction**, which we now prove.

2.1. THEOREM. *Suppose $S \subset \mathbb{N}$ is a set with the property that $1 \in S$ and that $x \in S$ implies $x' \in S$. Then $S = \mathbb{N}$.*

PROOF. Suppose S is a proper subset of $\mathbb{N}$ that satisfies the hypotheses of the theorem. Then $\mathbb{N} \setminus S$ is a nonempty set and therefore by (i) above has a first element x. Note that $x \neq 1$, since $1 \in S$. From (ii) we see that x has an immediate predecessor, y. Since $y \in S$, we have $y' \in S$. Since $x = y'$, we have $x \in S$, contradicting the choice of x as the first element of $\mathbb{N} \setminus S$.

Also, we have $x \in S$, since $x = y'$. By definition, x is the first element of $\mathbb{N} - S$, thus producing a contradiction. Hence, $S = \mathbb{N}$. □

The rational numbers $\mathbb{Q}$ may be constructed in a formal way from the natural numbers. This is accomplished by first defining the integers, both negative and positive, so that subtraction can be performed. Then the rationals are defined using the properties of the integers. We will not go into this construction but instead leave it to the reader to consult another source for this development. We list below the basic properties of the rational numbers.

The rational numbers are endowed with the operations of addition and multiplication that satisfy the following conditions:

(i) For every $r, s \in \mathbb{Q}$, $r + s \in \mathbb{Q}$, and $rs \in \mathbb{Q}$.

(ii) Both operations are commutative and associative, i.e., $r + s = s + r$, $rs = sr$, $(r + s) + t = r + (s + t)$, and $(rs)t = r(st)$.

(iii) The operations of addition and multiplication have identity elements 0 and 1 respectively, i.e., for each $r \in \mathbb{Q}$, we have

$$0 + r = r \qquad \text{and} \qquad 1 \cdot r = r.$$

(iv) The distributive law is valid:

$$r(s + t) = rs + rt$$

whenever r, s, and t are elements of $\mathbb{Q}$.

(v) The equation $r + x = s$ has a solution for every $r, s \in \mathbb{Q}$. The solution is denoted by $s - r$.

(vi) The equation $rx = s$ has a solution for every $r, s \in \mathbb{Q}$ with $r \neq 0$. This solution is denoted by s/r. A set containing at least two elements and satisfying the six conditions above is called a **field**; in particular, the rational numbers form a field. The set $\mathbb{Q}$ can also be endowed with a linear ordering. The order relation is related to the operations of addition and multiplication as follows:

(vii) If $r \geq s$, then for every $t \in \mathbb{Q}$, $r + t \geq s + t$.

(viii) $0 < 1$.

(ix) If $r \geq s$ and $t \geq 0$, then $rt \geq st$.

The set of rational numbers thus provides an example of an **ordered field**. The proof of the following is elementary and is left to the reader; see Exercise 6 at the end of this section.

2.2. THEOREM. *Every ordered field F contains an isomorphic image of $\mathbb{Q}$, and the isomorphism can be taken as order-preserving.*

In view of this result, we may view $\mathbb{Q}$ as a subset of F. Consequently, the following definition is meaningful.

2.3. DEFINITION. An ordered field F is called an Archimedean ordered field if for each $a \in F$ and each positive $b \in \mathbb{Q}$, there exists a positive integer n such that $nb > a$. Intuitively, this means that no matter how large a is and how small b, successive repetitions of b will eventually exceed a.

Although the rational numbers form a rich algebraic system, they are inadequate for the purposes of analysis, because they are, in a sense, incomplete. For example, a negative rational number does not have a rational square root, and not every positive rational number has a rational square root. We now proceed to construct the real numbers assuming knowledge of the integers and rational numbers. This is basically an assumption concerning the algebraic structure of the real numbers.

The linear order structure of a field permits us to define the notion of the absolute value of an element of that field. That is, the **absolute value** of x is defined by

$$|x| = \begin{cases} x & \text{if } x \geq 0, \\ -x & \text{if } x < 0. \end{cases}$$

We will freely use properties of the absolute value such as the triangle inequality in our development.

The following two definitions are undoubtedly well known to the reader; we state them only to emphasize that at this stage of the development, we assume knowledge of only the rational numbers.

2.4. DEFINITION. A sequence of rational numbers $\{r_i\}$ is **Cauchy** if for each rational $\varepsilon > 0$, there exists a positive integer $N(\varepsilon)$ such that $|r_i - r_k| < \varepsilon$ whenever $i, k \geq N(\varepsilon)$.

2.5. DEFINITION. A rational number r is said to be the limit of a sequence of rational numbers $\{r_i\}$ if for each rational $\varepsilon > 0$, there exists a positive integer $N(\varepsilon)$ such that

$$|r_i - r| < \varepsilon$$

for $i \geq N(\varepsilon)$. This is written as

$$\lim_{i \to \infty} r_i = r$$

and we say that $\{r_i\}$ **converges** to r.

We leave the proof of the following proposition to the reader.

2.6. PROPOSITION. *A sequence of rational numbers that converges to a rational number is Cauchy.*

2.7. PROPOSITION. *A Cauchy sequence of rational numbers, $\{r_i\}$, is bounded. That is, there exists a rational number M such that $|r_i| \leq M$ for $i = 1, 2, \ldots$.*

PROOF. Choose $\varepsilon = 1$. Since the sequence $\{r_i\}$ is Cauchy, there exists a positive integer N such that

$$|r_i - r_j| < 1 \quad \text{whenever} \quad i, j \geq N.$$

In particular, $|r_i - r_N| < 1$ whenever $i \geq N$. By the triangle inequality, $|r_i| - |r_N| \leq |r_i - r_N|$, and therefore,

$$|r_i| < |r_N| + 1 \quad \text{for all} \quad i \geq N.$$

If we define

$$M = \mathrm{Max}\{|r_1|, |r_2|, \ldots, |r_{N-1}|, |r_N| + 1\},$$

then $|r_i| \leq M$ for all $i \geq 1$. □

The reader can easily provide a proof of the following.

2.8. PROPOSITION. *Every Cauchy sequence of rational numbers has at most one limit.*

The fact that some Cauchy sequences in $\mathbb{Q}$ do not have a limit (in $\mathbb{Q}$) is what makes $\mathbb{Q}$ incomplete. We will construct the completion by means of equivalence classes of Cauchy sequences.

2.9. DEFINITION. Two Cauchy sequences of rational numbers $\{r_i\}$ and $\{s_i\}$ are said to be equivalent if

$$\lim_{i \to \infty} (r_i - s_i) = 0.$$

We write $\{r_i\} \sim \{s_i\}$ when $\{r_i\}$ and $\{s_i\}$ are equivalent. It is easy to show that this, in fact, is an **equivalence relation**. That is,

(i) $\{r_i\} \sim \{r_i\}$, (reflexivity)
(ii) $\{r_i\} \sim \{s_i\}$ if and only if $\{s_i\} \sim \{r_i\}$, (symmetry)
(iii) if $\{r_i\} \sim \{s_i\}$ and $\{s_i\} \sim \{t_i\}$, then $\{r_i\} \sim \{t_i\}$. (transitivity)

The set of all Cauchy sequences of rational numbers equivalent to a fixed Cauchy sequence is called an **equivalence class of Cauchy sequences.** The fact that we are dealing with an equivalence relation implies that the set of all Cauchy sequences of rational numbers is partitioned into mutually disjoint equivalence classes. For each rational number r, the sequence each of whose values is r (i.e., the constant sequence) will be denoted by $\bar{r}$. Hence, $\bar{0}$ is the constant sequence whose values are 0. This brings us to the definition of a real number.

2.10. DEFINITION. An equivalence class of Cauchy sequences of rational numbers is termed a **real number**. In this section, we will usually denote real numbers by ρ, σ, etc. With this convention, a real number ρ designates an equivalence class of Cauchy sequences, and if this equivalence class contains the sequence $\{r_i\}$, we will write

$$\rho = \overline{\{r_i\}}$$

and say that ρ is represented by $\{r_i\}$. Note that $\{1/i\}_{i=1}^{\infty} \sim \bar{0}$ and that every ρ has a representative $\{r_i\}_{i=1}^{\infty}$ with $r_i \neq 0$ for every i.

In order to define the sum and product of real numbers, we invoke the corresponding operations on Cauchy sequences of rational numbers. This will require the next two elementary propositions, whose proofs are left to the reader.

2.11. PROPOSITION. *If $\{r_i\}$ and $\{s_i\}$ are Cauchy sequences of rational numbers, then $\{r_i \pm s_i\}$ and $\{r_i \cdot s_i\}$ are Cauchy sequences. The sequence $\{r_i/s_i\}$ is also Cauchy, provided $s_i \neq 0$ for every i and $\{s_i\}_{i=1}^{\infty} \not\sim \bar{0}$.*

2.12. PROPOSITION. *If $\{r_i\} \sim \{r_i'\}$ and $\{s_i\} \sim \{s_i'\}$, then $\{r_i \pm s_i\} \sim \{r_i' \pm s_i'\}$ and $\{r_i \cdot s_i\} \sim \{r_i' \cdot s_i'\}$. Similarly, $\{r_i/s_i\} \sim \{r_i'/s_i'\}$, provided $\{s_i\} \not\sim \bar{0}$, and $s_i \neq 0$ and $s_i' \neq 0$ for every i.*

2.13. DEFINITION. If ρ and σ are represented by $\{r_i\}$ and $\{s_i\}$ respectively, then $\rho \pm \sigma$ is defined by the equivalence class containing $\{r_i \pm s_i\}$, and $\rho \cdot \sigma$ by $\{r_i \cdot s_i\}$. The class ρ/σ is defined to be the equivalence class containing $\{r_i/s_i'\}$, where $\{s_i\} \sim \{s_i'\}$ and $s_i' \neq 0$ for all i, provided $\{s_i\} \not\sim \bar{0}$.

Reference to Propositions 2.11 and 2.12 shows that these operations are well defined. That is, if ρ' and σ' are represented by $\{r_i'\}$ and $\{s_i'\}$, where $\{r_i\} \sim \{r_i'\}$ and $\{s_i\} \sim \{s_i'\}$, then $\rho + \sigma = \rho' + \sigma'$, and similarly for the other operations.

Since the rational numbers form a field, it is clear that the real numbers also form a field. However, we wish to show that they actually form an Archimedean ordered field. For this we first must define an ordering on the real numbers that is compatible with the field structure; this will be accomplished by the following theorem.

2.14. THEOREM. *If $\{r_i\}$ and $\{s_i\}$ are Cauchy, then one (and only one) of the following occurs:*

(i) $\{r_i\} \sim \{s_i\}$.
(ii) *There exist a positive integer N and a positive rational number k such that $r_i > s_i + k$ for $i \geq N$.*
(iii) *There exist a positive integer N and positive rational number k such that $s_i > r_i + k$ for $i \geq N$.*

PROOF. Suppose that (i) does not hold. Then there exists a rational number $k > 0$ with the property that for every positive integer N there exists an integer $i \geq N$ such that

$$|r_i - s_i| > 2k.$$

This is equivalent to saying that

$$|r_i - s_i| > 2k \quad \text{for infinitely many} \quad i \geq 1.$$

Since $\{r_i\}$ is Cauchy, there exists a positive integer N such that

$$|r_i - r_j| < k/2 \quad \text{for all} \quad i, j \geq N_1.$$

Likewise, there exists a positive integer N_2 such that

$$|s_i - s_j| < k/2 \quad \text{for all} \quad i, j \geq N_2.$$

Let $N^* \geq \max\{N_1, N_2\}$ be an integer with the property that

$$|r_{N^*} - s_{N^*}| > 2k.$$

Either $r_{N^*} > s_{N^*}$ or $s_{N^*} > r_{N^*}$. We will show that the first possibility leads to conclusion (ii) of the theorem. The proof that the second possibility leads to (iii) is similar and will be omitted. Assuming now that $r_{N^*} > s_{N^*}$, we have

$$r_{N^*} > s_{N^*} + 2k.$$

It follows from (2.1) and (2.4) that

$$|r_{N^*} - r_i| < k/2 \quad \text{and} \quad |s_{N^*} - s_i| < k/2 \quad \text{for all} \quad i \geq N^*.$$

From this and (2.6) we have that

$$r_i > r_{N^*} - k/2 > s_{N^*} + 2k - k/2 = s_{N^*} + 3k/2 \quad \text{for} \quad i \geq N^*.$$

But $s_{N^*} > s_i - k/2$ for $i \geq N^*$, and consequently,

$$r_i > s_i + k \quad \text{for} \quad i \geq N^*.$$

□

2.15. DEFINITION. If $\rho = \overline{\{r_i\}}$ and $\sigma = \overline{\{s_i\}}$, then we say that $\rho < \sigma$ if there exist rational numbers q_1 and q_2 with $q_1 < q_2$ and a positive integer N such that $r_i < q_1 < q_2 < s_i$ for all i with $i \geq N$. Note that q_1 and q_2 can be chosen to be independent of the representative Cauchy sequences of rational numbers that determine ρ and σ.

In view of this definition, Theorem 2.14 implies that the real numbers are **comparable**, which we state in the following corollary.

2.16. COROLLARY. *If ρ and σ are real numbers, then one (and only one) of the following must hold:*

(1) $\rho = \sigma$,
(2) $\rho < \sigma$,
(3) $\rho > \sigma$.

Moreover, $\mathbb{R}$ is an Archimedean ordered field.

The compatibility of $\leq$ with the field structure of $\mathbb{R}$ follows from Theorem 2.14. That $\mathbb{R}$ is Archimedean follows from Theorem 2.14 and the fact that $\mathbb{Q}$ is Archimedean. Note that the absolute value of a real number can thus be defined analogously to that of a rational number.

2.17. DEFINITION. If $\{\rho_i\}_{i=1}^{\infty}$ is a sequence in $\mathbb{R}$ and $\rho \in \mathbb{R}$, we define

$$\lim_{i\to\infty} \rho_i = \rho$$

to mean that for every real number $\varepsilon > 0$ there is a positive integer N such that

$$|\rho_i - \rho| < \varepsilon \quad \text{whenever} \quad i \geq N.$$

2.18. REMARK. Having shown that $\mathbb{R}$ is an Archimedean ordered field, we now know that $\mathbb{Q}$ has a natural injection into $\mathbb{R}$ by way of the constant sequences. That is, if $r \in \mathbb{Q}$, then the constant sequence $\bar{r}$ gives its corresponding equivalence class in $\mathbb{R}$. Consequently, we shall consider $\mathbb{Q}$ to be a subset of $\mathbb{R}$, that is, we do not distinguish between r and its corresponding equivalence class. Moreover, if ρ_1 and ρ_2 are in $\mathbb{R}$ with $\rho_1 < \rho_2$, then there is a rational number r such that $\rho_1 < r < \rho_2$.

The next proposition provides a connection between Cauchy sequences in $\mathbb{Q}$ with convergent sequences in $\mathbb{R}$.

2.19. THEOREM. *If* $\rho = \overline{\{r_i\}}$, *then*

$$\lim_{i\to\infty} r_i = \rho.$$

PROOF. Given $\varepsilon > 0$, we must show the existence of a positive integer N such that $|r_i - \rho| < \varepsilon$ whenever $i \geq N$. Let ε be represented by the rational sequence $\{\varepsilon_i\}$. Since $\varepsilon > 0$, we know from Theorem (2.14), (ii), that there exist a positive rational number k and an integer N_1 such that $\varepsilon_i > k$ for all $i \geq N_1$. Because the sequence $\{r_i\}$ is Cauchy, we know that there exists a positive integer N_2 such that $|r_i - r_j| < k/2$ whenever $i, j \geq N_2$. Fix an integer $i \geq N_2$ and let r_i be determined by the constant sequence $\{r_i, r_i, \ldots\}$. Then the real number $\rho - r_i$ is determined by the Cauchy sequence $\{r_j - r_i\}$, that is,

$$\rho - r_i = \overline{\{r_j - r_i\}}.$$

If $j \geq N_2$, then $|r_j - r_i| < k/2$. Note that the real number $|\rho - r_i|$ is determined by the sequence $\{|r_j - r_i|\}$. Now, the sequence $\{|r_j - r_i|\}$ has the property that $|r_j - r_i| < k/2 < k < \varepsilon_j$ for $j \geq \max(N_1, N_2)$. Hence, by Definition (2.15), $|\rho - r_i| < \varepsilon$. The proof is concluded by taking $N = \max(N_1, N_2)$. □

2.20. THEOREM. *The set of real numbers is complete; that is, every Cauchy sequence of real numbers converges to a real number.*

PROOF. Let $\{\rho_i\}$ be a Cauchy sequence of real numbers and let each ρ_i be determined by the Cauchy sequence of rational numbers, $\{r_{i,k}\}_{k=1}^{\infty}$. By the previous proposition,

$$\lim_{k\to\infty} r_{i,k} = \rho_i.$$

Thus, for each positive integer i, there exists k_i such that

$$|r_{i,k_i} - \rho_i| < \frac{1}{i}. \tag{2.1}$$

Let $s_i = r_{i,k_i}$. The sequence $\{s_i\}$ is Cauchy because

$$\begin{aligned}|s_i - s_j| &\le |s_i - \rho_i| + |\rho_i - \rho_j| + |\rho_j - s_j| \\ &\le 1/i + |\rho_i - \rho_j| + 1/j.\end{aligned}$$

Indeed, for $\varepsilon > 0$, there exists a positive integer $N > 4/\varepsilon$ such that $i, j \ge N$ implies $|\rho_i - \rho_j| < \varepsilon/2$. This, along with (2.1), shows that $|s_i - s_j| < \varepsilon$ for $i, j \ge N$. Moreover, if ρ is the real number determined by $\{s_i\}$, then

$$\begin{aligned}|\rho - \rho_i| &\le |\rho - s_i| + |s_i - \rho_i| \\ &\le |\rho - s_i| + 1/i.\end{aligned}$$

For $\varepsilon > 0$, we invoke Theorem 2.19 for the existence of $N > 2/\varepsilon$ such that the first term is less than $\varepsilon/2$ for $i \ge N$. For all such i, the second term is also less than $\varepsilon/2$. □

The completeness of the real numbers leads to another property that is of basic importance.

2.21. DEFINITION. A number M is called an **upper bound for** for a set $A \subset \mathbb{R}$ if $a \le M$ for all $a \in A$. An upper bound b for A is called a **least upper bound for** A if b is less than all other upper bounds for A. The term **supremum of** A is used interchangeably with least upper bound and is written $\sup A$. The terms **lower bound, greatest lower bound,** and **infimum** are defined analogously.

2.22. THEOREM. *Let $A \subset \mathbb{R}$ be a nonempty set that is bounded above (below). Then* $\sup A$ $(\inf A)$ *exists.*

PROOF. Let $b \in \mathbb{R}$ be any upper bound for A and let $a \in A$ be an arbitrary element. Further, using the Archimedean property of $\mathbb{R}$, let M and $-m$ be positive integers such that $M > b$ and $-m > -a$, so that we have $m < a \le b < M$. For each positive integer p let

$$I_p = \left\{k : k \text{ an integer and } \frac{k}{2^p} \text{ is an upper bound for } A\right\}.$$

Since A is bounded above, it follows that I_p is not empty. Furthermore, if $a \in A$ is an arbitrary element, there is an integer j that is less than a. If k is an integer such that $k \le 2^p j$, then k is not an element of I_p, thus showing that I_p is bounded below. Therefore, since I_p consists only of integers, it follows that I_p has a first element, call it k_p. Because

$$\frac{2k_p}{2^{p+1}} = \frac{k_p}{2^p},$$

the definition of k_{p+1} implies that $k_{p+1} \leq 2k_p$. But

$$\frac{2k_p - 2}{2^{p+1}} = \frac{k_p - 1}{2^p}$$

is not an upper bound for A, which implies that $k_{p+1} \neq 2k_p - 2$. In fact, it follows that $k_{p+1} > 2k_p - 2$. Therefore, either

$$k_{p+1} = 2k_p \quad \text{or} \quad k_{p+1} = 2k_p - 1.$$

Defining $a_p = \dfrac{k_p}{2^p}$, we have either

$$a_{p+1} = \frac{2k_p}{2^{p+1}} = a_p \quad \text{or} \quad a_{p+1} = \frac{2k_p - 1}{2^{p+1}} = a_p - \frac{1}{2^{p+1}},$$

and hence

$$a_{p+1} \leq a_p \quad \text{with} \quad a_p - a_{p+1} \leq \frac{1}{2^{p+1}}$$

for each positive integer p. If $q > p \geq 1$, then

$$\begin{aligned}
0 \leq a_p - a_q &= (a_p - a_{p+1}) + (a_{p+1} - a_{p+2}) + \cdots + (a_{q-1} - a_q) \\
&\leq \frac{1}{2^{p+1}} + \frac{1}{2^{p+2}} + \cdots + \frac{1}{2^q} \\
&= \frac{1}{2^{p+1}}\left(1 + \frac{1}{2} + \cdots + \frac{1}{2^{q-p-1}}\right) \\
&< \frac{1}{2^{p+1}}(2) = \frac{1}{2^p}.
\end{aligned}$$

Thus, whenever $q > p \geq 1$, we have $|a_p - a_q| < \frac{1}{2^p}$, which implies that $\{a_p\}$ is a Cauchy sequence. By the completeness of the real numbers, Theorem 2.20, there is a real number c to which the sequence converges.

We will show that c is the supremum of A. First, observe that c is an upper bound for A, since it is the limit of a decreasing sequence of upper bounds. Second, it must be the least upper bound, for otherwise, there would be an upper bound c' with $c' < c$. Choose an integer p such that $1/2^p < c - c'$. Then

$$a_p - \frac{1}{2^p} \geq c - \frac{1}{2^p} > c + c' - c = c',$$

which shows that $a_p - \frac{1}{2^p}$ is an upper bound for A. But the definition of a_p implies that

$$a_p - \frac{1}{2^p} = \frac{k_p - 1}{2^p},$$

a contradiction, since $\dfrac{k_p - 1}{2^p}$ is not an upper bound for A.

The existence of $\inf A$ if A is bounded below follows by an analogous argument. □

A linearly ordered field is said to have the **least upper bound** property if each nonempty subset that has an upper bound has a least upper bound (in the field). Hence, $\mathbb{R}$ has the least upper bound property. It can be shown that every linearly ordered field with the least upper bound property is a complete Archimedean ordered field. We will not prove this assertion.

Exercises for Section 2.1

1. Use the fact that

$$\mathbb{N} = \{n : n = 2k \text{ for some } k \in \mathbb{N}\} \cup \{n : n = 2k+1 \text{ for some } k \in \mathbb{N}\}$$

to prove $c \cdot c = c$. Consequently, card $(\mathbb{R}^n) = c$ for each $n \in \mathbb{N}$.

2. Prove that the set of numbers whose dyadic expansions are not unique is countable.
3. Prove that the equation $x^2 - 2 = 0$ has no solutions in the field $\mathbb{Q}$.
4. Prove: If $\{x_n\}_{n=1}^{\infty}$ is a bounded increasing sequence in an Archimedean ordered field, then the sequence is Cauchy.
5. Prove that each Archimedean ordered field contains a "copy" of $\mathbb{Q}$. Moreover, for each pair r_1 and r_2 of the field with $r_1 < r_2$, there exists a rational number r such that $r_1 < r < r_2$.
6. Consider the set $\{r+q\sqrt{2} : r \in \mathbb{Q}, q \in \mathbb{Q}\}$. Prove that it is an Archimedean ordered field.
7. Let F be the field of all rational polynomials with coefficients in $\mathbb{Q}$. Thus, a typical element of F has the form $\dfrac{P(x)}{Q(x)}$, where $P(x) = \displaystyle\sum_{k=0}^{n} a_k x^k$ and $Q(x) = \sum_{j=0}^{m} b_j x^j$, where the a_k and b_j are in $\mathbb{Q}$ with $a_n \neq 0$ and $b_m \neq 0$. We order F by saying that $\dfrac{P(x)}{Q(x)}$ is positive if $a_n b_m$ is a positive rational number. Prove that F is an ordered field that is not Archimedean.
8. Consider the set $\{0, 1\}$ with $+$ and $\times$ given by the following tables:

$+$	0	1
0	0	1
1	1	0

$\times$	0	1
0	0	0
1	0	1

Prove that $\{0, 1\}$ is a field and that there can be no ordering on $\{0, 1\}$ that results in a linearly ordered field.

9. Prove: For real numbers a and b,
   (a) $|a+b| \leq |a| + |b|$,
   (b) $||a| - |b|| \leq |a-b|$,
   (c) $|ab| = |a|\,|b|$.

2.2. Cardinal Numbers

There are many ways to determine the "size" of a set, the most basic being the enumeration of its elements when the set is finite. When the set is infinite, another means must be employed; the one that we use is not far from the enumeration concept.

2.23. DEFINITION. Two sets A and B are said to be **equivalent** if there exists a bijection $f\colon A \to B$, and then we write $A \sim B$. In other words, there is a one-to-one correspondence between A and B. It is not difficult to show that this notion of equivalence defines an equivalence relation as described in Definition 1.1, and therefore sets are partitioned into equivalence classes. Two sets in the same equivalence class are said to have the same **cardinal number** or to be of the same cardinality. The cardinal number of a set A is denoted by $\operatorname{card} A$; that is, $\operatorname{card} A$ is the symbol we attach to the equivalence class containing A. There are some sets so frequently encountered that we use special symbols for their cardinal numbers. For example, the cardinal number of the set $\{1, 2, \ldots, n\}$ is denoted by n, $\operatorname{card} \mathbb{N} = \aleph_0$, and $\operatorname{card} \mathbb{R} = c$.

2.24. DEFINITION. Let A be a nonempty set. If $\operatorname{card} A = n$, for some nonnegative integer n, then we say that A is a **finite** set. If A is not finite, then we say that it is an **infinite** set. If A is equivalent to the positive integers, then A is **denumerable.** If A is either finite or denumerable, then it is called **countable**; otherwise, it is called **uncountable**.

One of the first observations concerning cardinality is that it is possible for two sets to have the same cardinality even though one is a proper subset of the other. For example, the formula $y = 2x$, $x \in [0, 1]$, defines a bijection between the closed intervals $[0, 1]$ and $[0, 2]$. This also can be seen with the help of the figure below.

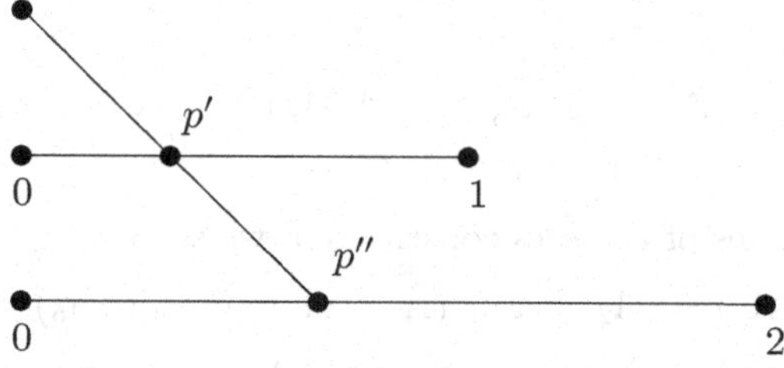

Another example, utilizing a two-step process, establishes the equivalence between points x of $(-1, 1)$ and y of $\mathbb{R}$. The semicircle with endpoints omitted serves as an intermediary.

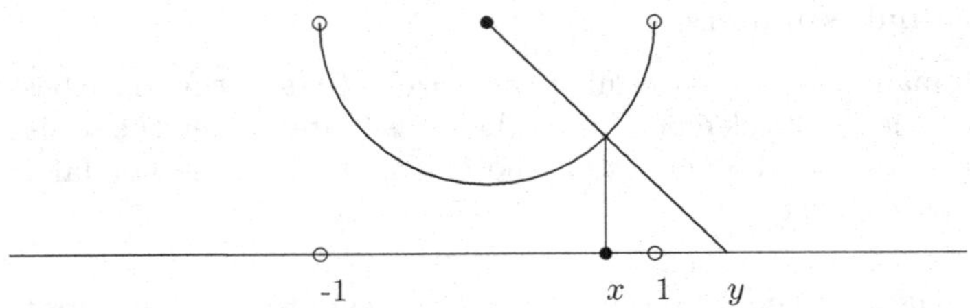

A bijection could also be explicitly given by $y = \frac{2x-1}{1-(2x-1)^2}$, $x \in (0,1)$.

Pursuing other examples, it should be true that $(0,1) \sim [0,1]$, although in this case, exhibiting the bijection is not immediately obvious (but not very difficult; see Exercise 7 at the end of this section). Aside from actually exhibiting the bijection, the facts that $(0,1)$ is equivalent to a subset of $[0,1]$ and that $[0,1]$ is equivalent to a subset of $(0,1)$ offer compelling evidence that $(0,1) \sim [0,1]$. The next two results make this rigorous.

2.25. THEOREM. *If $A \supset A_1 \supset A_2$ and $A \sim A_2$, then $A \sim A_1$.*

PROOF. Let $f\colon A \to A_2$ denote the bijection that determines the equivalence between A and A_2. The restriction of f to A_1, $f \llcorner A_1$, determines a set A_3 (actually, $A_3 = f(A_1)$) such that $A_1 \sim A_3$, where $A_3 \subset A_2$. Now we have sets $A_1 \supset A_2 \supset A_3$ such that $A_1 \sim A_3$. Repeating the argument, there exists a set A_4, $A_4 \subset A_3$, such that $A_2 \sim A_4$. Continue in this way to obtain a sequence of sets such that

$$A \sim A_2 \sim A_4 \sim \cdots \sim A_{2i} \sim \cdots$$

and

$$A_1 \sim A_3 \sim A_5 \sim \cdots \sim A_{2i+1} \sim \cdots .$$

For notational convenience, we take $A_0 = A$. Then we have

$$A_0 = (A_0 - A_1) \cup (A_1 - A_2) \cup (A_2 - A_3) \cup \cdots \cup (A_0 \cap A_1 \cap A_2 \cap \cdots) \tag{2.2}$$

and

$$A_1 = (A_1 - A_2) \cup (A_2 - A_3) \cup (A_3 - A_4) \cup \cdots \cup (A_1 \cap A_2 \cap A_3 \cap \cdots). \tag{2.3}$$

By the properties of the sets constructed, we see that

$$(A_0 - A_1) \sim (A_2 - A_3),\ (A_2 - A_3) \sim (A_4 - A_5), \ldots . \tag{2.4}$$

In fact, the bijection between $(A_0 - A_1)$ and $(A_2 - A_3)$ is given by f restricted to $A_0 - A_1$. Likewise, f restricted to $A_2 - A_3$ provides a bijection onto $A_4 - A_5$, and similarly for the remaining sets in the sequence. Moreover, since $A_0 \supset A_1 \supset A_2 \supset \cdots$, we have

$$(A_0 \cap A_1 \cap A_2 \cap \cdots) = (A_1 \cap A_2 \cap A_3 \cap \cdots).$$

The sets A_0 and A_1 are represented by a disjoint union of sets in (2.2) and (2.3). With the help of (2.4), note that the union of the first two sets that appear in the expressions for A and A_1 are equivalent; that is,

$$(A_0 - A_1) \cup (A_1 - A_2) \sim (A_1 - A_2) \cup (A_2 - A_3).$$

Likewise,

$$(A_2 - A_3) \cup (A_4 - A_5) \sim (A_3 - A_4) \cup (A_5 - A_6),$$

and similarly for the remaining sets. Thus, it is easy to see that $A \sim A_1$. □

2.26. THEOREM. (Schröder–Bernstein) *If $A \supset A_1$, $B \supset B_1$, $A \sim B_1$, and $B \sim A_1$, then $A \sim B$.*

PROOF. Denoting by f the bijection that determines the similarity between A and B_1, let $B_2 = f(A_1)$ to obtain $A_1 \sim B_2$ with $B_2 \subset B_1$. However, by hypothesis, we have $A_1 \sim B$ and therefore $B \sim B_2$. Now invoke Theorem 2.25 to conclude that $B \sim B_1$. But $A \sim B_1$ by hypothesis, and consequently, $A \sim B$. □

It is instructive to recast all of the information in this section in terms of cardinality. First, we introduce the concept of comparability of cardinal numbers.

2.27. DEFINITION. If α and β are the cardinal numbers of the sets A and B, respectively, we say that $\alpha \leq \beta$ if there exists a set $B_1 \subset B$ such that $A \sim B_1$. In addition, we say that $\alpha < \beta$ if there exists no set $A_1 \subset A$ such that $A_1 \sim B$.

With this terminology, the Schröder–Bernstein theorem states that

$$\alpha \leq \beta \quad \text{and} \quad \beta \leq \alpha \quad \text{implies} \quad \alpha = \beta.$$

The next definition introduces arithmetic operations on the cardinal numbers.

2.28. DEFINITION. Using the notation of Definition 2.27, we define

$$\begin{aligned} \alpha + \beta &= \operatorname{card}(A \cup B) \quad \text{where} \quad A \cap B = \emptyset \\ \alpha \cdot \beta &= \operatorname{card}(A \times B) \\ \alpha^\beta &= \operatorname{card} F, \end{aligned}$$

where F is the family of all functions $f\colon B \to A$.

Let us examine the last definition in the special case $\alpha = 2$. If we take the corresponding set A as $A = \{0, 1\}$, it is easy to see that F is equivalent to the class of all subsets of B. Indeed, the bijection can be defined by

$$f \to f^{-1}\{1\},$$

where $f \in F$. This bijection is nothing more than a correspondence between subsets of B and their associated characteristic functions. Thus, 2^β is the cardinality of all subsets of B, which agrees with what we already know in the case that β is finite. Also, from previous discussions in this section, we have

$$\aleph_0 + \aleph_0 = \aleph_0, \; \aleph_0 \cdot \aleph_0 = \aleph_0 \qquad \text{and} \qquad c + c = c.$$

In addition, we see that the customary basic arithmetic properties are preserved.

2.29. THEOREM. *If α, β, and γ are cardinal numbers, then*

(i) $\alpha + (\beta + \gamma) = (\alpha + \beta) + \gamma$,
(ii) $\alpha(\beta\gamma) = (\alpha\beta)\gamma$,
(iii) $\alpha + \beta = \beta + \alpha$,
(iv) $\alpha^{(\beta+\gamma)} = \alpha^\beta \alpha^\gamma$,
(v) $\alpha^\gamma \beta^\gamma = (\alpha\beta)^\gamma$,
(vi) $(\alpha^\beta)^\gamma = \alpha^{\beta\gamma}$.

The proofs of these properties are quite easy. We give an example by proving (vi):

PROOF OF (VI). Assume that sets A, B, and C respectively represent the cardinal numbers α, β, and γ. Recall that $(\alpha^\beta)^\gamma$ is represented by the family $\mathcal{F}$ of all mappings f defined on C, where $f(c)\colon B \to A$. Thus, $f(c)(b) \in A$. On the other hand, $\alpha^{\beta\gamma}$ is represented by the family $\mathcal{G}$ of all mappings $g\colon B \times C \to A$. Define

$$\varphi\colon \mathcal{F} \to \mathcal{G}$$

as $\varphi(f) = g$, where

$$g(b, c) := f(c)(b);$$

that is,

$$\varphi(f)(b, c) = f(c)(b) = g(b, c).$$

Clearly, φ is surjective. To show that φ is univalent, let $f_1, f_2 \in \mathcal{F}$ be such that $f_1 \neq f_2$. For this to be true, there exists $c_0 \in C$ such that

$$f_1(c_0) \neq f_2(c_0).$$

This, in turn, implies the existence of $b_0 \in B$ such that

$$f_1(c_0)(b_0) \neq f_2(c_0)(b_0),$$

and this means that $\varphi(f_1)$ and $\varphi(f_2)$ are different mappings, as desired. □

In addition to these arithmetic identities, we have the following theorems, which deserve special attention.

2.30. THEOREM. $2^{\aleph_0} = c$.

PROOF. First, to prove the inequality $2^{\aleph_0} \geq c$, observe that each real number r is uniquely associated with the subset $Q_r := \{q : q \in \mathbb{Q},\ q < r\}$ of $\mathbb{Q}$. Thus mapping $r \mapsto Q_r$ is an injection from $\mathbb{R}$ into $\mathcal{P}(\mathbb{Q})$. Hence,

$$c = \operatorname{card} \mathbb{R} \leq \operatorname{card}[\mathcal{P}(\mathbb{Q})] = \operatorname{card}[\mathcal{P}(\mathbb{N})] = 2^{\aleph_0},$$

because $\mathbb{Q} \sim \mathbb{N}$.

To prove the opposite inequality, consider the set $\mathcal{S}$ of all sequences of the form $\{x_k\}$, where x_k is either 0 or 1. Referring to the definition of a sequence (Definition 1.2), it is immediate that the cardinality of $\mathcal{S}$ is $2^{\aleph_0}$. We will see below (Corollary 2.36) that each number $x \in [0,1]$ has a decimal representation of the form

$$x = 0.x_1 x_2 \ldots, \; x_i \in \{0,1\}.$$

Of course, such representations do not uniquely represent x. For example,

$$\frac{1}{2} = 0.10000\ldots = 0.01111\ldots.$$

Accordingly, the mapping from $\mathcal{S}$ into $\mathbb{R}$ defined by

$$f(\{x_k\}) = \begin{cases} \displaystyle\sum_{k=1}^{\infty} \frac{x_k}{2^k} & \text{if } x_k \neq 0 \text{ for all but finitely many } k, \\ \displaystyle\sum_{k=1}^{\infty} \frac{x_k}{2^k} + 1 & \text{if } x_k = 0 \text{ for infinitely many } k, \end{cases}$$

is clearly an injection, thus proving that $2^{\aleph_0} \leq c$. Now apply the Schröder–Bernstein theorem to obtain our result. □

The previous result implies, in particular, that $2^{\aleph_0} > \aleph_0$; the next result is a generalization of this.

2.31. THEOREM. *For every cardinal number α, $2^\alpha > \alpha$.*

PROOF. If A has cardinal number α, it follows that $2^\alpha \geq \alpha$, since each element of A determines a singleton that is a subset of A. Proceeding by contradiction, suppose $2^\alpha = \alpha$. Then there exists a one-to-one correspondence between elements x and sets S_x, where $x \in A$ and $S_x \subset A$. Let $D = \{x \in A : x \notin S_x\}$. By assumption there exists $x_0 \in A$ such that x_0 is related to the set D under the one-to-one correspondence (i.e., $D = S_{x_0}$). However, this leads to a contradiction; consider the following two possibilities:

(1) If $x_0 \in D$, then $x_0 \notin S_{x_0}$ by the definition of D. But then, $x_0 \notin D$, a contradiction.

(2) If $x_0 \notin D$, similar reasoning leads to the conclusion that $x_0 \in D$. □

The next proposition, whose proof is left to the reader, shows that $\aleph_0$ is the smallest infinite cardinal.

2.32. PROPOSITION. *Every infinite set S contains a denumerable subset.*

An immediate consequence of the proposition is the following characterization of infinite sets.

2.33. THEOREM. *A nonempty set S is infinite if and only if for each $x \in S$ the sets S and $S - \{x\}$ are equivalent.*

By means of the Schröder–Bernstein theorem, it is now easy to show that the **rational numbers are denumerable**. In fact, we show a bit more.

2.34. PROPOSITION. (i) *The set of rational numbers is denumerable,* (ii) *If A_i is denumerable for $i \in \mathbb{N}$, then $A := \bigcup_{i \in \mathbb{N}} A_i$ is denumerable.*

PROOF. Case (i) is subsumed by (ii). Since the sets A_i are denumerable, their elements can be enumerated by $\{a_{i,1}, a_{i,2}, \ldots\}$. For each $a \in A$, let (k_a, j_a) be the unique pair in $\mathbb{N} \times \mathbb{N}$ such that

$$k_a = \min\{k : a = a_{k,j}\}$$

and

$$j_a = \min\{j : a = a_{k_a,j}\}.$$

(Be aware that a could be present more than once in A. If we visualize A as an infinite matrix, then (k_a, j_a) represents the position of a that is farthest to the "northwest" in the matrix.) Consequently, A is equivalent to a subset of $\mathbb{N} \times \mathbb{N}$. Further, observe that there is an injection of $\mathbb{N} \times \mathbb{N}$ into $\mathbb{N}$ given by

$$(i, j) \to 2^i 3^j.$$

Indeed, if this were not an injection, we would have

$$2^{i-i'} 3^{j-j'} = 1$$

for some distinct positive integers i, i', j, and j', which is impossible. Thus, it follows that A is equivalent to a subset of $\mathbb{N}$ and is therefore equivalent to a subset of A_1 because $\mathbb{N} \sim A_1$. Since $A_1 \subset A$, we can now appeal to the Schröder-Bernstein theorem to arrive at the desired conclusion. □

It is natural to ask whether the real numbers are also denumerable. This turns out to be false, as the following two results indicate. It was G. Cantor who first proved this fact.

2.35. THEOREM. *If $I_1 \supset I_2 \supset I_3 \supset \ldots$ are closed intervals with the property that length $I_i \to 0$, then*

$$\bigcap_{i=1}^{\infty} I_i = \{x_0\}$$

for some point $x_0 \in \mathbb{R}$.

PROOF. Let $I_i = [a_i, b_i]$ and choose $x_i \in I_i$. Then $\{x_i\}$ is a Cauchy sequence of real numbers, since $|x_i - x_j| \le \max[\text{length} I_i, \text{length} I_j]$. Since $\mathbb{R}$ is complete (Theorem 2.20), there exists $x_0 \in \mathbb{R}$ such that

$$\lim_{i \to \infty} x_i = x_0. \tag{2.5}$$

We claim that

$$x_0 \in \bigcap_{i=1}^{\infty} I_i, \tag{2.6}$$

for if not, there would be some positive integer i_0 for which $x_0 \notin I_{i_0}$. Therefore, since I_{i_0} is closed, there would be an $\eta > 0$ such that $|x_0 - y| > \eta$ for each $y \in I_{i_0}$. Since the intervals are nested, it would follow that $x_0 \notin I_i$ for all $i \geq i_0$ and thus $|x_0 - x_i| > \eta$ for all $i \geq i_0$. This would contradict (2.5), thus establishing (2.6). We leave it to the reader to verify that x_0 is the only point with this property. □

2.36. COROLLARY. *Every real number has a decimal representation relative to any basis.*

2.37. THEOREM. *The real numbers are uncountable.*

PROOF. The proof proceeds by contradiction. Thus, we assume that the real numbers can be enumerated as $a_1, a_2, \ldots, a_i, \ldots$. Let I_1 be a closed interval of positive length less than 1 such that $a_1 \notin I_1$. Let $I_2 \subset I_1$ be a closed interval of positive length less than $1/2$ such that $a_2 \notin I_2$. Continue in this way to produce a nested sequence of intervals $\{I_i\}$ of positive length less than $1/i$ with $a_i \notin I_i$. By Lemma 2.35, we have the existence of a point

$$x_0 \in \bigcap_{i=1}^{\infty} I_i.$$

Observe that $x_0 \neq a_i$ for every i, contradicting the assumption that all real numbers are among the a_i's. □

Exercises for Section 2.2

1. Suppose α, β, and δ are cardinal numbers. Prove that
$$\delta^{\alpha+\beta} = \delta^{\alpha} \cdot \delta^{\beta}.$$
2. Show that an arbitrary function $\mathbb{R} \xrightarrow{f} \mathbb{R}$ has at most a countable number of removable discontinuities; that is, prove that
$$A := \{a \in \mathbb{R} : \lim_{x\to a} f(x) \text{ exists and } \lim_{x\to a} f(x) \neq f(a)\}$$
is at most countable.
3. Show that an arbitrary function $\mathbb{R} \xrightarrow{f} \mathbb{R}$ has at most a countable number of jump discontinuities; that is, let
$$f^+(a) := \lim_{x\to a^+} f(x)$$
and
$$f^-(a) := \lim_{x\to a^-} f(x).$$
Show that the set $\{a \in \mathbb{R} : f^+(a) \neq f^-(a)\}$ is at most countable.
4. Prove: If A is the union of a countable collection of countable sets, then A is a countable set.
5. Prove Proposition 2.33.
6. Let B be a countable subset of an uncountable set A. Show that A is equivalent to $A \setminus B$.
7. Prove that a set $A \subset \mathbb{N}$ is finite if and only if A has an upper bound.

8. Exhibit an explicit bijection between $(0, 1)$ and $[0, 1]$.
9. If you are working in Zermelo–Fraenkel set theory without the axiom of choice, can you choose an element from ...
   a finite set?
   an infinite set?
   each member of an infinite set of singletons (i.e., one-element sets)?
   each member of an infinite set of pairs of shoes?
   each member of an infinite set of pairs of socks?
   each member of a finite set of sets if each of the members is infinite?
   each member of an infinite set of sets if each of the members is infinite?
   each member of a denumerable set of sets if each of the members is infinite?
   each member of an infinite set of sets of rationals?
   each member of a denumerable set of sets if each of the members is denumerable?
   each member of an infinite set of sets if each of the members is finite?
   each member of an infinite set of finite sets of reals?
   each member of an infinite set of sets of reals?
   each member of an infinite set of two-element sets whose members are sets of reals?

2.3. Ordinal Numbers

Here we construct the ordinal numbers and extend the familiar ordering of the natural numbers. The construction is based on the notion of a well-ordered set.

2.38. DEFINITION. Suppose W is a well-ordered set with respect to the ordering $\leq$. We will use the notation $<$ in its familiar sense; we write $x < y$ to indicate that both $x \leq y$ and $x \neq y$. Also, in this case, we will agree to say that x is **less than** y and that y is **greater than** x.

For $x \in W$ we define

$$W(x) = \{y \in W : y < x\}$$

and refer to $W(x)$ as the **initial segment** of W determined by x.

The following is the **principle of transfinite induction.**

2.39. THEOREM. *Let W be a well-ordered set and let $S \subset W$ be defined as*

$$S := \{x : W(x) \subset S \textit{ implies } x \in S\}.$$

Then $S = W$.

PROOF. If $S \neq W$ then $W - S$ is a nonempty subset of W and thus has a least element x_0. Then $W(x_0) \subset S$, which by hypothesis implies that $x_0 \in S$, contradicting the fact that $x_0 \in W - S$. □

When applied to the well-ordered set $\mathbb{Z}$ of natural numbers, the hypothesis of Theorem 2.39 appears to differ in two ways from that of the principle of finite induction, Theorem 2.1. First, it is not assumed that $1 \in S$, and second, in order to conclude that $x \in S$ we need to know that **every** predecessor of x is in S and not just its immediate predecessor. The first difference is illusory, for suppose a is the least element of W. Then $W(a) = \emptyset \subset S$ and thus $a \in S$. The second difference is more significant, because in contrast to the case of $\mathbb{N}$, an element of an arbitrary well-ordered set may not have an immediate predecessor.

2.40. DEFINITION. A mapping φ from a well-ordered set V into a well-ordered set W is order-preserving if $\varphi(v_1) \leq \varphi(v_2)$ whenever $v_1, v_2 \in V$ and $v_1 \leq v_2$. If, in addition, φ is a bijection, we will refer to it as an (order-preserving) **isomorphism**. Note that in this case, $v_1 < v_2$ implies $\varphi(v_1) < \varphi(v_2)$; in other words, an order-preserving isomorphism is strictly order-preserving.

Note: We have slightly abused the notation by using the same symbol $\leq$ to indicate the ordering in both V and W above. But this should cause no confusion.

2.41. LEMMA. *If φ is an order-preserving injection of a well-ordered set W into itself, then*

$$w \leq \varphi(w)$$

for each $w \in W$.

PROOF. Set

$$S = \{w \in W : \varphi(w) < w\}.$$

If S is not empty, then it has a least element, say a. Thus $\varphi(a) < a$, and consequently $\varphi(\varphi(a)) < \varphi(a)$, since φ is an order-preserving injection; moreover, $\varphi(a) \notin S$, since a is the least element of S. By the definition of S, this implies $\varphi(a) \leq \varphi(\varphi(a))$, which is a contradiction. □

2.42. COROLLARY. *If V and W are two well-ordered sets, then there is at most one order-preserving isomorphism of V onto W.*

PROOF. Suppose f and g are isomorphisms of V onto W. Then $g^{-1} \circ f$ is an isomorphism of V onto itself, and hence $v \leq g^{-1} \circ f(v)$ for each $v \in V$. This implies that $g(v) \leq f(v)$ for each $v \in V$. Since the same argument is valid with the roles of f and g interchanged, we see that $f = g$. □

2.43. COROLLARY. *If W is a well-ordered set, then W is not isomorphic to an initial segment of itself.*

PROOF. Suppose $a \in W$ and $W \xrightarrow{f} W(a)$ is an isomorphism. Since $w \leq f(w)$ for each $w \in W$, in particular we have $a \leq f(a)$. Hence $f(a) \notin W(a)$, a contradiction. □

2.44. COROLLARY. *No two distinct initial segments of a well-ordered set W are isomorphic.*

PROOF. Since one of the initial segments must be an initial segment of the other, the conclusion follows from the previous result. □

2.45. DEFINITION. We define an **ordinal number** as an equivalence class of well-ordered sets with respect to order-preserving isomorphisms. If W is a well-ordered set, we denote the corresponding ordinal number by $\mathrm{ord}(W)$. We define a linear ordering on the class of ordinal numbers as follows: if $\mathfrak{v} = \mathrm{ord}(V)$ and $\mathfrak{w} = \mathrm{ord}(W)$, then $\mathfrak{v} < \mathfrak{w}$ if and only if V is isomorphic to an initial segment of W. The fact that this defines a linear ordering follows from the next result.

2.46. THEOREM. *If $\mathfrak{v}$ and $\mathfrak{w}$ are ordinal numbers, then precisely one of the following holds:*
(i) $\mathfrak{v} = \mathfrak{w}$,
(ii) $\mathfrak{v} < \mathfrak{w}$,
(iii) $\mathfrak{v} > \mathfrak{w}$.

PROOF. Let V and W be well-ordered sets representing $\mathfrak{v}$, $\mathfrak{w}$ respectively and let $\mathcal{F}$ denote the family of all order isomorphisms from an initial segment of V (or V itself) onto either an initial segment of W (or W itself). Recall that a mapping from a subset of V into W is a subset of $V \times W$. We may assume that $V \neq \emptyset \neq W$. If v and w are the least elements of V and W respectively, then $\{(v, w)\} \in \mathcal{F}$, and so $\mathcal{F}$ is not empty. Ordering $\mathcal{F}$ by inclusion, we see that every linearly ordered subset S of $\mathcal{F}$ has an upper bound; indeed, the union of the subsets of $V \times W$ corresponding to the elements of S is easily seen to be an order isomorphism and thus an upper bound for S. Therefore, we may employ Zorn's lemma to conclude that $\mathcal{F}$ has a maximal element, say h. Since $h \in \mathcal{F}$, it is an order isomorphism and $h \subset V \times W$. If domain h and range h were initial segments, say V_x and W_y of V and W, then $h^* := h \cup \{(x, y)\}$ would contradict the maximality of h unless domain $h = V$ or range $h = W$. If domain $h = V$, then either range $h = W$ (i.e., $\mathfrak{v} < \mathfrak{w}$) or range h is an initial segment of W, (i.e., $\mathfrak{v} = \mathfrak{w}$). If domain $h \neq V$, then domain h is an initial segment of V and range $h = W$, and the existence of h^{-1} in this case establishes $\mathfrak{v} > \mathfrak{w}$. □

2.47. THEOREM. *The class of ordinal numbers is well ordered.*

PROOF. Let S be a nonempty set of ordinal numbers. Let $\alpha \in S$ and set

$$T = \{\beta \in S : \beta < \alpha\}.$$

If $T = \emptyset$, then α is the least element of S. If $T \neq \emptyset$, let W be a well-ordered set such that $\alpha = \mathrm{ord}(W)$. For each $\beta \in T$ there is a well-ordered set W_β such that $\beta = \mathrm{ord}(W_\beta)$, and there is a unique $x_\beta \in W$ such that W_β is isomorphic to the initial segment $W(x_\beta)$ of W. The nonempty subset $\{x_\beta : \beta \in T\}$ of

W has a least element x_{β_0}. The element $\beta_0 \in T$ is the least element of T and thus the least element of S. □

2.48. COROLLARY. *The cardinal numbers are comparable.*

PROOF. Suppose a is a cardinal number. Then, the set of all ordinals whose cardinal number is a forms a well-ordered set that has a least element, call it $\alpha(a)$. The ordinal $\alpha(a)$ is called the **initial ordinal** of a. Suppose b is another cardinal number and let $W(a)$ and $W(b)$ be the well-ordered sets whose ordinal numbers are $\alpha(a)$ and $\alpha(b)$, respectively. Either $W(a)$ or $W(b)$ is isomorphic to an initial segment of the other if a and b are not of the same cardinality. Thus, one of the sets $W(a)$ and $W(b)$ is equivalent to a subset of the other. □

2.49. COROLLARY. *Suppose α is an ordinal number. Then*

$$\alpha = \text{ord}(\{\beta : \beta \textit{ is an ordinal number and } \beta < \alpha\}).$$

PROOF. Let W be a well-ordered set such that $\alpha = \text{ord}(W)$. Let $\beta < \alpha$ and let $W(\beta)$ be the initial segment of W whose ordinal number is β. It is easy to verify that this establishes an isomorphism between the elements of W and the set of ordinals less than α. □

We may view the positive integers $\mathbb{N}$ as ordinal numbers in the following way. Set

$$\begin{aligned}
1 &= \text{ord}(\{1\}),\\
2 &= \text{ord}(\{1,2\}),\\
3 &= \text{ord}(\{1,2,3\}),\\
&\vdots\\
\omega &= \text{ord}(\mathbb{N}).
\end{aligned}$$

We see that

$$n < \omega \text{ for each } n \in \mathbb{N}. \tag{2.7}$$

If $\beta = \text{ord}(W) < \omega$, then W must be isomorphic to an initial segment of $\mathbb{N}$, i.e., $\beta = n$ for some $n \in \mathbb{N}$. Thus ω is the first ordinal number such that (2.7) holds and is thus the first infinite ordinal.

Consider the set of all ordinal numbers that have either finite or denumerable cardinal numbers and observe that this forms a well-ordered set. We denote the ordinal number of this set by Ω. It can be shown that Ω is the first nondenumerable ordinal number; see Exercise 2 at the end of this section. The cardinal number of Ω is designated by $\aleph_1$. We have shown that $2^{\aleph_0} > \aleph_0$ and that $2^{\aleph_0} = c$. A fundamental question that remains open is whether $2^{\aleph_0} = \aleph_1$. The assertion that this equality holds is known as the **continuum hypothesis**. The work of Gödel [32] and Cohen [16], [17] shows that the continuum hypothesis and its negation are both consistent with the standard axioms of set theory.

At this point we acknowledge the inadequacy of the intuitive approach that we have taken to set theory. In the statement of Theorem 2.47 we were careful to refer to the class of ordinal numbers. This is because the ordinal numbers must not be a set! Suppose, for a moment, that the ordinal numbers formed a set, say $\mathcal{O}$. Then according to Theorem 2.47, $\mathcal{O}$ would be a well-ordered set. Let $\sigma = \text{ord}(\mathcal{O})$. Since $\sigma \in \mathcal{O}$ we must conclude that $\mathcal{O}$ is isomorphic to an initial segment of itself, contradicting Corollary 2.43. For an enlightening discussion of this situation see the book by P.R. Halmos [34].

Exercises for Section 2.3

1. If E is a set of ordinal numbers, prove that that there is an ordinal number α such that $\alpha > \beta$ for each $\beta \in E$.
2. Prove that Ω is the smallest nondenumerable ordinal.
3. Prove that the cardinality of all open sets in $\mathbb{R}^n$ is c.
4. Prove that the cardinality of all countable intersections of open sets in $\mathbb{R}^n$ is c.
5. Prove that the cardinality of all sequences of real numbers is c.
6. Prove that there are uncountably many subsets of an infinite set that are infinite.

CHAPTER 3

Elements of Topology

3.1. Topological Spaces

The purpose of this short chapter is to provide enough point set topology for the development of the subsequent material in real analysis. An in-depth treatment is not intended. In this section, we begin with basic concepts and properties of topological spaces.

Here, instead of the word "set," the word "space" appears for the first time. Often the word "space" is used to designate a set that has been endowed with a special structure. For example, a vector space is a set, such as $\mathbb{R}^n$, that has been endowed with an algebraic structure. Let us now turn to a short discussion of topological spaces.

3.1. DEFINITION. The pair $(X, \mathcal{T})$ is called a **topological space** if X is a nonempty set and $\mathcal{T}$ is a family of subsets of X satisfying the following three conditions:

(i) The empty set $\emptyset$ and the whole space X are elements of $\mathcal{T}$.

(ii) If $\mathcal{S}$ is an arbitrary subcollection of $\mathcal{T}$, then

$$\bigcup\{U : U \in \mathcal{S}\} \in \mathcal{T}.$$

(iii) If $\mathcal{S}$ is any finite subcollection of $\mathcal{T}$, then

$$\bigcap\{U : U \in \mathcal{S}\} \in \mathcal{T}.$$

The collection $\mathcal{T}$ is called a **topology** for the space X, and the elements of $\mathcal{T}$ are called the **open sets** of X. An open set containing a point $x \in X$ is called a **neighborhood** of x. The **interior** of an arbitrary set $A \subset X$ is the union of all open sets contained in A and is denoted by A°. Note that A° is an open set and that it is possible for some sets to have an empty interior. A set $A \subset X$ is called **closed** if $X \setminus A\colon = \tilde{A}$ is open. The **closure** of a set $A \subset X$, denoted by $\overline{A}$, is

$$\overline{A} = X \cap \{x : U \cap A \neq \emptyset \text{ for each open set } U \text{ containing } x\}$$

and the **boundary** of A is $\partial A = \overline{A} \setminus A^{\circ}$. Note that $A \subset \overline{A}$.

These definitions are fundamental and will be used extensively throughout this text.

W.P. Ziemer, *Modern Real Analysis*, Graduate Texts in Mathematics 278,
https://doi.org/10.1007/978-3-319-64629-9_3

3.2. DEFINITION. A point x_0 is called a **limit point** of a set $A \subset X$ if $A \cap U$ contains a point of A different from x_0 whenever U is an open set containing x_0. The definition does not require x_0 to be an element of A. We will use the notation A^* to denote the set of limit points of A.

3.3. EXAMPLES. (i) If X is any set and $\mathcal{T}$ the family of all subsets of X, then $\mathcal{T}$ is called the **discrete topology**. It is the largest topology (in the sense of inclusion) that X can possess. In this topology, all subsets of X are open.

(ii) The **indiscrete** is the topology in which $\mathcal{T}$ comprises only the empty set $\emptyset$ and X itself; it is obviously the smallest topology on X. In this topology, the only open sets are X and $\emptyset$.

(iii) Let $X = \mathbb{R}^n$ and let $\mathcal{T}$ consist of all sets U satisfying the following property: for each point $x \in U$ there exists a number $r > 0$ such that $B(x,r) \subset U$. Here, $B(x,r)$ denotes the ball of radius r centered at x; that is,

$$B(x,r) = \{y : |x - y| < r\}.$$

It is easy to verify that $\mathcal{T}$ is a topology. Note that $B(x,r)$ itself is an open set. This is true because if $y \in B(x,r)$ and $t = r - |y - x|$, then an application of the triangle inequality shows that $B(y,t) \subset B(x,r)$. Of course, for $n = 1$, we have that $B(x,r)$ is an open interval in $\mathbb{R}$.

(iv) Let $X = [0,1] \cup (1,2)$ and let $\mathcal{T}$ consist of $\{0\}$ and $\{1\}$ along with all open sets (open relative to $\mathbb{R}$) in $(0,1) \cup (1,2)$. Then the open sets in this topology contain, in particular, $[0,1]$ and $[1,2)$.

3.4. DEFINITION. Suppose $Y \subset X$ and $\mathcal{T}$ is a topology for X. Then it is easy to see that the family $\mathcal{S}$ of sets of the form $Y \cap U$, where U ranges over all elements of $\mathcal{T}$, satisfies the conditions for a topology on Y. The topology formed in this way is called the **induced topology**, or equivalently, the **relative topology** on Y. The space Y is said to inherit the topology from its parent space X.

3.5. EXAMPLE. Let $X = \mathbb{R}^2$ and let $\mathcal{T}$ be the topology described in (iii) above. Let $Y = \mathbb{R}^2 \cap \{x = (x_1, x_2) : x_2 \geq 0\} \cup \{x = (x_1, x_2) : x_1 = 0\}$. Thus, Y is the upper half-plane of $\mathbb{R}^2$ along with both the horizontal and vertical axes. All intervals I of the form $I = \{x = (x_1, x_2) : x_1 = 0,\ a < x_2 < b < 0\}$, where a and b are arbitrary negative real numbers, are open in the induced topology on Y, but none of them is open in the topology on X. However, all intervals J of the form $J = \{x = (x_1, x_2) : x_1 = 0, a \leq x_2 \leq b\}$ are closed in both the relative topology and the topology on X.

3.6. THEOREM. *Let $(X, \mathcal{T})$ be a topological space. Then*

(i) *The union of an arbitrary collection of open sets is open.*

(ii) *The intersection of a finite number of open sets is open.*

(iii) *The union of a finite number of closed sets is closed.*

(iv) *The intersection of an arbitrary collection of closed sets is closed.*

(v) $\overline{A \cup B} = \overline{A} \cup \overline{B}$ *whenever* $A, B \subset X$.

(vi) *If* $\{A_\alpha\}$ *is an arbitrary collection of subsets of* X, *then*

$$\bigcup_\alpha \overline{A_\alpha} \subset \overline{\bigcup_\alpha A_\alpha}.$$

(vii) $\overline{A \cap B} \subset \overline{A} \cap \overline{B}$ *whenever* $A, B \subset X$.

(viii) *A set* $A \subset X$ *is closed if and only if* $A = \overline{A}$.

(ix) $\overline{A} = A \cup A^*$.

PROOF. Parts (i) and (ii) constitute a restatement of the definition of a topological space. Parts (iii) and (iv) follow from (i) and (ii) and de Morgan's laws, (1.5).

(v) Since $A \subset A \cup B$, we have $\overline{A} \subset \overline{A \cup B}$. Similarly, $\overline{B} \subset \overline{A \cup B}$, thus proving $\overline{A \cup B} \supset \overline{A} \cup \overline{B}$. By contradiction, suppose the converse is not true. Then there exists $x \in \overline{A \cup B}$ with $x \notin \overline{A} \cup \overline{B}$, and therefore there exist open sets U and V containing x such that $U \cap A = \emptyset = V \cap B$. However, since $U \cap V$ is an open set containing x, it follows that

$$\emptyset \neq (U \cap V) \cap (A \cup B) \subset (U \cap A) \cup (V \cap B) = \emptyset,$$

a contradiction.

(vi) This follows from the same reasoning used to establish the first part of (v).

(vii) This is immediate from the definitions.

(viii) If $A = \overline{A}$, then $\tilde{A}$ is open (and thus A is closed) because $x \notin \overline{A}$ implies that there exists an open set U containing x with $U \cap A = \emptyset$; that is, $U \subset \tilde{A}$. Conversely, if A is closed and $x \in \tilde{A}$, then x belongs to some open set U with $U \subset \tilde{A}$. Thus, $U \cap A = \emptyset$ and therefore $x \notin \overline{A}$. This proves $\tilde{A} \subset (\overline{A})^\sim$ or $\overline{A} \subset A$. But always $A \subset \overline{A}$, and hence $A = \overline{A}$.

(ix) This is left as Exercise 2, Section 3.1. □

3.7. DEFINITION. Let $(X, \mathcal{T})$ be a topological space and $\{x_i\}_{i=1}^\infty$ a sequence in X. The sequence is said to **converge to** $x_0 \in X$ if for each neighborhood U of x_0 there is a positive integer N such that $x_i \in U$ whenever $i \geq N$.

It is important to observe that the structure of a topological space is so general that a sequence could possibly have more than one limit. For example, every sequence in the space with the indiscrete topology (Example 3.3 (ii)) converges to every point in X. This cannot happen if an additional restriction is placed on the topological structure, as in the following definition. (Also note that the only sequences that converge in the discrete topology are those that are eventually constant.)

3.8. DEFINITION. A topological space X is said to be a **Hausdorff space** if for each pair of distinct points $x_1, x_2 \in X$ there exist disjoint open sets U_1

and U_2 containing x_1 and x_2 respectively. That is, two distinct points can be **separated** by disjoint open sets.

3.9. DEFINITION. Suppose $(X, \mathcal{T})$ and $(Y, \mathcal{S})$ are topological spaces. A function $f: X \to Y$ is said to be **continuous at** $x_0 \in X$ if for each neighborhood V containing $f(x_0)$ there is a neighborhood U of x_0 such that $f(U) \subset V$. The function f is said to be **continuous on** X if it is continuous at each point $x \in X$.

The proof of the next result is given as Exercise 4, Section 3.1.

3.10. THEOREM. *Let $(X, \mathcal{T})$ and $(Y, \mathcal{S})$ be topological spaces. Then for a function $f: X \to Y$, the following statements are equivalent:*
(i) *f is continuous.*
(ii) *$f^{-1}(V)$ is open in X for each open set V in Y.*
(iii) *$f^{-1}(K)$ is closed in X for each closed set K in Y.*

3.11. DEFINITION. A collection of open sets, $\mathcal{F}$, in a topological space X is said to be an **open cover** of a set $A \subset X$ if

$$A \subset \bigcup_{U \in \mathcal{F}} U.$$

The family $\mathcal{F}$ is said to admit a **subcover**, $\mathcal{G}$, of A if $\mathcal{G} \subset \mathcal{F}$ and $\mathcal{G}$ is a cover of A. A subset $K \subset X$ is called **compact** if each open cover of K possesses a finite subcover of K. A space X is said to be **locally compact** if each point of X is contained in some open set whose closure is compact.

It is easy to give illustrations of sets that are not compact. For example, it is readily seen that the set $A = (0, 1]$ in $\mathbb{R}$ is not compact, since the collection of open intervals of the form $(1/i, 2)$, $i = 1, 2, \ldots$, provides an open cover of A that admits no finite subcover. On the other hand, it is true that $[0, 1]$ is compact, but the proof is not obvious. The reason for this is that the definition of compactness is usually not easy to employ directly. Later, in the context of metric spaces (Section 3.3), we will find other ways of dealing with compactness.

The following two propositions reveal some basic connections between closed and compact subsets.

3.12. PROPOSITION. *Let $(X, \mathcal{T})$ be a topological space. If A and K are respectively closed and compact subsets of X with $A \subset K$, then A is compact.*

PROOF. If $\mathcal{F}$ is an open cover of A, then the elements of $\mathcal{F}$ along with $X \setminus A$ form an open cover of K. This open cover has a finite subcover, $\mathcal{G}$, of K, since K is compact. The set $X \setminus A$ may possibly be an element of $\mathcal{G}$. If $X \setminus A$ is not a member of $\mathcal{G}$, then $\mathcal{G}$ is a finite subcover of A; if $X \setminus A$ is a member of $\mathcal{G}$, then $\mathcal{G}$ with $X \setminus A$ omitted is a finite subcover of A. □

3.13. PROPOSITION. *A compact subset of a Hausdorff space $(X, \mathcal{T})$ is closed.*

PROOF. We will show that $X \setminus K$ is open, where $K \subset X$ is compact. Choose a fixed $x_0 \in X \setminus K$ and for each $y \in K$, let V_y and U_y denote disjoint neighborhoods of y and x_0 respectively. The family

$$\mathcal{F} = \{V_y : y \in K\}$$

forms an open cover of K. Hence, $\mathcal{F}$ possesses a finite subcover, say $\{V_{y_i} : i = 1, 2, \dots, N\}$. Since $V_{y_i} \cap U_{y_i} = \emptyset$, $i = 1, 2, \dots, N$, it follows that $\bigcap_{i=1}^{N} U_{y_i} \cap \bigcup_{i=1}^{N} V_{y_i} = \emptyset$. Since $K \subset \bigcup_{i=1}^{N} V_{y_i}$, it follows that $\bigcap_{i=1}^{N} V_{y_i}$ is an open set containing x_0 that does not intersect K. Thus, $X \setminus K$ is an open set, as desired. □

The characteristic property of a Hausdorff space is that two distinct points can be separated by disjoint open sets. The next result shows that a stronger property holds, namely, that a compact set and a point not in that compact set can be separated by disjoint open sets.

3.14. PROPOSITION. *Suppose K is a compact subset of a Hausdorff space X and assume $x_0 \notin K$. Then there exist disjoint open sets U and V containing x_0 and K respectively.*

PROOF. This follows immediately from the preceding proof by taking

$$U = \bigcap_{i=1}^{N} U_{y_i} \quad \text{and} \quad V = \bigcup_{i=1}^{N} V_{y_i}.$$

□

3.15. DEFINITION. A family $\{E_\alpha : \alpha \in I\}$ of subsets of a set X is said to have the **finite intersection property** if for each finite subset $F \subset I$, one has

$$\bigcap_{\alpha \in F} E_\alpha \neq \emptyset.$$

3.16. LEMMA. *A topological space X is compact if and only if every family of closed subsets of X having the finite intersection property has a nonempty intersection.*

PROOF. First assume that X is compact and let $\{C_\alpha\}$ be a family of closed sets with the finite intersection property. Then $\{U_\alpha\} := \{X \setminus C_\alpha\}$ is a family, $\mathcal{F}$, of open sets. If $\bigcap_\alpha C_\alpha$ were empty, then $\mathcal{F}$ would form an open covering of X, and therefore the compactness of X would imply that $\mathcal{F}$ had a finite subcover. This would imply that $\{C_\alpha\}$ had a finite subfamily with an empty intersection, contradicting the fact that $\{C_\alpha\}$ has the finite intersection property.

For the converse, let $\{U_\alpha\}$ be an open covering of X and let $\{C_\alpha\} := \{X \setminus U_\alpha\}$. If $\{U_\alpha\}$ had no finite subcover of X, then $\{C_\alpha\}$ would have the finite intersection property, and therefore, $\bigcap_\alpha C_\alpha$ would be nonempty, thus contradicting the assumption that $\{U_\alpha\}$ is a covering of X. □

3.17. REMARK. An equivalent way of stating the previous result is as follows: a topological space X is compact if and only if every family of closed subsets of X whose intersection is empty has a finite subfamily whose intersection is also empty.

3.18. THEOREM. *Suppose $K \subset U$ are respectively compact and open sets in a locally compact Hausdorff space X. Then there is an open set V whose closure is compact such that*

$$K \subset V \subset \overline{V} \subset U.$$

PROOF. Since each point of K is contained in an open set whose closure is compact, and since K can be covered by finitely many such open sets, it follows that the union of these open sets, call it G, is an open set containing K with compact closure. Thus if $U = X$, the proof is compete.

Now consider the case $U \neq X$. Proposition 3.14 states that for each $x \in \widetilde{U}$ there is an open set V_x such that $K \subset V_x$ and $x \notin \overline{V}_x$. Let $\mathcal{F}$ be the family of compact sets defined by

$$\mathcal{F} := \{\widetilde{U} \cap \overline{G} \cap \overline{V}_x : x \in \widetilde{U}\}$$

and observe that the intersection of all sets in $\mathcal{F}$ is empty, for otherwise, we would be faced with the impossibility of some $x_0 \in \widetilde{U} \cap \overline{G}$ that also belongs to $\overline{V}_{x_0}$. Lemma 3.16 (or Remark 3.17) implies there is some finite subfamily of $\mathcal{F}$ that has an empty intersection. That is, there exist points $x_1, x_2, \ldots, x_k \in \widetilde{U}$ such that

$$\widetilde{U} \cap \overline{G} \cap \overline{V}_{x_1} \cap \cdots \cap \overline{V}_{x_k} = \emptyset.$$

The set

$$V = G \cap V_{x_1} \cap \cdots \cap V_{x_k}$$

satisfies the conclusion of our theorem, since

$$K \subset V \subset \overline{V} \subset \overline{G} \cap \overline{V}_{x_1} \cap \cdots \cap \overline{V}_{x_k} \subset U.$$ □

Exercises for Section 3.1

1. In a topological space $(X, \mathcal{T})$, prove that $\overline{A} = \overline{\overline{A}}$ whenever $A \subset X$.
2. Prove (ix) of Theorem 3.6.
3. Prove that A^* is a closed set.
4. Prove Theorem 3.10.

3.2. Bases for a Topology

Often a topology is described in terms of a primitive family of sets, called a basis. We will give a brief description of this concept.

3.19. DEFINITION. A collection $\mathcal{B}$ of open sets in a topological space $(X, \mathcal{T})$ is called a **basis** for the topology $\mathcal{T}$ if $\mathcal{B}$ is a subfamily of $\mathcal{T}$ with the property that for each $U \in \mathcal{T}$ and each $x \in U$, there exists $B \in \mathcal{B}$ such that $x \in B \subset U$. A collection $\mathcal{B}$ of open sets containing a point x is said to be a **basis** at x if for each open set U containing x there is a $B \in \mathcal{B}$ such that $x \in B \subset U$. Observe that a collection $\mathcal{B}$ forms a basis for a topology if and only if it contains a basis at each point $x \in X$. For example, the collection of all sets $B(x,r)$, $r > 0$, $x \in \mathbb{R}^n$, provides a basis for the topology on $\mathbb{R}^n$ as described in (iii) of Example 3.3.

The following is a useful tool for generating a topology on a space X.

3.20. PROPOSITION. *Let X be an arbitrary space. A collection $\mathcal{B}$ of subsets of X is a basis for some topology on X if and only if each $x \in X$ is contained in some $B \in \mathcal{B}$ and if $x \in B_1 \cap B_2$, then there exists $B_3 \in \mathcal{B}$ such that $x \in B_3 \subset B_1 \cap B_2$.*

PROOF. It is easy to verify that the conditions specified in the proposition are necessary. To show that they are sufficient, let $\mathcal{T}$ be the collection of sets U with the property that for each $x \in U$, there exists $B \in \mathcal{B}$ such that $x \in B \subset U$. It is easy to verify that $\mathcal{T}$ is closed under arbitrary unions. To show that it is closed under finite intersections, it is sufficient to consider the case of two sets. Thus, suppose $x \in U_1 \cap U_2$, where U_1 and U_2 are elements of $\mathcal{T}$. There exist $B_1, B_2 \in \mathcal{B}$ such that $x \in B_1 \subset U_1$ and $x \in B_2 \subset U_2$. We are given that there is $B_3 \in \mathcal{B}$ such that $x \in B_3 \subset B_1 \cap B_2$, thus showing that $U_1 \cap U_2 \in \mathcal{T}$. □

3.21. DEFINITION. A topological space $(X, \mathcal{T})$ is said to satisfy the **first axiom of countability** if each point $x \in X$ has a countable basis $\mathcal{B}_x$ at x. It is said to satisfy the **second axiom of countability** if the space $(X, \mathcal{T})$ has a countable basis $\mathcal{B}$.

The second axiom of countability obviously implies the first axiom of countability. The usual topology on $\mathbb{R}^n$, for example, satisfies the second axiom of countability.

3.22. DEFINITION. A family S of subsets of a topological space $(X, \mathcal{T})$ is called a **subbasis** for the topology $\mathcal{T}$ if the family consisting of all finite intersections of members of S forms a basis for the topology $\mathcal{T}$.

In view of Proposition 3.20, every nonempty family of subsets of X is a subbasis for some topology on X. This leads to the concept of the **product topology**.

3.23. DEFINITION. Given an index set A, consider the Cartesian product $\prod_{\alpha \in A} X_\alpha$, where each $(X_\alpha, \mathcal{T}_\alpha)$ is a topological space. For each $\beta \in A$ there is a natural projection

$$P_\beta \colon \prod_{\alpha \in A} X_\alpha \to X_\beta$$

defined by $P_\beta(x) = x_\beta$, where x_β is the βth coordinate of x (see (1.13) and its following remarks). Consider the collection S of subsets of $\prod_{\alpha \in A} X_\alpha$ given by

$$P_\alpha^{-1}(V_\alpha),$$

where $V_\alpha \in \mathcal{T}_\alpha$ and $\alpha \in A$. The topology formed by the subbasis S is called the **product topology** on $\prod_{\alpha \in A} X_\alpha$. In this topology, the projection maps P_β are continuous.

It is easily seen that a function f from a topological space $(Y, \mathcal{T})$ into a product space $\prod_{\alpha \in A} X_\alpha$ is continuous if and only if $(P_\alpha \circ f)$ is continuous for each $\alpha \in A$. Moreover, a sequence $\{x_i\}_{i=1}^\infty$ in a product space $\prod_{\alpha \in A} X_\alpha$ converges to a point x_0 of the product space if and only if the sequence $\{P_\alpha(x_i)\}_{i=1}^\infty$ converges to $P_\alpha(x_0)$ for each $\alpha \in A$. See Exercises 4 and 5 at the end of this section.

Exercises for Section 3.2

1. Prove that the product topology on $\mathbb{R}^n$ agrees with the Euclidean topology on $\mathbb{R}^n$.
2. Suppose that X_i, $i = 1, 2$, satisfy the second axiom of countability. Prove that the product space $X_1 \times X_2$ also satisfies the second axiom of countability.
3. Let $(X, \mathcal{T})$ be a topological space and let $f\colon X \to \mathbb{R}$ and $g\colon X \to \mathbb{R}$ be continuous functions. Define $F\colon X \to \mathbb{R} \times \mathbb{R}$ by
$$F(x) = (f(x), g(x)), \quad x \in X.$$
Prove that F is continuous.
4. Show that a function f from a topological space $(X, \mathcal{T})$ into a product space $\prod_{\alpha \in A} X_\alpha$ is continuous if and only if $(P_\alpha \circ f)$ is continuous for each $\alpha \in A$.
5. Prove that a sequence $\{x_i\}_{i=1}^\infty$ in a product space $\prod_{\alpha \in A} X_\alpha$ converges to a point x_0 of the product space if and only if the sequence $\{P_\alpha(x_i)\}_{i=1}^\infty$ converges to $P_\alpha(x_0)$ for each $\alpha \in A$.

3.3. Metric Spaces

Metric spaces are used extensively throughout analysis. The main purpose of this section is to introduce basic definitions.

We already have mentioned two structures placed on sets that deserve the designation "space," namely that of a vector space and that of a topological space. We now come to our third structure, that of a metric space.

3.24. Definition. A metric space is an arbitrary set X endowed with a **metric** $\rho\colon X \times X \to [0, \infty)$ that satisfies the following properties for all x, y, and z in X:

(i) $\rho(x, y) = 0$ if and only if $x = y$,
(ii) $\rho(x, y) = \rho(y, x)$,

(iii) $\rho(x,y) \le \rho(x,z) + \rho(z,y)$.

We will write (X,ρ) to denote the metric space X endowed with a metric ρ. Often the metric ρ is called the distance function, and a reasonable name for property (iii) is the **triangle inequality**. If $Y \subset X$, then the metric space $(Y, \rho \llcorner (Y \times Y))$ is called the **subspace induced** by (X,ρ).

The following are easily seen to be metric spaces.

3.25. EXAMPLE.

(i) Let $X = \mathbb{R}^n$ and with $x = (x_1, \ldots, x_n)$, $y = (y_1, \ldots, y_n) \in \mathbb{R}^n$, define

$$\rho(x,y) = \left(\sum_{i=1}^{n} |x_i - y_i|^2 \right)^{1/2} .$$

(ii) Let $X = \mathbb{R}^n$ and with $x = (x_1, \ldots, x_n)$, $y = (y_1, \ldots, y_n) \in \mathbb{R}^n$, define

$$\rho(x,y) = \max\{|x_i - y_i| : i = 1, 2, \ldots, n\}.$$

(iii) The **discrete metric** on an arbitrary set X is defined as follows: for $x, y \in X$,

$$\rho(x,y) = \begin{cases} 1 & \text{if } x \neq y, \\ 0 & \text{if } x = y\,. \end{cases}$$

(iv) Let X denote the space of all continuous functions defined on $[0,1]$ and for $f, g \in C(X)$, let

$$\rho(f,g) = \int_0^1 |f(t) - g(t)| \; dt.$$

(v) Let X denote the space of all continuous functions defined on $[0,1]$ and for $f, g \in C(X)$, let

$$\rho(f,g) = \max\{|f(x) - g(x)| : x \in [0,1]\}.$$

3.26. DEFINITION. If X is a metric space with metric ρ, the **open ball** centered at $x \in X$ with radius $r > 0$ is defined as

$$B(x,r) = X \cap \{y : \rho(x,y) < r\}.$$

The **closed ball** is defined as

$$\overline{B}(x,r) := X \cap \{y : \rho(x,y) \le r\}.$$

In view of the triangle inequality, the family $S = \{B(x,r) : x \in X, r > 0\}$ forms a basis for a topology $\mathcal{T}$ on X, called the **topology induced** by ρ. The two metrics in $\mathbb{R}^n$ defined in Examples 3.25, (i) and (ii) induce the same topology on $\mathbb{R}^n$. Two metrics on a set X are said to be **topologically equivalent** if they induce the same topology on X.

3.27. DEFINITION. Using the notion of convergence given in Definition 3.7, p. 35, the reader can easily verify that the convergence of a sequence $\{x_i\}_{i=1}^{\infty}$ in a metric space (X, ρ) becomes the following:

$$\lim_{i\to\infty} x_i = x_0$$

if and only if for each positive number ε there is a positive integer N such that

$$\rho(x_i, x_0) < \varepsilon \quad \text{whenever} \quad i \geq N.$$

We often write $x_i \to x_0$ for $\lim_{i\to\infty} x_i = x_0$.

The notion of a **fundamental sequence**, or **Cauchy sequence**, is not a topological one and requires a separate definition:

3.28. DEFINITION. A sequence $\{x_i\}_{i=1}^{\infty}$ is called **Cauchy** if for every $\varepsilon > 0$, there exists a positive integer N such that $\rho(x_i, x_j) < \varepsilon$ whenever $i, j \geq N$. The notation for this is

$$\lim_{i,j\to\infty} \rho(x_i, x_j) = 0.$$

Recall the definition of continuity given in Definition 3.9. In a metric space, it is convenient to have the following characterization, whose proof is left as an exercise.

3.29. THEOREM. *If (X, ρ) and (Y, σ) are metric spaces, then a mapping $f \colon X \to Y$ is continuous at $x \in X$ if for each $\varepsilon > 0$, there exists $\delta > 0$ such that $\sigma[f(x), f(y)] < \varepsilon$ whenever $\rho(x, y) < \delta$.*

3.30. DEFINITION. If X and Y are topological spaces and if $f \colon X \to Y$ is a bijection with the property that both f and f^{-1} are continuous, then f is called a **homeomorphism**, and the spaces X and Y are said to be **homeomorphic**. A substantial part of topology is devoted to the investigation of properties that remain unchanged under the action of a homeomorphism. For example, in view of Exercise 12 at the end of this section, it follows that if $U \subset X$ is open, then so is $f(U)$ whenever $f \colon X \to Y$ is a homeomorphism; that is, the property of being open is a **topological invariant.** Consequently, so is closedness. But of course, not all properties are topological invariants. For example, the distance between two points might be changed under a homeomorphism. A mapping that preserves distances, that is, one for which

$$\sigma[f(x), f(y)] = \rho(x, y)$$

for all $x, y \in X$ is called an **isometry**. In particular, it is a homeomorphism. The spaces X and Y are called **isometric** if there exists a surjection $f \colon X \to Y$ that is an isometry. In the context of metric space topology, isometric spaces can be regarded as identical.

It is easy to verify that a convergent sequence in a metric space is Cauchy, but the converse need not be true. For example, the metric space $\mathbb{Q}$, consisting of the rational numbers endowed with the usual metric on $\mathbb{R}$, possesses

Cauchy sequences that do not converge to elements in $\mathbb{Q}$. If a metric space has the property that every Cauchy sequence converges (to an element of the space), the space is said to be **complete**. Thus, the metric space of rational numbers is not complete, whereas the real numbers are complete. However, we can apply the technique that was employed in the construction of the real numbers (see Section 2.1, p. 11) to complete an arbitrary metric space. A precise statement of this is incorporated in the following theorem, whose proof is left as Exercise 2, Section 3.4.

3.31. THEOREM. *If (X, ρ) is a metric space, there exists a complete metric space (X^*, ρ^*) in which X is isometrically embedded as a dense subset.*

In the statement, the notion of a dense set is used. This notion is a topological one. In a topological space $(X, \mathcal{T})$, a subset A of X is said to be a **dense** subset of X if $X = \overline{A}$.

Exercises for Section 3.3

1. In a metric space, prove that $B(x, \rho)$ is an open set and that $\overline{B}(x, \rho)$ is closed. Is it true that $\overline{B}(x, \rho) = \overline{B(x, \rho)}$?
2. Suppose X is a complete metric space. Show that if $F_1 \supset F_2 \supset \ldots$ are nonempty closed subsets of X with diameter $F_i \to 0$, then there exists $x \in X$ such that
$$\bigcap_{i=1}^{\infty} F_i = \{x\}.$$
3. Suppose (X, ρ) and (Y, σ) are metric spaces with X compact and Y complete. Let $C(X, Y)$ denote the space of all continuous mappings $f\colon X \to Y$. Define a metric on $C(X, Y)$ by
$$d(f, g) = \sup\{\sigma(f(x), g(x)) : x \in X\}.$$
Prove that $C(X, Y)$ is a complete metric space.
4. Let (X_1, ρ_1) and (X_2, ρ_2) be metric spaces and define metrics on $X_1 \times X_2$ as follows: For $\boldsymbol{x} = (x_1, x_2)$, $\boldsymbol{y} := (y_1, y_2) \in X_1 \times X_2$, let
   (a) $d_1(\boldsymbol{x}, \boldsymbol{y}) := \rho_1(x_1, y_1) + \rho_2(x_2, y_2)$,
   (b) $d_2(\boldsymbol{x}, \boldsymbol{y}) := \sqrt{(\rho_1(x_1, y_1))^2 + (\rho_2(x_2, y_2))^2}$.
   (i) Prove that d_1 and d_2 define identical topologies.
   (ii) Prove that $(X_1 \times X_2, d_1)$ is complete if and only if X_1 and X_2 are complete.
   (iii) Prove that $(X_1 \times X_2, d_1)$ is compact if and only if X_1 and X_2 are compact.
5. Suppose A is a subset of a metric space X. Prove that a point $x_0 \notin A$ is a limit point of A if and only if there is a sequence $\{x_i\}$ in A such that $x_i \to x_0$.

6. Prove that a closed subset of a complete metric space is a complete metric space.
7. A mapping $f\colon X \to X$ with the property that there exists a number $0 < K < 1$ such that $\rho(f(x), f(y)) < K\rho(x, y)$ for all $x \neq y$ is called a **contraction**. Prove that a contraction on a complete metric space has a unique fixed point.
8. Suppose (X, ρ) is metric space and consider a mapping $f\colon X \to X$ from X into itself. A point $x_0 \in X$ is called a **fixed point** for f if $f(x_0) = x_0$. Prove that if X is compact and f has the property that $\rho(f(x), f(y)) < \rho(x, y)$ for all $x \neq y$, then f has a unique fixed point.
9. As on p. 276, a mapping $f\colon X \to X$ with the property that $\rho(f(x), f(y)) = \rho(x, y)$ for all $x, y \in X$ is called an **isometry**. If X is compact, prove that an isometry is a surjection. Is compactness necessary?
10. Show that a metric space X is compact if and only if every continuous real-valued function on X attains a maximum value.
11. If X and Y are topological spaces, prove that $f\colon X \to Y$ is continuous if and only if $f^{-1}(U)$ is open whenever $U \subset Y$ is open.
12. Suppose $f\colon X \to Y$ is surjective and a homeomorphism. Prove that if $U \subset X$ is open, then so is $f(U)$.
13. If X and Y are topological spaces, show that if $f\colon X \to Y$ is continuous, then $f(x_i) \to f(x_0)$ whenever $\{x_i\}$ is a sequence that converges to x_0. Show that the converse is true if X and Y are metric spaces.
14. Prove that a subset C of a metric space X is closed if and only if every convergent sequence $\{x_i\}$ in C converges to a point in C.
15. Prove that $C[0, 1]$ is not a complete space when endowed with the metric given in (iv) of Example 3.25, p. 41.
16. Prove that in a topological space $(X, \mathcal{T})$, if A is dense in B and B is dense in C, then A is dense in C.
17. A metric space is said to be **separable** if it has a countable dense subset.
    (i) Show that $\mathbb{R}^n$ with its usual topology is separable.
    (ii) Prove that a metric space is separable if and only if it satisfies the second axiom of countability.
    (iii) Prove that a subspace of a separable metric space is separable.
    (iv) Prove that if a metric space X is separable, then card $X \leq c$.
18. Let (X, ϱ) be a metric space, $Y \subset X$, and let $(Y, \varrho \llcorner (Y \times Y))$ be the induced subspace. Prove that if $E \subset Y$, then the closure of E in the subspace Y is the same as the closure of E in the space X intersected with Y.
19. Prove that the discrete metric on X induces the discrete topology on X.

3.4. Meager Sets in Topology

Throughout this book, we will encounter several ways of describing the "size" of a set. In Chapter 2, the size of a set was described in terms of its cardinality. Later, we will discuss other methods. The notion of a nowhere dense set and its related concept, that of a set being of the first category, are ways of saying that a set is "meager" in the topological sense. In this section we shall prove one of the main results involving these concepts, the Baire category theorem, which asserts that a complete metric space is not meager.

Recall Definition 3.24, in which a subset S of a metric space (X, ρ) is endowed with the induced topology. The metric placed on S is obtained by restricting the metric ρ to $S \times S$. Thus, the distance between any two points $x, y \in S$ is defined as $\rho(x, y)$, which is the distance between x, y as points of X.

As a result of the definition, a subset $U \subset S$ is open in S if for each $x \in U$, there exists $r > 0$ such that if $y \in S$ and $\rho(x, y) < r$, then $y \in U$. In other words, $B(x, r) \cap S \subset U$, where $B(x, r)$ is taken as the ball in X. Thus, it is easy to see that U is open in S if and only if there exists an open set V in X such that $U = V \cap S$. Consequently, a set $F \subset S$ is closed relative to S if and only if $F = C \cap S$ for some closed set C in S. Moreover, the closure of a set E relative to S is $\overline{E} \cap S$, where $\overline{E}$ denotes the closure of E in X. This is true because if a point x is in the closure of E in X, then it is a point in the closure of E in S if it belongs to S.

3.32. DEFINITIONS. A subset E of a metric space X is said to be **dense in an open set** U if $\overline{E} \supset U$. Also, a set E is defined to be **nowhere dense** if it is not dense in any open subset U of X. Alternatively, we could say that E is nowhere dense if $\overline{E}$ does not contain any open set. For example, the set of integers is a nowhere dense set in $\mathbb{R}$, whereas the set $Q \cap [0, 1]$ is not nowhere dense in $\mathbb{R}$. A set E is said to be of **first category** in X if it is the union of a countable collection of nowhere dense sets. A set that is not of the first category is said to be of the **second category** in X.

We now proceed to investigate a fundamental result related to these concepts.

3.33. THEOREM (Baire category theorem). *A complete metric space X is not the union of a countable collection of nowhere dense sets. That is, a complete metric space is of the second category.*

Before going on, it is important to examine the statement of the theorem in various contexts. For example, let X be the set of integers endowed with the metric induced from $\mathbb{R}$. Thus, X is a complete metric space, and therefore, by the Baire category theorem, it is of the second category. At first, this may seem counterintuitive, since X is the union of a countable collection of points. But remember that a point in this space is an open set, and therefore is not nowhere dense. However, if X is viewed as a subset of $\mathbb{R}$ and not as a space in itself, then indeed, X is the union of a countable number of nowhere dense sets.

PROOF. Assume, to obtain a contradiction, that X is of the first category. Then there exists a countable collection of nowhere dense sets $\{E_i\}$ such that

$$X = \bigcup_{i=1}^{\infty} E_i.$$

Let $B(x_1, r_1)$ be an open ball with radius $r_1 < 1$. Since E_1 is not dense in any open set, it follows that $B(x_1, r_1) \setminus \overline{E_1} \neq \emptyset$. This is a nonempty open set, and therefore there exists a ball $B(x_2, r_2) \subset B(x_1, r_1) \setminus \overline{E_1}$ with $r_2 < \frac{1}{2} r_1$. In fact, by also choosing r_2 smaller than $r_1 - \rho(x_1, x_2)$, we may assume that $\overline{B}(x_2, r_2) \subset B(x_1, r_1) \setminus \overline{E_1}$. Similarly, since E_2 is not dense in any open set, we have that $B(x_2, r_2) \setminus \overline{E_2}$ is a nonempty open set. As before, we can find a closed ball with center x_3 and radius $r_3 < \frac{1}{2} r_2 < \frac{1}{2^2} r_1$:

$$\begin{aligned}\overline{B}(x_3, r_3) &\subset B(x_2, r_2) \setminus \overline{E_2} \\ &\subset \Big(B(x_1, r_1) \setminus \overline{E_1} \Big) \setminus \overline{E_2} \\ &= B(x_1, r_1) \setminus \bigcup_{j=1}^{2} \overline{E_j}.\end{aligned}$$

Proceeding inductively, we obtain a nested sequence $\overline{B}(x_1, r_1) \supset B(x_1, r_1) \supset \overline{B}(x_2, r_2) \supset B(x_2, r_2) \dots$ with $r_i < \frac{1}{2^i} r_1 \to 0$ such that

$$\overline{B}(x_{i+1}, r_{i+1}) \subset B(x_1, r_1) \setminus \bigcup_{j=1}^{i} \overline{E_j} \tag{3.1}$$

for each i. Now, for $i, j > N$, we have $x_i, x_j \in B(x_N, r_N)$ and therefore $\rho(x_i, x_j) \leq 2r_N$. Thus, the sequence $\{x_i\}$ is Cauchy in X. Since X is assumed to be complete, it follows that $x_i \to x$ for some $x \in X$. For each positive integer N, $x_i \in B(x_N, r_N)$ for $i \geq N$. Hence, $x \in \overline{B}(x_N, r_N)$ for each positive integer N. For each positive integer i it follows from (3.1) that

$$x \in \overline{B}(x_{i+1}, r_{i+1}) \subset B(x_1, r_1) \setminus \bigcup_{j=1}^{i} \overline{E_j}.$$

In particular, for each $i \in \mathbb{N}$,

$$x \notin \bigcup_{j=1}^{i} \overline{E_j},$$

and therefore

$$x \notin \bigcup_{j=1}^{\infty} \overline{E_j} = X,$$

a contradiction. □

3.34. DEFINITION. A function $f: X \to Y$, where (X, ρ) and (Y, σ) are metric spaces, is said to **bounded** if there exists $0 < M < \infty$ such that $\sigma(f(x), f(y)) \leq M$ for all $x, y \in X$. A family $\mathcal{F}$ of functions $f: X \to Y$ is called **uniformly bounded** if $\sigma(f(x), f(y)) \leq M$ for all $x, y \in X$ and for all $f \in \mathcal{F}$.

An immediate consequence of the Baire category theorem is the following result, which is known as the **uniform boundedness principle.** We will encounter this result again in the framework of functional analysis, Theorem 8.21. It states that if the upper envelope of a family of continuous functions on a complete metric space is finite everywhere, then the upper envelope is bounded above by some constant on some nonempty open subset. In other words, the family is uniformly bounded on some open set. Of course, there is no estimate of how large the open set is, but in some applications just the knowledge that such an open set exists, no matter how small, is of great importance.

3.35. THEOREM. *Let $\mathcal{F}$ be a family of real-valued continuous functions defined on a complete metric space X and suppose*

$$f^*(x) := \sup_{f \in \mathcal{F}} |f(x)| < \infty \tag{3.2}$$

for each $x \in X$. That is, for each $x \in X$, there is a constant M_x such that

$$f(x) \leq M_x \quad \text{for all } f \in \mathcal{F}.$$

Then there exist a nonempty open set $U \subset X$ and a constant M such that $|f(x)| \leq M$ for all $x \in U$ and all $f \in \mathcal{F}$.

3.36. REMARK. Condition (3.2) states that the family $\mathcal{F}$ is bounded at each point $x \in X$; that is, the family is pointwise bounded by M_x. In applications, a difficulty arises from the possibility that $\sup_{x \in X} M_x = \infty$. The main thrust of the theorem is that there exist $M > 0$ and an open set U such that $\sup_{x \in U} M_x \leq M$.

PROOF. For each positive integer i, let

$$E_{i,f} = \{x : |f(x)| \leq i\}, \; E_i = \bigcap_{f \in \mathcal{F}} E_{i,f}.$$

Note that $E_{i,f}$ is closed, and therefore so is E_i, since f is continuous. From the hypothesis, it follows that

$$X = \bigcup_{i=1}^{\infty} E_i.$$

Since X is a complete metric space, the Baire category theorem implies that there is some set, say E_M, that is not nowhere dense. Because E_M is closed, it must contain an open set U. Now for each $x \in U$, we have $|f(x)| \leq M$ for all $f \in \mathcal{F}$, which is the desired conclusion. □

3.37. EXAMPLE. Here is a simple example that illustrates this result. Define a sequence of functions $f_k\colon [0,1] \to \mathbb{R}$ by

$$f_k(x) = \begin{cases} k^2x, & 0 \le x \le 1/k, \\ -k^2x + 2k, & 1/k \le x \le 2/k, \\ 0, & 2/k \le x \le 1. \end{cases}$$

Thus, $f_k(x) \le k$ on $[0,1]$ and $f^*(x) \le k$ on $[1/k, 1]$, and so $f^*(x) < \infty$ for all $0 \le x \le 1$. The sequence $\{f_k\}$ is not uniformly bounded on $[0,1]$, but it is uniformly bounded on some open set $U \subset [0,1]$. Indeed, in this example, the open set U can be taken as any interval (a,b) where $0 < a < b < 1$, because the sequence $\{f_k\}$is bounded by $1/k$ on $(2/k, 1)$.

Exercises for Section 3.4

1. Prove that a set E in a metric space is nowhere dense if and only if for each open set U, there is a nonempty open set $V \subset U$ such that $V \cap E = \emptyset$.
2. If (X, ρ) is a metric space, prove that there exists a complete metric space (X^*, ρ^*) in which X is isometrically embedded as a dense subset.
3. Prove that the boundary of an open set (or closed set) is nowhere dense in a topological space.

3.5. Compactness in Metric Spaces

In topology there are various notions related to compactness including sequential compactness and the Bolzano–Weierstrass property. The main objective of this section is to show that these concepts are equivalent in a metric space.

The concept of completeness in a metric space is very useful, but it is limited to only those sequences that are Cauchy. A stronger notion called sequential compactness allows consideration of sequences that are not Cauchy. This notion is more general in the sense that it is topological, whereas completeness is meaningful only in the setting of a metric space.

There is an abundant supply of sets that are not compact. For example, the set $A\colon = (0,1]$ in $\mathbb{R}$ is not compact since the collection of open intervals of the form $(1/i, 2]$, $i = 1, 2, \ldots$, provides an open cover of A that admits no finite subcover. On the other hand, while it is true that $[0,1]$ is compact, the proof is not obvious. The reason for this is that the definition of compactness usually is not easy to employ directly. It is best to first determine how it intertwines with other related concepts.

3.38. DEFINITION. If (X, ρ) is a metric space, a set $A \subset X$ is called **totally bounded** if for every $\varepsilon > 0$, A can be covered by finitely many balls of radius ε. A set A is **bounded** if there is a positive number M such that $\rho(x,y) \le M$ for all $x, y \in A$. While it is true that a totally bounded set is bounded (Exercise 3.1), the converse is easily seen to be false; consider (iii) of Example 3.25.

3.39. DEFINITION. A set $A \subset X$ is said to be **sequentially compact** if every sequence in A has a subsequence that converges to a point in A. Also, A is said to have the **Bolzano–Weierstrass property** if every infinite subset of A has a limit point that belongs to A.

3.40. THEOREM. *If A is a subset of a metric space (X, ρ), the following are equivalent:*

(i) *A is compact.*

(ii) *A is sequentially compact.*

(iii) *A is complete and totally bounded.*

(iv) *A has the Bolzano–Weierstrass property.*

PROOF. Beginning with (i), we shall prove that each statement implies its successor.

(i) implies (ii): Let $\{x_i\}$ be a sequence in A; that is, there is a function f defined on the positive integers such that $f(i) = x_i$ for $i = 1, 2, \ldots$. Let E denote the range of f. If E has only finitely many elements, then some member of the sequence must be repeated an infinite number of times, thus showing that the sequence has a convergent subsequence.

Assuming now that E is infinite, we proceed by contradiction and thus suppose that $\{x_i\}$ has no convergent subsequence. If that were the case, then each element of E would be **isolated**. That is, for each $x \in E$ there would exist $r = r_x > 0$ such that $B(x, r_x) \cap E = \{x\}$. This would imply that E has no limit points; thus, Theorem 3.6 (viii) and (ix), p. 35) would imply that E is closed and therefore compact by Proposition 3.12. However, this would lead to a contradiction, since the family $\{B(x, r_x) : x \in E\}$ is an open cover of E that possesses no finite subcover; this is impossible, since E consists of infinitely many points.

(ii) implies (iii): The denial of (iii) leads to two possibilities: Either A is not complete or it is not totally bounded. If A were not complete, there would exist a fundamental sequence $\{x_i\}$ in A that did not converge to any point in A. Hence, no subsequence converges, for otherwise the whole sequence would converge, thus contradicting the sequential compactness of A.

On the other hand, suppose A is not totally bounded; then there exists $\varepsilon > 0$ such that A cannot be covered by finitely many balls of radius ε. In particular, we conclude that A has infinitely many elements. Now inductively choose a sequence $\{x_i\}$ in A as follows: select $x_1 \in A$. Then, since $A \setminus B(x_1, \varepsilon) \neq \emptyset$, we can choose $x_2 \in A \setminus B(x_1, \varepsilon)$. Similarly, $A \setminus [B(x_1, \varepsilon) \cup B(x_2, \varepsilon)] \neq \emptyset$ and $\rho(x_1, x_2) \geq \varepsilon$. Assuming that $x_1, x_2, \ldots, x_{i-1}$ have been chosen so that $\rho(x_k, x_j) \geq \varepsilon$ when $1 \leq k < j \leq i-1$, select

$$x_i \in A \setminus \bigcup_{j=1}^{i-1} B(x_j, \varepsilon),$$

thus producing a sequence $\{x_i\}$ with $\rho(x_i, x_j) \geq \varepsilon$ whenever $i \neq j$. Clearly, $\{x_i\}$ has no convergent subsequence.

(iii) implies (iv): We may as well assume that A has an infinite number of elements. Under the assumptions of (iii), A can be covered by a finite number of balls of radius 1, and therefore, at least one of them, call it B_1, contains infinitely many points of A. Let x_1 be one of these points. By a similar argument, there is a ball B_2 of radius $1/2$ such that $A \cap B_1 \cap B_2$ has infinitely many elements, and thus it contains an element $x_2 \neq x_1$. Continuing in this way, we find a sequence of balls $\{B_i\}$ with B_i of radius $1/i$ and mutually distinct points x_i such that

$$\bigcap_{i=1}^{k} A \bigcap B_i \tag{3.3}$$

is infinite for each $k = 1, 2, \ldots$ and therefore contains a point x_k distinct from $\{x_1, x_2, \ldots, x_{k-1}\}$. Observe that $0 < \rho(x_k, x_l) < 2/k$ whenever $l \geq k$, thus implying that $\{x_k\}$ is a Cauchy sequence, which, by assumption, converges to some $x_0 \in A$. It is easy to verify that x_0 is a limit point of A.

(iv) implies (i): Let $\{U_\alpha\}$ be an arbitrary open cover of A. First, we claim that there exist $\lambda > 0$ and a countable number of balls, call them $B_1, B_2, \ldots$, such that each has radius λ, A is contained in their union, and each B_k is contained in some U_α. To establish our claim, suppose that for each positive integer i, there is a ball, B_i, of radius $1/i$ such that

$$\begin{gathered} B_i \cap A \neq \emptyset, \\ B_i \text{ is not contained in any } U_\alpha. \end{gathered} \tag{3.4}$$

For each positive integer i, select $x_i \in B_i \cap A$. Since A satisfies the Bolzano–Weierstrass property, the sequence $\{x_i\}$ possesses a limit point, and therefore it has a subsequence $\{x_{i_j}\}$ that converges to some $x \in A$. Now $x \in U_\alpha$ for some α. Since U_α is open, there exists $\varepsilon > 0$ such that $B(x, \varepsilon) \subset U_\alpha$. If i_j is chosen so large that $\rho(x_{i_j}, x) < \frac{\varepsilon}{2}$ and $\frac{1}{i_j} < \frac{\varepsilon}{4}$, then for $y \in B_{i_j}$ we have

$$\rho(y, x) \leq \rho(y, x_{i_j}) + \rho(x_{i_j}, x) < 2\frac{\varepsilon}{4} + \frac{\varepsilon}{2} = \varepsilon,$$

which shows that $B_{i_j} \subset B(x, \varepsilon) \subset U_\alpha$, contradicting (3.4). Thus, our claim is established.

In view of our claim, A can be covered by a family $\mathcal{F}$ of balls of radius λ such that each ball belongs to some U_α. A finite number of these balls also covers A, for if not, we could proceed exactly as in the proof above of (ii) implies (iii) to construct a sequence of points $\{x_i\}$ in A with $\rho(x_i, x_j) \geq \lambda$ whenever $i \neq j$. This leads to a contradiction, since the Bolzano–Wierstrass condition on A implies that $\{x_i\}$ possesses a limit point $x_0 \in A$. Thus, a finite number of balls covers A, say $B_1, \ldots B_k$. Each B_i is contained in some U_α, say U_{a_i}, and therefore we have

$$A \subset \bigcup_{i=1}^{k} B_i \subset \bigcup_{i=1}^{k} U_{\alpha_i},$$

which proves that a finite number of the U_α covers A. □

3.41. COROLLARY. *A set $A \subset \mathbb{R}^n$ is compact if and only if A is closed and bounded.*

PROOF. Clearly, A is bounded if it is compact. Proposition 3.13 shows that it is also closed.

Conversely, if A is closed, it is complete (see Exercise 6, Section 3.3); it thus suffices to show that every bounded subset of $\mathbb{R}^n$ is totally bounded. (Recall that bounded sets in an arbitrary metric space are not generally totally bounded; see Exercise 1, Section 3.5.) Since every bounded set is contained in some cube

$$Q = [-a, a]^n = \{x \in \mathbb{R}^n : \max(|x_1|, \ldots, |x_n| \leq a)\},$$

it is sufficient to show that Q is totally bounded. For this purpose, choose $\varepsilon > 0$ and let k be an integer such that $k > \sqrt{n}a/\varepsilon$. Then Q can be expressed as the union of k^n congruent subcubes by dividing the interval $[-a, a]$ into k equal pieces. The side length of each of these subcubes is $2a/k$, and hence the diameter of each cube is $2\sqrt{n}a/k < 2\varepsilon$. Therefore, each cube is contained in a ball of radius ε about its center. □

Exercises for Section 3.5

1. Prove that a totally bounded set in a metric space is bounded.
2. Prove that a subset E of a metric space is totally bounded if and only if $\overline{E}$ is totally bounded.
3. Prove that a totally bounded metric space is separable.
4. The proof that (iv) implies (i) in Theorem 3.40 utilizes a result that needs to be emphasized. Prove: For each open cover $\mathcal{F}$ of a compact set in a metric space, there is a number $\eta > 0$ with the property that if x, y are any two points in X with $\rho(x, y) < \eta$, then there is an open set $V \in \mathcal{F}$ such that both x, y belong to V. The number η is called a **Lebesgue number** for the covering $\mathcal{F}$.
5. Let $\varrho \colon \mathbb{R} \times \mathbb{R} \to \mathbb{R}$ be defined by

$$\varrho(x, y) = \min\{|x - y|, 1\} \quad \text{for} \quad (x, y) \in \mathbb{R} \times \mathbb{R}.$$

Prove that ϱ is a metric on $\mathbb{R}$. Show that closed, bounded subsets of $(\mathbb{R}, \varrho)$ need not be compact. Hint: This metric is topologically equivalent to the Euclidean metric.

3.6. Compactness of Product Spaces

In this section we prove Tychonoff's theorem, which states that the product of an arbitrary number of compact topological spaces is compact. This is one of the most important theorems in general topology, in particular for its applications to functional analysis.

Let $\{X_\alpha : \alpha \in A\}$ be a family of topological spaces and set $X = \prod_{\alpha\in A} X_\alpha$. Let $P_\alpha : X \to X_\alpha$ denote the projection of X onto X_α for each α. Recall that the family of subsets of X of the form $P_\alpha^{-1}(U)$, where U is an open subset of X_α and $\alpha \in A$, is a subbasis for the product topology on X.

The proof of Tychonoff's theorem will utilize the finite intersection property introduced in Definition 3.15 and Lemma 3.16.

In the following proof, we use the Hausdorff maximal principle; see p. 7.

3.42. LEMMA. *Let $\mathcal{A}$ be a family of subsets of a set Y having the finite intersection property and suppose $\mathcal{A}$ is maximal with respect to the finite intersection property, i.e., no family of subsets of Y that properly contains $\mathcal{A}$ has the finite intersection property. Then*

(i) *$\mathcal{A}$ contains all finite intersections of members of $\mathcal{A}$.*

(ii) *If $S \subset Y$ and $S \cap A \neq \emptyset$ for each $A \in \mathcal{A}$, then $S \in \mathcal{A}$.*

PROOF. To prove (i) let $\mathcal{B}$ denote the family of all finite intersections of members of $\mathcal{A}$. Then $\mathcal{A} \subset \mathcal{B}$, and $\mathcal{B}$ has the finite intersection property. Thus by the maximality of $\mathcal{A}$, it is clear that $\mathcal{A} = \mathcal{B}$.

To prove (ii), suppose $S \cap A \neq \emptyset$ for each $A \in \mathcal{A}$. Set $\mathcal{C} = \mathcal{A} \cup \{S\}$. Then, since $\mathcal{C}$ has the finite intersection property, the maximality of $\mathcal{A}$ implies that $\mathcal{C} = \mathcal{A}$. □

We can now prove Tychonoff's theorem.

3.43. THEOREM (Tychonoff's product theorem). *If $\{X_\alpha : \alpha \in A\}$ is a family of compact topological spaces and $X = \prod_{\alpha\in A} X_\alpha$ with the product topology, then X is compact.*

PROOF. Suppose $\mathcal{C}$ is a family of closed subsets of X having the finite intersection property and let $\mathcal{E}$ denote the collection of all families of subsets of X such that each family contains $\mathcal{C}$ and has the finite intersection property. Then $\mathcal{E}$ satisfies the conditions of the Hausdorff maximal principle, and hence there is a maximal element $\mathcal{B}$ of $\mathcal{E}$ in the sense that $\mathcal{B}$ is not a subset of any other member of $\mathcal{E}$.

For each α the family $\{P_\alpha(B) : B \in \mathcal{B}\}$ of subsets of X_α has the finite intersection property. Since X_α is compact, there is a point $x_\alpha \in X_\alpha$ such that

$$x_\alpha \in \bigcap_{B\in\mathcal{B}} \overline{P_\alpha(B)}.$$

For $\alpha \in A$, let U_α be an open subset of X_α containing x_α. Then

$$B \bigcap P_\alpha^{-1}(U_\alpha) \neq \emptyset$$

for each $B \in \mathcal{B}$. In view of Lemma 3.42 (ii) we see that $P_\alpha^{-1}(U_\alpha) \in \mathcal{B}$. Thus by Lemma 3.42 (i), every finite intersection of sets of this form is a member of $\mathcal{B}$. It follows that every open subset of X containing x has a nonempty intersection with each member of $\mathcal{B}$. Since $\mathcal{C} \subset \mathcal{B}$ and each member of $\mathcal{C}$ is closed, it follows that $x \in C$ for each $C \in \mathcal{C}$. □

Exercises for Section 3.6

1. The set of all sequences $\{x_i\}_{i=1}^{\infty}$ in $[0,1]$ can be written as $[0,1]^{\mathbb{N}}$. Tychonoff's theorem asserts that $[0,1]^{\mathbb{N}}$ with the product topology is compact. Prove that the function ϱ defined by

$$\varrho(\{x_i\},\{y_i\}) = \sum_{i=1}^{\infty} \frac{1}{2^i} |x_i - y_i| \quad \text{for} \quad \{x_i\},\{y_i\} \in [0,1]^{\mathbb{N}}$$

is a metric on $[0,1]^{\mathbb{N}}$ and that this metric induces the product topology on $[0,1]^{\mathbb{N}}$. Prove that every sequence of sequences in $[0,1]$ has a convergent subsequence in the metric space $([0,1]^{\mathbb{N}}, \varrho)$. This space is sometimes called the **Hilbert cube.**

3.7. The Space of Continuous Functions

In this section we investigate an important metric space, $C(X)$, the space of continuous functions on a metric space X. It is shown that this space is complete. More importantly, necessary and sufficient conditions for the compactness of subsets of $C(X)$ are given.

Recall the discussion of continuity given in Theorems 3.10 and 3.29. Our discussion will be carried out in the context of functions $f: X \to Y$, where (X, ρ) and (Y, σ) are metric spaces. Continuity of f at x_0 requires that points near x_0 be mapped into points near $f(x_0)$. We introduce the concept of "oscillation" to assist in making this idea precise.

3.44. Definition. If $f: X \to Y$ is an arbitrary mapping, then the **oscillation** of f on a ball $B(x_0)$ is defined by

$$\operatorname{osc}[f, B(x_0, r)] = \sup\{\sigma[f(x), f(y)] : x, y \in B(x_0, r)\}.$$

Thus, the oscillation of f on a ball $B(x_0, r)$ is nothing more than the diameter of the set $f(B(x_0, r))$ in Y. The **diameter** of an arbitrary set E is defined as $\sup\{\sigma(x, y) : x, y \in E\}$. It may possibly assume the value $+\infty$. Note that $\operatorname{osc}[f, B(x_0, r)]$ is a nondecreasing function of r for each point x_0.

We leave it to the reader to supply the proof of the following assertion.

3.45. Proposition. *A function $f: X \to Y$ is continuous at $x_0 \in X$ if and only if*

$$\lim_{r \to 0} osc[f, B(x_0, r)] = 0.$$

The concept of oscillation is useful in providing information concerning the set on which an arbitrary function is continuous.
For this we need the following definitions.

3.46. Definition. A subset E of a topological space is called a G_δ set if E can be written as the countable intersection of open sets, and it is an F_σ set if it can be written as the countable union of closed sets.

3.47. THEOREM. *Let $f: X \to Y$ be an arbitrary function. Then the set of points at which f is continuous is a G_δ set.*

PROOF. For each integer i, let

$$G_i = X \cap \{x : \inf_{r>0} \operatorname{osc}[f, B(x,r)] < 1/i\}.$$

From the proposition above, we know that f is continuous at x if and only if $\lim_{r\to 0} \operatorname{osc}[f, B(x,r)] = 0$. Therefore, the set of points at which f is continuous is given by

$$A = \bigcap_{i=1}^{\infty} G_i.$$

To complete the proof we need only show that each G_i is open. For this, observe that if $x \in G_i$, then there exists $r > 0$ such that $\operatorname{osc}[f, B(x,r)] < 1/i$. Now for each $y \in B(x,r)$, there exists $t > 0$ such that $B(y,t) \subset B(x,r)$, and consequently,

$$\operatorname{osc}[f, B(y,t)] \leq \operatorname{osc}[f, B(x,r)] < 1/i.$$

This implies that each point y of $B(x,r)$ is an element of G_i. That is, $B(x,r) \subset G_i$, and since x is an arbitrary point of G_i, it follows that G_i is open. □

3.48. THEOREM. *Let f be an arbitrary function defined on $[0,1]$ and let $E := \{x \in [0,1] : f \text{ is continuous at } x\}$. Then E cannot be the set of rational numbers in $[0,1]$.*

PROOF. It suffices to show that the rationals in $[0,1]$ do not constitute a G_δ set. If this were false, the irrationals in $[0,1]$ would be an F_σ set and thus would be the union of a countable number of closed sets, each having an empty interior. Since the rationals are a countable union of closed sets (singletons, with no interiors), it would follow that $[0,1]$ is also of the first category, contrary to the Baire category theorem. Thus, the rationals cannot be a G_δ set. □

Since continuity is such a fundamental notion, it is useful to know the properties that remain invariant under a continuous transformation. The following result shows that compactness is a continuous invariant.

3.49. THEOREM. *Suppose X and Y are topological spaces and $f: X \to Y$ is a continuous mapping. If $K \subset X$ is a compact set, then $f(K)$ is a compact subset of Y.*

PROOF. Let $\mathcal{F}$ be an open cover of $f(K)$; that is, the elements of $\mathcal{F}$ are open sets whose union contains $f(K)$. The continuity of f implies that each $f^{-1}(U)$ is an open subset of X for each $U \in \mathcal{F}$. Moreover, the collection $\{f^{-1}(U) : U \in \mathcal{F}\}$ provides an open cover of K. Indeed, if $x \in K$, then $f(x) \in f(K)$, and therefore $f(x) \in U$ for some $U \in \mathcal{F}$. This implies that $x \in f^{-1}(U)$. Since K is compact, $\mathcal{F}$ possesses a finite subcover for K, say $\{f^{-1}(U_1), \ldots, f^{-1}(U_k)\}$. From this it easily follows that the corresponding collection $\{U_1, \ldots, U_k\}$ is an open cover of $f(K)$, thus proving that $f(K)$ is compact. □

3.50. COROLLARY. *Assume that X is a compact topological space and suppose $f\colon X \to \mathbb{R}$ is continuous. Then f attains its maximum and minimum on X; that is, there are points $x_1, x_2 \in X$ such that $f(x_1) \le f(x) \le f(x_2)$ for all $x \in X$.*

PROOF. From the preceding result and Corollary 3.41, it follows that $f(X)$ is a closed and bounded subset of $\mathbb{R}$. Consequently, by Theorem 2.22, $f(X)$ has a least upper bound, say y_0, that belongs to $f(X)$, since $f(X)$ is closed. Thus there is a point $x_2 \in X$ such that $f(x_2) = y_0$. Then $f(x) \le f(x_2)$ for all $x \in X$. Similarly, there is a point x_1 at which f attains a minimum. □

We proceed to examine yet another implication of continuous mappings defined on compact spaces. The next definition sets the stage.

3.51. DEFINITION. Suppose X and Y are metric spaces. A mapping $f\colon X \to Y$ is said to be **uniformly continuous on** X if for each $\varepsilon > 0$ there exists $\delta > 0$ such that $\sigma[f(x), f(y)] < \varepsilon$ whenever x and y are points in X with $\rho(x, y) < \delta$. The important distinction between continuity and uniform continuity is that in the latter concept, the number δ depends only on ε and not on ε and x as in continuity. An equivalent formulation of uniform continuity can be stated in terms of oscillation, which was defined in Definition 3.44. For each number $r > 0$, let

$$\omega_f(r) \colon = \sup_{x \in X} \operatorname{osc}[f, B(x, r)].$$

The function ω_f is called the **modulus of continuity of** f. It is not difficult to show that f is uniformly continuous on X if

$$\lim_{r \to 0} \omega_f(r) = 0.$$

3.52. THEOREM. *Let $f\colon X \to Y$ be a continuous mapping. If X is compact, then f is uniformly continuous on X.*

PROOF. Choose $\varepsilon > 0$. Then the collection

$$\mathcal{F} = \{f^{-1}(B(y, \varepsilon)) : y \in Y\}$$

is an open cover of X. Let η denote a Lebesgue number of this open cover (see Exercise 4, Section 3.5). Thus, for every $x \in X$, we have that $B(x, \eta/2)$ is contained in $f^{-1}(B(y, \varepsilon))$ for some $y \in Y$. This implies $\omega_f(\eta/2) \le \varepsilon$. □

3.53. DEFINITION. For (X, ρ) a metric space, let

$$d(f, g) \colon = \sup(|f(x) - g(x)| : x \in X) \tag{3.5}$$

denote the distance between two bounded real-valued functions f and g defined on X. This metric is related to the notion of **uniform convergence**. Indeed, a sequence of bounded functions $\{f_i\}$ defined on X is said to **converge uniformly** to a bounded function f on X if $d(f_i, f) \to 0$ as $i \to \infty$. We denote by

$$C(X)$$

the space of bounded real-valued continuous functions on X.

3.54. THEOREM. *The space $C(X)$ is complete.*

PROOF. Let $\{f_i\}$ be a Cauchy sequence in $C(X)$. Since

$$|f_i(x) - f_j(x)| \leq d(f_i, f_j)$$

for all $x \in X$, it follows that for each $x \in X$, the sequence $\{f_i(x)\}$ is a Cauchy sequence of real numbers. Therefore, $\{f_i(x)\}$ converges to a number that depends on x and is denoted by $f(x)$. In this way, we define a function f on X. In order to complete the proof, we need to show that f is an element of $C(X)$ and that the sequence $\{f_i\}$ converges to f in the metric of (3.5). First, observe that f is a bounded function on X, because for every $\varepsilon > 0$, there exists an integer N such that

$$|f_i(x) - f_j(x)| < \varepsilon$$

whenever $x \in X$ and $i, j \geq N$. Therefore,

$$|f(x)| \leq |f_N(x)| + \varepsilon$$

for all $x \in X$, thus showing that f is bounded, since f_N is.

Next, we show that

$$\lim_{i \to \infty} d(f, f_i) = 0. \tag{3.6}$$

For this, let $\varepsilon > 0$. Since $\{f_i\}$ is a Cauchy sequence in $C(X)$, there exists $N > 0$ such that $d(f_i, f_j) < \varepsilon$ whenever $i, j \geq N$. That is, $|f_i(x) - f_j(x)| < \varepsilon$ for all $i, j \geq N$ and for all $x \in X$. Thus,

$$|f(x) - f_i(x)| = \lim_{j \to \infty} |f_i(x) - f_j(x)| < \varepsilon,$$

for each $x \in X$ and $i > N$. This implies that $d(f, f_i) < \varepsilon$ for $i > N$, which establishes (3.6), as required.

Finally, it will be shown that f is continuous on X. For this, let $x_0 \in X$ and $\varepsilon > 0$ be given. Let f_i be a member of the sequence such that $d(f, f_i) < \varepsilon/3$. Since f_i is continuous at x_0, there is a $\delta > 0$ such that $|f_i(x_0) - f_i(y)| < \varepsilon/3$ when $\rho(x_0, y) < \delta$. Then for all y with $\rho(x_0, y) < \delta$, we have

$$\begin{aligned} |f(x_0) - f(y)| &\leq |f(x_0) - f_i(x_0)| + |f_i(x_0) - f_i(y)| + |f_i(y) - f(y)| \\ &< d(f, f_i) + \frac{\varepsilon}{3} + d(f_i, f) < \varepsilon. \end{aligned}$$

This shows that f is continuous at x_0, and the proof is complete. □

3.55. COROLLARY. *The uniform limit of a sequence of continuous functions is continuous.*

Now that we have shown that $C(X)$ is complete, it is natural to inquire about other topological properties it may possess. We will close this section with an investigation of its compactness properties. We begin by examining the consequences of uniform convergence on a compact space.

3.56. THEOREM. *Let $\{f_i\}$ be a sequence of continuous functions defined on a compact metric space X that converges uniformly to a function f. Then for each $\varepsilon > 0$, there exists $\delta > 0$ such that $\omega_{f_i}(r) < \varepsilon$ for all positive integers i and for $0 < r < \delta$.*

PROOF. We know from Corollary 3.55 that f is continuous, and Theorem 3.52 asserts that f is uniformly continuous, as is each f_i as well. Thus, for each i, we know that

$$\lim_{r\to 0} \omega_{f_i}(r) = 0.$$

That is, for each $\varepsilon > 0$ and for each i, there exists $\delta_i > 0$ such that

$$\omega_{f_i}(r) < \varepsilon \quad \text{for} \quad r < \delta_i. \tag{3.7}$$

However, since f_i converges uniformly to f, we claim that there exists $\delta > 0$ independent of f_i such that (3.7) holds with δ_i replaced by δ. To see this, observe that since f is uniformly continuous, there exists $\delta' > 0$ such that $|f(y) - f(x)| < \varepsilon/3$ whenever $x, y \in X$ and $\rho(x, y) < \delta'$. Furthermore, there exists an integer N such that $|f_i(z) - f(z)| < \varepsilon/3$ for $i \geq N$ and for all $z \in X$. Therefore, by the triangle inequality, for each $i \geq N$, we have

$$\begin{aligned} |f_i(x) - f_i(y)| &\leq |f_i(x) - f(x)| + |f(x) - f(y)| + |f(y) - f_i(y)| \\ &< \frac{\varepsilon}{3} + \frac{\varepsilon}{3} + \frac{\varepsilon}{3} = \varepsilon \end{aligned} \tag{3.8}$$

whenever $x, y \in X$ with $\rho(x, y) < \delta'$. Consequently, if we let

$$\delta = \min\{\delta_1, \dots, \delta_{N-1}, \delta'\},$$

it follows from (3.7) and (3.8) that for each positive integer i,

$$|f_i(x) - f_i(y)| < \varepsilon$$

whenever $\rho(x, y) < \delta$, thus establishing our claim. □

This argument shows not only that the functions f_i uniformly continuous, but that the modulus of continuity of each function tends to 0 with r, uniformly with respect to i. We use this to formulate the following definition.

3.57. DEFINITION. A family, $\mathcal{F}$, of functions defined on X is called **equicontinuous** if for each $\varepsilon > 0$ there exists $\delta > 0$ such that for each $f \in \mathcal{F}$, $|f(x) - f(y)| < \varepsilon$ whenever $\rho(x, y) < \delta$. Alternatively, $\mathcal{F}$ is equicontinuous if for each $f \in \mathcal{F}$, $\omega_f(r) < \varepsilon$ whenever $0 < r < \delta$. Sometimes equicontinuous families are defined pointwise; see Exercise 3.14.

We are now in a position to give a characterization of compact subsets of $C(X)$ when X is a compact metric space.

3.58. THEOREM (Arzelà–Ascoli). *Suppose (X, ρ) is a compact metric space. Then a set $\mathcal{F} \subset C(X)$ is compact if and only if $\mathcal{F}$ is closed, bounded, and equicontinuous.*

PROOF. **Sufficiency:** It suffices to show that $\mathcal{F}$ is sequentially compact. Thus, it suffices to show that an arbitrary sequence $\{f_i\}$ in $\mathcal{F}$ has a convergent subsequence. Since X is compact, it is totally bounded, and therefore separable. Let $D = \{x_1, x_2, \ldots\}$ denote a countable dense subset. The boundedness of $\mathcal{F}$ implies that there is a number M' such that $d(f, g) < M'$ for all $f, g \in \mathcal{F}$. In particular, if we fix an arbitrary element $f_0 \in \mathcal{F}$, then $d(f_0, f_i) < M'$ for all positive integers i. Since $|f_0(x)| < M''$ for some $M'' > 0$ and for all $x \in X$, it follows that $|f_i(x)| < M' + M''$ for all i and for all x.

Our first objective is to construct a sequence of functions, $\{g_i\}$, that is a subsequence of $\{f_i\}$ and that converges at each point of D. As a first step toward this end, observe that $\{f_i(x_1)\}$ is a sequence of real numbers that is contained in the compact interval $[-M, M]$, where $M := M' + M''$. It follows that this sequence of numbers has a convergent subsequence, denoted by $\{f_{1i}(x_1)\}$. Note that the point x_1 determines a subsequence of functions that converges at x_1. For example, the subsequence of $\{f_i\}$ that converges at the point x_1 might be $f_1(x_1), f_3(x_1), f_5(x_1), \ldots$, in which case $f_{11} = f_1, f_{12} = f_3, f_{13} = f_5, \ldots$. Since the subsequence $\{f_{1i}\}$ is a uniformly bounded sequence of functions, we proceed exactly as in the previous step with f_{1i} replacing f_i. Thus, since $\{f_{1i}(x_2)\}$ is a bounded sequence of real numbers, it too has a convergent subsequence, which we denote by $\{f_{2i}(x_2)\}$. Similarly to the first step, we see that f_{2i} is a sequence of functions that is a subsequence of $\{f_{1i}\}$, which, in turn, is a subsequence of f_i. Continuing this process, we see that the sequence $\{f_{2i}(x_3)\}$ also has a convergent subsequence, denoted by $\{f_{3i}(x_3)\}$. We proceed in this way and then set $g_i = f_{ii}$, so that g_i is the ith function occurring in the ith subsequence. We have the following situation:

$$
\begin{array}{cccccccl}
f_{11} & f_{12} & f_{13} & \cdots & f_{1i} & \cdots & & \text{first subsequence} \\
f_{21} & f_{22} & f_{23} & \cdots & f_{2i} & \cdots & & \text{subsequence of previous subsequence} \\
f_{31} & f_{32} & f_{33} & \cdots & f_{3i} & \cdots & & \text{subsequence of previous subsequence} \\
\vdots & \vdots & \vdots & \vdots & \vdots & \vdots & & \\
f_{i1} & f_{i2} & f_{i3} & \cdots & f_{ii} & \cdots & & i^{\text{th}} \text{ subsequence} \\
\vdots & \vdots & \vdots & \vdots & \vdots & \vdots & &
\end{array}
$$

Observe that the sequence of functions $\{g_i\}$ converges at each point of D. Indeed, g_i is an element of the jth row for $i \geq j$. In other words, the tail end of $\{g_i\}$ is a subsequence of $\{f_{ji}\}$ for every $j \in \mathbb{N}$, and so it will converge as $i \to \infty$ at every point for which $\{f_{ji}\}$ converges as $i \to \infty$, i.e., for each point of D.

We now proceed to show that $\{g_i\}$ converges at each point of X and that the convergence is, in fact, uniform on X. For this purpose, choose $\varepsilon > 0$ and let $\delta > 0$ be the number obtained from the definition of equicontinuity. Since X is compact, it is totally bounded, and therefore there is a finite number of balls of radius $\delta/2$, say k of them, whose union covers X: $X = \bigcup_{i=1}^{k} B_i(\delta/2)$. Then selecting any $y_i \in B_i(\delta/2) \cap D$, it follows that

$$X = \bigcup_{i=1}^{k} B(y_i, \delta).$$

Let $D' := \{y_1, y_2, \ldots, y_k\}$ and note that $D' \subset D$. Therefore, each of the k sequences

$$\{g_i(y_1)\}, \{g_i(y_2)\}, \ldots, \{g_i(y_k)\}$$

converges, and so there is an integer $N \in \mathbb{N}$ such that if $i, j \geq N$, then

$$|g_i(y_m) - g_j(y_m)| < \varepsilon \text{ for } m = 1, 2, \ldots, k.$$

For each $x \in X$, there exists $y_m \in D'$ such that $|x - y_m| < \delta$. Thus, by equicontinuity, it follows that

$$|g_i(x) - g_i(y_m)| < \varepsilon$$

for all positive integers i. Therefore, we have

$$\begin{aligned} |g_i(x) - g_j(x)| &\leq |g_i(x) - g_i(y_m)| + |g_i(y_m) - g_j(y_m)| \\ &\quad + |g_j(y_m) - g_j(x)| \\ &< \varepsilon + \varepsilon + \varepsilon = 3\varepsilon, \end{aligned}$$

provided $i, j \geq N$. This shows that

$$d(g_i, g_j) < 3\varepsilon \quad \text{for} \quad i, j \geq N.$$

That is, $\{g_i\}$ is a Cauchy sequence in $\mathcal{F}$. Since $C(X)$ is complete (Theorem 3.54) and $\mathcal{F}$ is closed, it follows that $\{g_i\}$ converges to an element $g \in \mathcal{F}$. Since $\{g_i\}$ is a subsequence of the original sequence $\{f_i\}$, we have shown that $\mathcal{F}$ is sequentially compact, thus establishing the sufficiency argument.

Necessity: Note that $\mathcal{F}$ is closed, since $\mathcal{F}$ is assumed to be compact. Furthermore, the compactness of $\mathcal{F}$ implies that $\mathcal{F}$ is totally bounded and therefore bounded. For the proof that $\mathcal{F}$ is equicontinuous, note that $\mathcal{F}$ being totally bounded implies that for each $\varepsilon > 0$, there exists a finite number of elements in $\mathcal{F}$, say $f_1, \ldots, f_k$, such that every $f \in \mathcal{F}$ is within $\varepsilon/3$ of f_i, for some $i \in \{1, \ldots, k\}$. Consequently, by Exercise 3.5, we have

$$\omega_f(r) \leq \omega_{f_i}(r) + 2d(f, f_i) < \omega_{f_i}(r) + 2\varepsilon/3. \tag{3.9}$$

Since X is compact, each f_i is uniformly continuous on X. Thus, for each i, $i = 1, \ldots, k$, there exists $\delta_i > 0$ such that $\omega_{f_i}(r) < \varepsilon/3$ for $r < \delta_i$. Now let $\delta = \min\{\delta_1, \ldots, \delta_k\}$. By (3.9) it follows that $\omega_f(r) < \varepsilon$ whenever $r < \delta$, which proves that $\mathcal{F}$ is equicontinuous. □

In many applications, it is not of great interest to know whether $\mathcal{F}$ itself is compact, but whether a given sequence in $\mathcal{F}$ has a subsequence that converges uniformly to an element of $C(X)$, and not necessarily to an element of $\mathcal{F}$. In other words, the compactness of the closure of $\mathcal{F}$ is the critical question. It is easy to see that if $\mathcal{F}$ is equicontinuous, then so is $\bar{\mathcal{F}}$. This leads to the following corollary.

3.59. COROLLARY. *Suppose (X, ρ) is a compact metric space and suppose that $\mathcal{F} \subset C(X)$ is bounded and equicontinuous. Then $\bar{\mathcal{F}}$ is compact.*

PROOF. This follows immediately from the previous theorem, since $\bar{\mathcal{F}}$ is both bounded and equicontinuous. □

In particular, this corollary yields the following special result.

3.60. COROLLARY. *Let $\{f_i\}$ be an equicontinuous, uniformly bounded sequence of functions defined on $[0,1]$. Then there is a subsequence that converges uniformly to a continuous function on $[0,1]$.*

We close this section with a result that will be used frequently throughout the sequel.

3.61. THEOREM. *Suppose f is a bounded function on $[a,b]$ that is either nondecreasing or nonincreasing. Then f has at most a countable number of discontinuities.*

PROOF. We will give the proof only for f nondecreasing; the proof for f nonincreasing is essentially the same.

Since f is nondecreasing, it follows that the left- and right-hand limits exist at each point (see Exercise 25, Section 3.7), and the discontinuities of f occur precisely where these limits are not equal. Thus, setting

$$f(x^+) = \lim_{y \to x^+} f(y) \quad \text{and} \quad f(x^-) = \lim_{y \to x^-} f(y),$$

we have that the set D of discontinuities of f in (a,b) is given by

$$D = (a,b) \cap \left(\bigcup_{k=1}^{\infty} \{x : f(x^+) - f(x^-) > \frac{1}{k}\} \right).$$

For each k the set

$$\{x \; : \; f(x^+) - f(x^-) > \frac{1}{k}\}$$

is finite, since f is bounded and thus D is countable. □

Exercises for Section 3.7

1. Prove that the set of rational numbers on the real line is not a G_δ set.
2. Prove that the two definitions of uniform continuity given in Definition 3.51 are equivalent.

3. Assume that (X, ρ) is a metric space with the property that each function $f: X \to \mathbb{R}$ is uniformly continuous.
   (a) Show that X is a complete metric space.
   (b) Give an example of a space X with the above property that is not compact.
   (c) Prove that if X has only a finite number of isolated points, then X is compact. See p. 49 for the definition of isolated point.
4. Prove that a family of functions F is equicontinuous if there exists a nondecreasing real-valued function φ such that
$$\lim_{r \to 0} \varphi(r) = 0$$
and $\omega_f(r) \leq \varphi(r)$ for all $f \in F$.
5. Suppose f, g are two functions defined on a metric space. Prove that
$$\omega_f(r) \leq \omega_g(r) + 2d(f, g).$$
6. Prove that a Lipschitz function is uniformly continuous.
7. Prove: If F is a family of Lipschitz functions from a bounded metric space X into a metric space Y such that M is a Lipschitz constant for each member of F and $\{f(x_0) : f \in F\}$ is a bounded set in Y for some $x_0 \in X$, then F is a uniformly bounded, equicontinuous family.
8. Let (X, ϱ) and (Y, σ) be metric spaces and let $f: X \to Y$ be uniformly continuous. Prove that if X is totally bounded, then $f(X)$ is totally bounded.
9. Let (X, ϱ) and (Y, σ) be metric spaces and let $f: X \to Y$ be an arbitrary function. The **graph** of f is a subset of $X \times Y$ defined by
$$G_f := \{(x, y) : y = f(x)\}.$$
Let d be the metric d_1 on $X \times Y$ as defined in Exercise 4, Section 3.3. If Y is compact, show that f is continuous if and only if G_f is a closed subset of the metric space $(X \times Y, d)$. Can the compactness assumption on Y be dropped?
10. Let Y be a dense subset of a metric space (X, ϱ). Let $f: Y \to Z$ be a uniformly continuous function, where Z is a complete metric space. Show that there is a uniformly continuous function $g: X \to Z$ with the property that $f = g \llcorner Y$. Can the assumption of uniform continuity be relaxed to mere continuity?
11. Exhibit a bounded function that is continuous on $(0, 1)$ but not uniformly continuous.
12. Let $\{f_i\}$ be a sequence of real-valued, uniformly continuous functions on a metric space (X, ρ) with the property that for some $M > 0$, $|f_i(x) - f_j(x)| \leq M$ for all positive integers i, j and all $x \in X$. Suppose also that $d(f_i, f_j) \to 0$ as $i, j \to \infty$. Prove that there is a uniformly continuous function f on X such that $d(f_i, f) \to 0$ as $i \to \infty$.

13. Let $X \xrightarrow{f} Y$, where (X, ρ) and (Y, σ) are metric spaces and where f is continuous. Suppose f has the following property: for each $\varepsilon > 0$ there is a compact set $K_\varepsilon \subset X$ such that $\sigma(f(x), f(y)) < \varepsilon$ for all $x, y \in X \setminus K_e$. Prove that f is uniformly continuous on X.
14. A family $\mathcal{F}$ of functions defined on a metric space X is called **equicontinuous at** $x \in X$ if for every $\varepsilon > 0$ there exists $\delta > 0$ such that $|f(x) - f(y)| < \varepsilon$ for all y with $|x - y| < \delta$ and all $f \in \mathcal{F}$. Show that the Arzela–Ascoli Theorem remains valid with this definition of equicontinuity. That is, prove that if X is compact and $\mathcal{F}$ is closed, bounded, and equicontinuous at each $x \in X$, then $\mathcal{F}$ is compact.
15. Give an example of a sequence of real valued functions defined on $[a, b]$ that converges uniformly to a continuous function, but is not equicontinuous.
16. Let $\{f_i\}$ be a sequence of nonnegative, equicontinuous functions defined on a totally bounded metric space X such that

$$\limsup_{i \to \infty} f_i(x) < \infty \quad \text{for each } x \in X.$$

Prove that there is a subsequence that converges uniformly to a continuous function f.
17. Let $\{f_i\}$ be a sequence of nonnegative, equicontinuous functions defined on $[0, 1]$ with the property that

$$\limsup_{i \to \infty} f_i(x_0) < \infty \quad \text{for some } x_0 \in [0, 1].$$

Prove that there is a subsequence that converges uniformly to a continuous function f.
18. Let $\{f_i\}$ be a sequence of nonnegative, equicontinuous functions defined on a locally compact metric space X such that

$$\limsup_{i \to \infty} f_i(x) < \infty \quad \text{for each } x \in X.$$

Prove that there exist an open set U and a subsequence that converges uniformly on U to a continuous function f.
19. Let $\{f_i\}$ be a sequence of real-valued functions defined on a compact metric space X with the property that $x_k \to x$ implies $f_k(x_k) \to f(x)$, where f is a continuous function on X. Prove that $f_k \to f$ uniformly on X.
20. Let $\{f_i\}$ be a sequence of nondecreasing, real-valued (not necessarily continuous) functions defined on $[a, b]$ that converges pointwise to a continuous function f. Show that the convergence is necessarily uniform.
21. Let $\{f_i\}$ be a sequence of continuous, real valued functions defined on a compact metric space X that converges pointwise on some dense set to a continuous function on X. Prove that $f_i \to f$ uniformly on X.

22. Let $\{f_i\}$ be a uniformly bounded sequence in $C[a,b]$. For each $x \in [a,b]$, define
$$F_i(x) := \int_a^x f_i(t)\,dt.$$
Prove that there is a subsequence of $\{F_i\}$ that converges uniformly to some function $F \in C[a,b]$.

23. For each integer $k > 1$ let $\mathcal{F}_k$ be the family of continuous functions on $[0,1]$ with the property that for some $x \in [0, 1-1/k]$ we have
$$|f(x+h) - f(x)| \le kh \quad \text{whenever } 0 < h < \frac{1}{k}.$$
   (a) Prove that $\mathcal{F}_k$ is nowhere dense in the space $C[0,1]$ endowed with its usual metric of uniform convergence.
   (b) Using the Baire category theorem, prove that there exists $f \in C[0,1]$ that is not differentiable at any point of $(0,1)$.

24. The previous problem demonstrates the remarkable fact that functions that are nowhere differentiable are in great abundance, whereas functions that are well behaved are relatively scarce. The following are examples of functions that are continuous and nowhere differentiable.
   (a) For $x \in [0,1]$ let
$$f(x) := \sum_{n=0}^{\infty} \frac{[10^n x]}{10^n},$$
   where $[y]$ denotes the distance from the greatest integer in y.
   (b)
$$f(x) := \sum_{k=0}^{\infty} a^k \cos^b \pi x,$$
   where $1 < ab < b$. Weierstrass was the first to prove the existence of continuous nowhere differentiable functions by conceiving of this function and then proving that it is nowhere differentiable for certain values of a and b [50]. Later, Hardy proved the same result for all a and b [35].

25. Let f be a nondecreasing function on (a,b). Show that $f(x+)$ and $f(x-)$ exist at every point x of (a,b). Show also that if $a < x < y < b$, then $f(x+) \le f(y-)$.

3.8. Lower Semicontinuous Functions

In many applications in analysis, lower and upper semicontinuous functions play an important role. The purpose of this section is to introduce these functions and develop their basic properties.

Recall that a function f on a metric space is continuous at x_0 if for each $\varepsilon > 0$, there exists $r > 0$ such that

$$f(x_0) - \varepsilon < f(x) < f(x_0) + \varepsilon$$

whenever $x \in B(x_0, r)$. Semicontinuous functions require only one part of this inequality to hold.

3.62. DEFINITION. Suppose (X, ρ) is a metric space. A function f defined on X with possibly infinite values is said to be **lower semicontinuous** at $x_0 \in X$ if the following conditions hold. If $f(x_0) < \infty$, then for every $\varepsilon > 0$ there exists $r > 0$ such that $f(x) > f(x_0) - \varepsilon$ whenever $x \in B(x_0, r)$. If $f(x_0) = \infty$, then for every positive number M there exists $r > 0$ such that $f(x) \geq M$ for all $x \in B(x_0, r)$. The function f is called **lower semicontinuous** if it is lower semicontinuous at all $x \in X$. An **upper semicontinuous** function is defined analogously: if $f(x_0) > -\infty$, then $f(x) < f(x_0) + \varepsilon$ for all $x \in B(x_0, r)$. If $f(x_0) = -\infty$, then $f(x) < -M$ for all $x \in B(x_0, r)$.

Of course, a continuous function is both lower and upper semicontinuous. It is easy to see that the characteristic function of an open set is lower semicontinuous and that the characteristic function of a closed set is upper semicontinuous.

Semicontinuity can be reformulated in terms of the **lower limit** (also called **limit inferior**) and **upper limit of** (also called **limit superior**) f.

3.63. DEFINITION. We define

$$\liminf_{x \to x_0} f(x) = \lim_{r \to 0} m(r, x_0),$$

where $m(r, x_0) = \inf\{f(x) : 0 < \rho(x, x_0) < r\}$. Similarly,

$$\limsup_{x \to x_0} f(x) = \lim_{r \to 0} M(r, x_0),$$

where $M(r, x_0) = \sup\{f(x) : 0 < \rho(x, x_0) < r\}$.

One readily verifies that f is lower semicontinuous at a limit point x_0 of X if and only if

$$\liminf_{x \to x_0} f(x) \geq f(x_0),$$

and f is upper semicontinuous at x_0 if and only if

$$\limsup_{x \to x_0} f(x) \leq f(x_0).$$

In terms of sequences, these statements are equivalent, respectively, to the following:

$$\liminf_{k \to \infty} f(x_k) \geq f(x_0)$$

and

$$\limsup_{k \to \infty} f(x_k) \leq f(x_0)$$

whenever $\{x_k\}$ is a sequence converging to x_0. This leads immediately to the following.

3.64. THEOREM. *Suppose X is a compact metric space. Then a real-valued lower (upper) semicontinuous function on X assumes its minimum (maximum) on X.*

PROOF. We will give the proof for f lower semicontinuous, the proof for f upper semicontinuous being similar. Let

$$m = \inf\{f(x) : x \in X\}.$$

We will see that $m \neq -\infty$ and that there exists $x_0 \in X$ such that $f(x_0) = m$, thus establishing the result.

To see this, let $y_k \in f(X)$ be such that $\{y_k\} \to m$ as $k \to \infty$. At this point of the proof, we must allow the possibility that $m = -\infty$. Note that $m \neq +\infty$. Let $x_k \in X$ be such that $f(x_k) = y_k$. Since X is compact, there exist a point $x_0 \in X$ and a subsequence (still denoted by $\{x_k\}$) such that $\{x_k\} \to x_0$. Since f is lower semicontinuous, we obtain

$$m = \liminf_{k\to\infty} f(x_k) \geq f(x_0),$$

which implies that $f(x_0) = m$ and that $m \neq -\infty$. □

The following result will require the definition of a Lipschitz function.

3.65. DEFINITION. Suppose (X, ρ) and (Y, σ) are metric spaces. A mapping $f \colon X \to Y$ is called **Lipschitz** if there is a constant C_f such that

$$\sigma[f(x), f(y)] \leq C_f \rho(x, y) \tag{3.10}$$

for all $x, y \in X$. The smallest such constant C_f is called the **Lipschitz constant** of f.

3.66. THEOREM. *Suppose (X, ρ) is a metric space.*

(i) *f is lower semicontinuous on X if and only if $\{f > t\}$ is open for all $t \in \mathbb{R}$.*
(ii) *If both f and g are lower semicontinuous on X, then* $\min\{f, g\}$ *is lower semicontinuous.*
(iii) *The upper envelope of a collection of lower semicontinuous functions is lower semicontinuous.*
(iv) *Every nonnegative lower semicontinuous function on X is the upper envelope of a nondecreasing sequence of continuous (in fact, Lipschitz) functions.*

PROOF. To prove (i), choose $x_0 \in \{f > t\}$. Let $\varepsilon = f(x_0) - t$, and use the definition of lower semicontinuity to find a ball $B(x_0, r)$ such that $f(x) > f(x_0) - \varepsilon = t$ for all $x \in B(x_0, r)$. Thus, $B(x_0, r) \subset \{f > t\}$, which proves that $\{f > t\}$ is open. Conversely, choose $x_0 \in X$ and $\varepsilon > 0$ and let $t = f(x_0) - \varepsilon$. Then $x_0 \in \{f > t\}$, and since $\{f > t\}$ is open, there exists a ball $B(x_0, r) \subset \{f > t\}$. This implies that $f(x) > f(x_0) - \varepsilon$ whenever $x \in B(x_0, r)$, thus establishing lower semicontinuity.

(i) immediately implies (ii) and (iii). For (ii), let $h = \min(f, g)$ and observe that $\{h > t\} = \{f > t\} \cap \{g > t\}$, which is the intersection of two open sets.

Similarly, for (iii), let $\mathcal{F}$ be a family of lower semicontinuous functions and set

$$h(x) = \sup\{f(x) : f \in \mathcal{F}\} \quad \text{for} \quad x \in X.$$

Then for each real number t,

$$\{h > t\} = \bigcup_{f \in \mathcal{F}} \{f > t\},$$

which is open, since each set on the right is open.

Proof of (iv): For each positive integer k define

$$f_k(x) = \inf\{f(y) + k\rho(x, y) : y \in X\}.$$

Observe that $f_1 \leq f_2, \ldots, \leq f$. To show that each f_k is Lipschitz, it is sufficient to prove

$$f_k(x) \leq f_k(w) + k\rho(x, w) \quad \text{for all} \quad w \in X, \tag{3.11}$$

since the roles of x and w can be interchanged. To prove (3.11), observe that for each $\varepsilon > 0$, there exists $y \in X$ such that

$$f_k(w) \leq f(y) + k\rho(w, y) \leq f_k(w) + \varepsilon.$$

Now,

$$\begin{aligned} f_k(x) &\leq f(y) + k\rho(x, y) \\ &= f(y) + k\rho(w, y) + k\rho(x, y) - k\rho(w, y) \\ &\leq f_k(w) + \varepsilon + k\rho(x, w), \end{aligned}$$

where the triangle inequality has been used to obtain the last inequality. This implies (3.11), since ε is arbitrary.

Finally, to show that $f_k(x) \to f(x)$ for each $x \in X$, observe that for each $x \in X$ there is a sequence $\{x_k\} \subset X$ such that

$$f(x_k) + k\rho(x_k, x) \leq f_k(x) + \frac{1}{k} \leq f(x) + 1 < \infty.$$

As a consequence, we have that $\lim_{k\to\infty} \rho(x_k, x) = 0$. Given $\varepsilon > 0$, there exists $n \in \mathbb{N}$ such that

$$f_k(x) + \varepsilon \geq f_k(x) + \frac{1}{k} \geq f(x_k) \geq f(x) - \varepsilon$$

whenever $k \geq n$ and thus $f_k(x) \to f(x)$.

3.67. REMARK. Of course, the previous theorem has a companion that pertains to upper semicontinuous functions. Thus, the result analogous to (i) states that f is upper semicontinuous on X if and only if $\{f < t\}$ is open for all $t \in \mathbb{R}$. We leave it to the reader to formulate and prove the remaining three statements.

3.68. DEFINITION. Theorem 3.66 provides a means of defining upper and lower semicontinuity for functions defined merely on a topological space X.

Thus, $f\colon X \to \mathbb{R}$ is called **upper semicontinuous** (**lower semicontinuous**) if $\{f < t\}$ ($\{f > t\}$) is open for all $t \in \mathbb{R}$. It is easily verified that (ii) and (iii) of Theorem 3.66 remain true when X is assumed to be only a topological space.

Exercises for Section 3.8

1. Let $\{f_i\}$ be a decreasing sequence of upper semicontinuous functions defined on a compact metric space X such that $f_i(x) \to f(x)$, where f is lower semicontinuous. Prove that $f_i \to f$ uniformly.
2. Show that Theorem 3.66, (iv), remains true for lower semicontinuous functions that are bounded below. Show also that this assumption is necessary.

CHAPTER 4

Measure Theory

4.1. Outer Measure

An outer measure on an abstract set X is a monotone, countably subadditive function defined on all subsets of X. In this section, the notion of measurable set is introduced, and it is shown that the class of measurable sets forms a σ-algebra, i.e., measurable sets are closed under the operations of complementation and countable unions. It is also shown that an outer measure is countably additive on disjoint measurable sets.

In this section we introduce the concept of outer measure, which will underlie and motivate some of the most important concepts of abstract measure theory. The "length" of set in $\mathbb{R}$, the "area" of a set in $\mathbb{R}^2$, and the "volume" of a set in $\mathbb{R}^3$ are notions that can be developed from basic and strongly intuitive geometric principles, provided the sets are well behaved. If one wished to develop a concept of volume in $\mathbb{R}^3$, for example, that would allow the assignment of volume to any set, then one could hope for a function $\mathcal{V}$ that assigns to each subset $E \subset \mathbb{R}^3$ a number $\mathcal{V}(E) \in [0,\infty]$ having the following properties:

(i) If $\{E_i\}_{i=1}^k$ is any finite sequence of mutually disjoint sets, then

$$\mathcal{V}\left(\bigcup_{i=1}^{k} E_i\right) = \sum_{i=i}^{k} \mathcal{V}(E_i). \tag{4.1}$$

(ii) If two sets E and F are congruent, then $\mathcal{V}(E) = \mathcal{V}(F)$.

(iii) $\mathcal{V}(Q) = 1$, where Q is the cube of side length 1.

However, these three conditions are inconsistent. In 1924, Banach and Tarski [2] proved that it is possible to decompose a ball in $\mathbb{R}^3$ into six pieces that can be reassembled by rigid motions to form two balls, each the same size as the original. The sets in this decomposition are pathological and require the axiom of choice for their existence. If condition (i) is changed to require countable additivity rather than mere finite additivity, that is, to require that if $\{E_i\}_{i=1}^{\infty}$ is any infinite sequence of mutually disjoint sets, then

$$\sum_{i=i}^{\infty} \mathcal{V}(E_i) = \mathcal{V}\left(\bigcup_{i=1}^{\infty} E_i\right),$$

W.P. Ziemer, *Modern Real Analysis*, Graduate Texts in Mathematics 278,
https://doi.org/10.1007/978-3-319-64629-9_4

this too suffers from the same inconsistency, and thus we are led to the conclusion that there is no function $\mathcal{V}$ satisfying all three conditions above. Later, we will also see that if we restrict $\mathcal{V}$ to a large class of subsets of $\mathbb{R}^3$ that omits only the truly pathological sets, then it is possible to incorporate $\mathcal{V}$ into a satisfactory theory of volume.

We will proceed to find this large class of sets by considering a very general context and replace countable additivity by countable subadditivity.

4.1. Definition. A function φ defined for every subset A of an arbitrary set X is called an **outer measure** on X if the following conditions are satisfied:

(i) $\varphi(\emptyset) = 0$,

(ii) $0 \leq \varphi(A) \leq \infty$ whenever $A \subset X$,

(iii) $\varphi(A_1) \leq \varphi(A_2)$ whenever $A_1 \subset A_2$,

(iv) $\varphi(\bigcup\limits_{i=1}^{\infty} A_i) \leq \sum\limits_{i=1}^{\infty} \varphi(A_i)$ for every countable collection of sets $\{A_i\}$ in X.

Condition (iii) states that φ is **monotone**, while (iv) states that φ is **countably subadditive**. As we mentioned earlier, suitable additivity properties are necessary in measure theory; subadditivity, in general, will not suffice to produce a useful theory. We will now introduce the concept of a "measurable set" and show later that measurable sets enjoy a wide spectrum of additivity properties.

The term "outer measure" is derived from the way outer measures are constructed in practice. Often one uses a set function that is defined on some family of primitive sets (such as the family of intervals in $\mathbb{R}$) to approximate an arbitrary set from the "outside" to define its measure. Examples of this procedure will be given in Sections 4.3 and 4.4. First, consider some elementary examples of outer measures.

4.2. Examples. (i) In an arbitrary set X, define $\varphi(A) = 1$ if A is nonempty and $\varphi(\emptyset) = 0$.

(ii) Let $\varphi(A)$ be the number (possibly infinite) of points in A.

(iii) Let $\varphi(A) = \begin{cases} 0 & \text{if card } A \leq \aleph_0, \\ 1 & \text{if card } A > \aleph_0. \end{cases}$

(iv) If X is a metric space, fix $\varepsilon > 0$. Let $\varphi(A)$ be the smallest number of balls of radius ε that cover A.

(v) Select a fixed x_0 in an arbitrary set X, and let

$$\varphi(A) = \begin{cases} 0 & \text{if } x_0 \notin A, \\ 1 & \text{if } x_0 \in A; \end{cases}$$

φ is called the **Dirac measure** concentrated at x_0.

Notice that the domain of an outer measure φ is $\mathcal{P}(X)$, the collection of all subsets of X. In general, it may happen that the equality $\varphi(A \cup B) = \varphi(A)+\varphi(B)$ fails when $A \cap B = \emptyset$. This property and more generally, property (4.1), will require a more restrictive class of subsets of X, called measurable sets, which we now define.

4.3. DEFINITION. Let φ be an outer measure on a set X. A set $E \subset X$ is called φ-measurable if

$$\varphi(A) = \varphi(A \cap E) + \varphi(A - E)$$

for every set $A \subset X$. In view of property (iv) above, observe that φ-measurability requires only

$$\varphi(A) \geq \varphi(A \cap E) + \varphi(A - E). \tag{4.2}$$

This definition, while not very intuitive, says that a set is φ-measurable if it decomposes an arbitrary set into two parts for which φ is additive. We use this definition in deference to Carathéodory, who established this property as an alternative characterization of measurability in the special case of Lebesgue measure (see Definition 4.21 below). The following characterization of φ-measurability is perhaps more intuitively appealing.

4.4. LEMMA. *A set $E \subset X$ is φ-measurable if and only if*

$$\varphi(P \cup Q) = \varphi(P) + \varphi(Q)$$

for all sets P and Q such that $P \subset E$ and $Q \subset \tilde{E}$.

PROOF. Sufficiency: Let $A \subset X$. Then with $P: = A \cap E \subset E$ and $Q: = A - E \subset \tilde{E}$ we have $A = P \cup Q$ and therefore

$$\varphi(A) = \varphi(P \cup Q) = \varphi(P) + \varphi(Q) = \varphi(A \cap E) + \varphi(A - E).$$

Necessity: Let P and Q be arbitrary sets such that $P \subset E$ and $Q \subset \tilde{E}$. Then, by the definition of φ-measurability,

$$\begin{aligned} \varphi(P \cup Q) &= \varphi[(P \cup Q) \cap E] + \varphi[(P \cup Q) \cap \tilde{E}] \\ &= \varphi(P \cap E) + \varphi(Q \cap \tilde{E}) \\ &= \varphi(P) + \varphi(Q). \end{aligned}$$

□

4.5. REMARK. Recalling Examples 4.2, one verifies that only the empty set and X are measurable for (i), while all sets are measurable for (ii).

Now that we have an alternative definition of φ-measurability, we investigate the properties of φ-measurable sets. We begin with the following theorem, which is basic to the theory. A set function that satisfies property (iv) below on every sequence of disjoint sets is said to be **countably additive**.

4.6. THEOREM. *Suppose φ is an outer measure on an arbitrary set X. Then the following four statements hold:*

(i) *E is φ-measurable whenever $\varphi(E) = 0$.*

(ii) $\emptyset$ *and* X *are* φ*-measurable.*

(iii) $E_1 - E_2$ *is* φ*-measurable whenever* E_1 *and* E_2 *are* φ*-measurable.*

(iv) *If* $\{E_i\}$ *is a countable collection of disjoint* φ*-measurable sets, then* $\cup_{i=1}^{\infty} E_i$ *is* φ*-measurable and*

$$\varphi(\bigcup_{i=1}^{\infty} E_i) = \sum_{i=1}^{\infty} \varphi(E_i).$$

More generally, if $A \subset X$ *is an arbitrary set, then*

$$\varphi(A) = \sum_{i=1}^{\infty} \varphi(A \cap E_i) + \varphi(A \cap \tilde{S}),$$

where $S = \bigcup_{i=1}^{\infty} E_i$.

PROOF. (i) If $A \subset X$, then $\varphi(A \cap E) = 0$. Thus, $\varphi(A) \leq \varphi(A \cap E) + \varphi(A \cap \tilde{E}) = \varphi(A \cap \tilde{E}) \leq \varphi(A)$.

(ii) This follows immediately from Lemma 4.4.

(iii) We will use Lemma 4.4 to establish the φ-measurability of $E_1 - E_2$. Thus, let $P \subset E_1 - E_2$ and $Q \subset (E_1 - E_2)^{\sim} = \tilde{E}_1 \cup E_2$ and note that $Q = (Q \cap E_2) \cup (Q - E_2)$. The φ-measurability of E_2 implies

$$\begin{aligned} \varphi(P) + \varphi(Q) &= \varphi(P) + \varphi[(Q \cap E_2) \cup (Q - E_2)] \\ &= \varphi(P) + \varphi(Q \cap E_2) + \varphi(Q - E_2). \end{aligned} \tag{4.3}$$

But $P \subset E_1$, $Q - E_2 \subset \tilde{E}_1$ and the φ-measurability of E_1 imply

$$\begin{aligned} &\varphi(P) + \varphi(Q \cap E_2) + \varphi(Q - E_2) \\ &\qquad = \varphi(Q \cap E_2) + \varphi[P \cup (Q - E_2)]. \end{aligned} \tag{4.4}$$

Also, $Q \cap E_2 \subset E_2, P \cup (Q - E_2) \subset \tilde{E}_2$ and the φ-measurability of E_2 imply

$$\begin{aligned} &\varphi(Q \cap E_2) + \varphi[P \cup (Q - E_2)] \\ &\qquad = \varphi[(Q \cap E_2) \cup (P \cup (Q - E_2))] \\ &\qquad = \varphi(Q \cup P) = \varphi(P \cup Q). \end{aligned} \tag{4.5}$$

Hence, by (4.3), (4.4), and (4.5) we have

$$\varphi(P) + \varphi(Q) = \varphi(P \cup Q).$$

(iv) Let $S_k = \cup_{i=1}^{k} E_i$ and let A be an arbitrary subset of X. We proceed by finite induction and first note that the result is obviously true for $k = 1$. For $k > 1$ assume that S_k is φ-measurable and that

$$\varphi(A) \geq \sum_{i=1}^{k} \varphi(A \cap E_i) + \varphi(A \cap \tilde{S}_k), \tag{4.6}$$

for every set A. Then

$$\begin{aligned}
\varphi(A) &= \varphi(A \cap E_{k+1}) + \varphi(A \cap \tilde{E}_{k+1}) && \text{because } E_{k+1} \text{ is } \varphi\text{-measurable} \\
&= \varphi(A \cap E_{k+1}) + \varphi(A \cap \tilde{E}_{k+1} \cap S_k) \\
&\quad + \varphi(A \cap \tilde{E}_{k+1} \cap \tilde{S}_k) && \text{because } S_k \text{ is } \varphi\text{-measurable} \\
&= \varphi(A \cap E_{k+1}) + \varphi(A \cap S_k) \\
&\quad + \varphi(A \cap \tilde{S}_{k+1}) && \text{because } S_k \subset \tilde{E}_{k+1} \\
&\geq \sum_{i=1}^{k+1} \varphi(A \cap E_i) + \varphi(A \cap \tilde{S}_{k+1}) && \text{use (4.6) with } A \text{ replaced by } A \cap S_k.
\end{aligned}$$

By the countable subadditivity of φ, this shows that

$$\varphi(A) \geq \varphi(A \cap S_{k+1}) + \varphi(A \cap \tilde{S}_{k+1});$$

this, in turn, implies that S_{k+1} is φ-measurable. Since we now know that for every set $A \subset X$ and for all positive integers k

$$\varphi(A) \geq \sum_{i=1}^{k} \varphi(A \cap E_i) + \varphi(A \cap \tilde{S}_k)$$

and that $\tilde{S}_k \supset \tilde{S}$, we have

$$\begin{aligned}
\varphi(A) &\geq \sum_{i=1}^{\infty} \varphi(A \cap E_i) + \varphi(A \cap \tilde{S}) \\
&\geq \varphi(A \cap S) + \varphi(A \cap \tilde{S}).
\end{aligned} \tag{4.7}$$

Again, the countable subadditivity of φ was used to establish the last inequality. This implies that S is φ-measurable, which establishes the first part of (iv). For the second part of (iv), note that the countable subadditivity of φ yields

$$\begin{aligned}
\varphi(A) &\leq \varphi(A \cap S) + \varphi(A \cap \tilde{S}) \\
&\leq \sum_{i=1}^{\infty} \varphi(A \cap E_i) + \varphi(A \cap \tilde{S}).
\end{aligned}$$

This, along with (4.7), establishes the last part of (iv). □

The preceding result shows that φ-measurable sets are closed under the set-theoretic operations of taking complements and countable disjoint unions. Of course, it would be preferable if they were closed under countable unions, and not merely countable disjoint unions. The proposition below addresses this issue. But first, we will prove a lemma that will be frequently used throughout. It states that the union of a countable family of sets can be written as the union of a countable family of disjoint sets.

4.7. LEMMA. *Let $\{E_i\}$ be a sequence of arbitrary sets. Then there exists a sequence of disjoint sets $\{A_i\}$ such that each $A_i \subset E_i$ and*

$$\bigcup_{i=1}^{\infty} E_i = \bigcup_{i=1}^{\infty} A_i.$$

If each E_i is φ-measurable, and so is A_i.

PROOF. For each positive integer j, define $S_j = \cup_{i=1}^{j} E_i$. Note that

$$\bigcup_{i=1}^{\infty} E_i = S_1 \bigcup \Big(\bigcup_{k=1}^{\infty} (S_{k+1} \setminus S_k) \Big).$$

Now take $A_1 = S_1$ and $A_{i+1} = S_{i+1} \setminus S_i$ for all integers $i \geq 1$.

If each E_i is φ-measurable, and the same is true for each S_j. Indeed, referring to Theorem 4.6 (iii), we see that S_2 is φ-measurable because $S_2 = E_2 \cup (E_1 \setminus E_2)$ is the disjoint union of φ-measurable sets. Inductively, we see that $S_j = E_j \cup (S_{j-1} \setminus E_j)$ is the disjoint union of φ-measurable sets, and therefore the sets A_i are also φ-measurable. □

4.8. THEOREM. *If $\{E_i\}$ is a sequence of φ-measurable sets in X, then $\cup_{i=1}^{\infty} E_i$ and $\cap_{i=1}^{\infty} E_i$ are φ-measurable.*

PROOF. From the previous lemma, we have

$$\bigcup_{i=1}^{\infty} E_i = \bigcup_{i=1}^{\infty} A_i,$$

where each A_i is a φ-measurable subset of E_i and where the sequence $\{A_i\}$ is disjoint. Thus, it follows immediately from Theorem 4.6 (iv) that $\cup_{i=1}^{\infty} E_i$ is φ-measurable.

To establish the second claim, note that

$$X \setminus \Big(\bigcap_{i=1}^{\infty} E_i\Big) = \bigcup_{i=1}^{\infty} \tilde{E}_i.$$

The right side is φ-measurable in view of Theorem 4.6 (ii), (iii) and the first claim. A further appeal to Theorem 4.6 (iv) concludes the proof. □

Classes of sets that are closed under complementation and countable unions play an important role in measure theory and are therefore given a special name.

4.9. DEFINITION. A nonempty collection Σ of sets E satisfying the following two conditions is called a σ**−algebra**:

(i) if $E \in \Sigma$, then $\tilde{E} \in \Sigma$;

(ii) $\cup_{i=1}^{\infty} E_i \in \Sigma$ if each E_i is in Σ.

Note that it easily follows from the definition that a σ-algebra is closed under countable intersections and finite differences. Note also that the entire space and the empty set are elements of the σ-algebra, since $\emptyset = E \cap \tilde{E} \in \Sigma$.

4.10. Definition. In a topological space, the elements of the smallest σ-algebra that contains all open sets are called **Borel sets**. The term "smallest" is taken in the sense of inclusion, and it is left as an exercise (Exercise 1, Section 4.2) to show that such a smallest σ-algebra, denoted by $\mathcal{B}$, exists.

The following is an immediate consequence of Theorem 4.6 and Theorem 4.8.

4.11. Corollary. *If φ is an outer measure on an arbitrary set X, then the class of φ-measurable sets forms a σ-algebra.*

Next we state a result that exhibits the basic additivity and continuity properties of outer measure when restricted to its measurable sets. These properties follow almost immediately from Theorem 4.6.

4.12. Corollary. *Suppose φ is an outer measure on X and $\{E_i\}$ a countable collection of φ-measurable sets.*

(i) *If $E_1 \subset E_2$ with $\varphi(E_1) < \infty$, then*

$$\varphi(E_2 \setminus E_1) = \varphi(E_2) - \varphi(E_1).$$

(See Exercise 3, Section 4.1.)

(ii) *(Countable additivity) If $\{E_i\}$ is a disjoint sequence of sets, then*

$$\varphi(\bigcup_{i=1}^{\infty} E_i) = \sum_{i=1}^{\infty} \varphi(E_i).$$

(iii) *(Continuity from the left) If $\{E_i\}$ is an increasing sequence of sets, that is, if $E_i \subset E_{i+1}$ for each i, then*

$$\varphi(\bigcup_{i=1}^{\infty} E_i) = \varphi(\lim_{i\to\infty} E_i) = \lim_{i\to\infty} \varphi(E_i).$$

(iv) *(Continuity from the right) If $\{E_i\}$ is a decreasing sequence of sets, that is, if $E_i \supset E_{i+1}$ for each i, and if $\varphi(E_{i_0}) < \infty$ for some i_0, then*

$$\varphi(\bigcap_{i=1}^{\infty} E_i) = \varphi(\lim_{i\to\infty} E_i) = \lim_{i\to\infty} \varphi(E_i).$$

(v) *If $\{E_i\}$ is any sequence of φ-measurable sets, then*

$$\varphi(\liminf_{i\to\infty} E_i) \leq \liminf_{i\to\infty} \varphi(E_i).$$

(vi) *If*

$$\varphi(\bigcup_{i=i_0}^{\infty} E_i) < \infty$$

for some positive integer i_0, then

$$\varphi(\limsup_{i\to\infty} E_i) \geq \limsup_{i\to\infty} \varphi(E_i).$$

PROOF. We first observe that in view of Corollary 4.11, each of the sets that appears on the left side of (ii) through (vi) is φ-measurable. Consequently, all sets encountered in the proof will be φ-measurable.

(i): Observe that $\varphi(E_2) = \varphi(E_2 \setminus E_1) + \varphi(E_1)$, since $E_2 \setminus E_1$ is φ-measurable (Theorem 4.6 (iii)).

(ii): This is a restatement of Theorem 4.6 (iv).

(iii): We may assume that $\varphi(E_i) < \infty$ for each i, for otherwise, the result follows from the monotonicity of φ. Since the sets $E_1, E_2 \setminus E_1, \ldots, E_{i+1} \setminus E_i, \ldots$ are φ-measurable and disjoint, it follows that

$$\lim_{i\to\infty} E_i = \bigcup_{i=1}^{\infty} E_i = E_1 \cup \big[\bigcup_{i=1}^{\infty} (E_{i+1} \setminus E_i)\big]$$

and therefore, from (iv) of Theorem 4.6, that

$$\varphi(\lim_{i\to\infty} E_i) = \varphi(E_1) + \sum_{i=1}^{\infty} \varphi(E_{i+1} \setminus E_i).$$

Since the sets E_i and $E_{i+1} \setminus E_i$ are disjoint and φ-measurable, we have $\varphi(E_{i+1}) = \varphi(E_{i+1} \setminus E_i) + \varphi(E_i)$. Therefore, because $\varphi(E_i) < \infty$ for each i, we have from (4.1)

$$\begin{aligned} \varphi(\lim_{i\to\infty} E_i) &= \varphi(E_1) + \sum_{i=1}^{\infty} [\varphi(E_{i+1}) - \varphi(E_i)] \\ &= \lim_{i\to\infty} \varphi(E_{i+1}), \end{aligned}$$

which proves (iii).

(iv): By replacing E_i with $E_i \cap E_{i_0}$ if necessary, we may assume that $\varphi(E_1) < \infty$. Since $\{E_i\}$ is decreasing, the sequence $\{E_1 \setminus E_i\}$ is increasing, and therefore (iii) implies

$$\begin{aligned} \varphi(\bigcup_{i=1}^{\infty} (E_1 \setminus E_i)) &= \lim_{i\to\infty} \varphi(E_1 \setminus E_i) \\ &= \varphi(E_1) - \lim_{i\to\infty} \varphi(E_i). \end{aligned} \tag{4.8}$$

It is easy to verify that

$$\bigcup_{i=1}^{\infty} (E_1 \setminus E_i) = E_1 \setminus \bigcap_{i=1}^{\infty} E_i,$$

and therefore, from (i) of Corollary 4.12, we have

$$\varphi(\bigcup_{i=1}^{\infty} (E_1 \setminus E_i)) = \varphi(E_1) - \varphi(\bigcap_{i=1}^{\infty} E_i),$$

which, along with (4.8), yields

$$\varphi(E_1) - \varphi(\bigcap_{i=1}^{\infty} E_i) = \varphi(E_1) - \lim_{i\to\infty} \varphi(E_i).$$

The fact that $\varphi(E_1) < \infty$ allows us to conclude (iv).

(v): Let $A_j = \cap_{i=j}^{\infty} E_i$ for $j = 1, 2, \ldots$. Then A_j is an increasing sequence of φ-measurable sets with the property that $\lim_{j\to\infty} A_j = \liminf_{i\to\infty} E_i$, and therefore, by (iii),

$$\varphi(\liminf_{i\to\infty} E_i) = \lim_{j\to\infty} \varphi(A_j).$$

But since $A_j \subset E_j$, it follows that

$$\lim_{j\to\infty} \varphi(A_j) \leq \liminf_{j\to\infty} \varphi(E_j),$$

thus establishing (v).

The proof of (vi) is similar to that of (v) and is left as Exercise 1, Section 4.1. □

4.13. REMARK. We mentioned earlier that one of our major concerns is to determine whether there is a rich supply of measurable sets for a given outer measure φ. Although we have learned that the class of measurable sets constitutes a σ-algebra, this is not sufficient to guarantee that the measurable sets exist in great numbers. For example, suppose that X is an arbitrary set and φ is defined on X as $\varphi(E) = 1$ whenever $E \subset X$ is nonempty, while $\varphi(\emptyset) = 0$. Then it is easy to verify that X and $\emptyset$ are the only φ-measurable sets. In order to overcome this difficulty, it is necessary to impose an additivity condition on φ. This will be developed in the following section.

We will need the following definitions:

4.14. DEFINITIONS. An outer measure φ on a topological space X is called a **Borel outer measure** if all Borel sets are φ-measurable. A Borel outer measure is **finite** if $\varphi(X)$ is finite.

An outer measure φ on a set X is called **regular** if for each $A \subset X$ there exists a φ-measurable set $B \supset A$ such that $\varphi(B) = \varphi(A)$. A **Borel regular outer measure** is a Borel outer measure such that for each $A \subset X$, there exists a Borel set B such that $\varphi(B) = \varphi(A)$ (see Theorem 4.52 and Corollary 4.56 for regularity properties of Borel outer measures). A **Radon outer measure** is a Borel regular outer measure that is finite on compact sets.

We began this section with the concept of an outer measure on an arbitrary set X and proved that the family of φ-measurable sets forms a σ-algebra. We will see in Section 4.9 that the restriction of φ to this σ-algebra generates a measure space (see Definition 4.47). In the next few sections we will introduce important examples of outer measures, which in turn will provide measure spaces that appear in many areas of mathematics.

Exercises for Section 4.1

1. Prove (vi) of Corollary 4.12.
2. In Examples 4.2 (iv), let $\varphi_\varepsilon(A) := \varphi(A)$ denote the dependence on ε, and define

$$\psi(A) := \lim_{\varepsilon\to 0} \varphi_\varepsilon(A).$$

What is $\psi(A)$ and what are the corresponding ψ-measurable sets?

3. In (i) of Corollary 4.12, it was shown that $\varphi(E_2 \setminus E_1) = \varphi(E_2) - \varphi(E_1)$ if $E_1 \subset E_2$ are φ-measurable with $\varphi(E_1) < \infty$. Prove that this result remains true if E_2 is not assumed to be φ-measurable.

4.2. Carathéodory Outer Measure

In the previous section, we considered an outer measure φ on an arbitrary set X. We now restrict our attention to a metric space X and impose a further condition (an additivity condition) on the outer measure. This will allow us to conclude that all closed sets are measurable.

4.15. DEFINITION. An outer measure φ defined on a metric space (X, ρ) is called a **Carathéodory outer measure** if

$$\varphi(A \cup B) = \varphi(A) + \varphi(B) \tag{4.9}$$

whenever A, B are arbitrary subsets of X with $\boldsymbol{d}(A, B) > 0$. The notation $\boldsymbol{d}(A, B)$ denotes the distance between the sets A and B and is defined by

$$\boldsymbol{d}(A, B) := \inf\{\rho(a, b) : a \in A, b \in B\}.$$

4.16. THEOREM. *If φ is a Carathéodory outer measure on a metric space X, then all closed sets are φ-measurable.*

PROOF. We will verify the condition in Definition 4.3 whenever C is a closed set. Because φ is subadditive, it suffices to show that

$$\varphi(A) \geq \varphi(A \cap C) + \varphi(A \setminus C) \tag{4.10}$$

whenever $A \subset X$. In order to prove (4.10), consider $A \subset X$ with $\varphi(A) < \infty$ and for each positive integer i, let $C_i = \{x : \boldsymbol{d}(x, C) \leq 1/i\}$. Note that

$$\boldsymbol{d}(A \setminus C_i, A \cap C) \geq \frac{1}{i} > 0.$$

Since $A \supset (A \setminus C_i) \cup (A \setminus C)$, (4.15) implies

$$\varphi(A) \geq \varphi\big((A \setminus C_i) \cup (A \cap C)\big) = \varphi(A \setminus C_i) + \varphi(A \cap C). \tag{4.11}$$

Because of this inequality, the proof of (4.10) will be concluded if we can show that

$$\lim_{i \to \infty} \varphi(A \setminus C_i) = \varphi(A \setminus C). \tag{4.12}$$

For each positive integer i, let

$$T_i = A \cap \left\{x : \frac{1}{i+1} < \boldsymbol{d}(x, C) \leq \frac{1}{i}\right\}$$

and note that since C is closed, $x \notin C$ if and only if $\boldsymbol{d}(x, C) > 0$ and therefore that

$$A \setminus C = (A \setminus C_j) \cup \big(\bigcup_{i=j}^{\infty} T_i\big) \tag{4.13}$$

for each positive integer j. This, in turn, implies

$$\varphi(A \setminus C) \leq \varphi(A \setminus C_j) + \sum_{i=j}^{\infty} \varphi(T_i). \tag{4.14}$$

We now note that

$$\sum_{i=1}^{\infty} \varphi(T_i) < \infty. \tag{4.15}$$

To establish (4.15), first observe that $\boldsymbol{d}(T_i, T_j) > 0$ if $|i-j| \geq 2$. Thus, we obtain from (4.9) that for each positive integer m,

$$\sum_{i=1}^{m} \varphi(T_{2i}) = \varphi\big(\bigcup_{i=1}^{m} T_{2i}\big) \leq \varphi(A) < \infty,$$

$$\sum_{i=1}^{m} \varphi(T_{2i-1}) = \varphi\big(\bigcup_{i=1}^{m} T_{2i-1}\big) \leq \varphi(A) < \infty.$$

From (4.14) and since $A \setminus C_j \subset A \setminus C$ and $\sum_{i=1}^{\infty} \varphi(T_i) < \infty$, we have

$$\varphi(A \setminus C) - \sum_{i=j}^{\infty} \varphi(T_i) \leq \varphi(A \setminus C_j) \leq \varphi(A \setminus C).$$

Hence, by letting $j \to \infty$ and using $\lim_{j\to\infty} \sum_{i=j}^{\infty} \varphi(T_i) = 0$, we obtain the desired conclusion. □

The following proposition provides a useful description of the Borel sets.

4.17. THEOREM. *Suppose $\mathcal{F}$ is a family of subsets of a topological space X that contains all open and all closed subsets of X. Suppose also that $\mathcal{F}$ is closed under countable unions and countable intersections. Then $\mathcal{F}$ contains all Borel sets; that is, $\mathcal{B} \subset \mathcal{F}$.*

PROOF. Let

$$\mathcal{H} = \mathcal{F} \cap \{A : \tilde{A} \in \mathcal{F}\}.$$

Observe that $\mathcal{H}$ contains all closed sets. Moreover, it is easily seen that $\mathcal{H}$ is closed under complementation and countable unions. Thus, $\mathcal{H}$ is a σ-algebra that contains the open sets and therefore contains all Borel sets. □

As a direct result of Corollary 4.11 and Theorem 4.16 we have the main result of this section.

4.18. THEOREM. *If φ is a Carathéodory outer measure on a metric space X, then the Borel sets of X are φ-measurable.*

If $X = \mathbb{R}^n$, it follows that the cardinality of the Borel sets is at least as great as that of the closed sets. Since the Borel sets contain all singletons of $\mathbb{R}^n$, their cardinality is at least c. We thus have shown that not only do the

φ-measurable sets have nice additivity properties (they form a σ-algebra), but in addition, there is a plentiful supply of them if φ is a Carathéodory outer measure on $\mathbb{R}^n$. Thus, the difficulty that arises from the example in Remark 4.13 is avoided. In the next section we discuss a concrete illustration of such a measure.

Exercises for Section 4.2

1. Prove that in every topological space X, there exists a smallest σ-algebra that contains all open sets in X. That is, prove that there is a σ-algebra Σ that contains all open sets and has the property that if Σ_1 is another σ-algebra containing all open sets, then $\Sigma \subset \Sigma_1$. In particular, for $X = \mathbb{R}^n$, note that there is a smallest σ-algebra that contains all the closed sets in $\mathbb{R}^n$.
2. In a topological space X the family of Borel sets, $\mathcal{B}$, is by definition the σ-algebra generated by the closed sets. The method below is another way of describing the Borel sets using transfinite induction. You are to fill in the necessary steps:
   (a) For an arbitrary family $\mathcal{F}$ of sets, let
   $$\mathcal{F}^* = \{\bigcup_{k=1}^{\infty} E_k : \text{where either } E_i \in \mathcal{F} \text{ or } \widetilde{E}_i \in \mathcal{F} \text{ for all } i \in \mathbb{N}\}.$$
   Let Ω denote the smallest uncountable ordinal. We will use transfinite induction to define a family $\mathcal{E}_\alpha$ for each $\alpha < \Omega$.
   (b) Let $\mathcal{E}_0 :=$ all closed sets $:= \mathcal{K}$. Now choose $\alpha < \Omega$ and assume that $\mathcal{E}_\beta$ has been defined for each β such that $0 \leq \beta < \alpha$. Define
   $$\mathcal{E}_\alpha := \left(\bigcup_{0 \leq \beta < \alpha} \mathcal{E}_\beta \right)^*$$
   and define
   $$\mathcal{A} := \bigcup_{0 \leq \alpha < \Omega} \mathcal{E}_\alpha .$$
   (c) Show that each $\mathcal{E}_\alpha \in \mathcal{B}$.
   (d) Show that $\mathcal{A} \subset \mathcal{B}$.
   (e) Now show that $\mathcal{A}$ is a σ-algebra to conclude that $\mathcal{A} = \mathcal{B}$.
      (i) Show that $\emptyset, X \in \mathcal{A}$.
      (ii) Let $A \in \mathcal{A} \implies A \in \mathcal{E}_\alpha$ for some $\alpha < \Omega$. Show that this implies $\widetilde{A} \in \mathcal{E}_\alpha^* \subset \mathcal{E}_\beta$ for every $\beta > \alpha$.
      (iii) Conclude that $\widetilde{A} \in \mathcal{A}$ and thus conclude that $\mathcal{A}$ is closed under complementation.
      (iv) Now let $\{A_k\}$ be a sequence in $\mathcal{A}$. Show that $\bigcup_{k=1}^{\infty} A_k \in \mathcal{A}$. (Hint: Each A_k is in $\mathcal{A}_{\alpha_k}$ for some $\alpha_k < \Omega$. We know that there is $\beta < \Omega$ such that $\beta > \alpha_k$ for each $k \in \mathbb{N}$.)

3. Prove that the set function μ defined in (4.40) is an outer measure whose measurable sets include all open sets.
4. Prove that the set function ψ defined in (4.45) is an outer measure on X.
5. An outer measure φ on a space X is called σ-finite if there exists a countable number of sets A_i with $\varphi(A_i) < \infty$ such that $X \subset \cup_{i=1}^{\infty} A_i$. Assuming that φ is a σ-finite Borel regular outer measure on a metric space X, prove that $E \subset X$ is φ-measurable if and only if there exists an F_σ set $F \subset E$ such that $\varphi(E \setminus F) = 0$.
6. Let φ be an outer measure on a space X. Suppose $A \subset X$ is an arbitrary set with $\varphi(A) < \infty$ such that there exists a φ-measurable set $E \supset A$ with $\varphi(E) = \varphi(A)$. Prove that $\varphi(A \cap B) = \varphi(E \cap B)$ for every φ-measurable set B.
7. In $\mathbb{R}^2$, find two disjoint closed sets A and B such that $\boldsymbol{d}(A, B) = 0$. Show that this is not possible if one of the sets is compact.
8. Let φ be an outer Carathéodory measure on $\mathbb{R}$ and let $f(x) := \varphi(I_x)$, where I_x is an open interval of fixed length centered at x. Prove that f is lower semicontinuous. What can you say about f if I_x is taken as a closed interval? Prove the analogous result in $\mathbb{R}^n$; that is, let $f(x) := \varphi(B(x, a))$, where $B(x, a)$ is the open ball with fixed radius a centered at x.
9. In a metric space X, prove that dist $(A, B) = \boldsymbol{d}(\bar{A}, \bar{B})$ for arbitrary sets $A, B \in X$.
10. Let A be a non-Borel subset of $\mathbb{R}^n$ and define for each subset E,

$$\varphi(E) = \begin{cases} 0 & \text{if } E \subset A, \\ \infty & \text{if } E \setminus A \neq \emptyset. \end{cases}$$

Prove that φ is an outer measure that is not Borel regular.
11. Let $\mathcal{M}$ denote the class of φ-measurable sets of an outer measure φ defined on a set X. If $\varphi(X) < \infty$, prove that the family

$$\mathcal{F} := \{A \in \mathcal{M} : \varphi(A) > 0\}$$

is at most countable.

4.3. Lebesgue Measure

Lebesgue measure on $\mathbb{R}^n$ is perhaps the most important example of a Carathéodory outer measure. We will investigate the properties of this measure and show, among other things, that it agrees with the primitive notion of volume on sets such as n-dimensional "intervals."

For the purpose of defining Lebesgue outer measure on $\mathbb{R}^n$, we consider closed n-dimensional intervals

$$I = \{x : a_i \leq x_i \leq b_i, i = 1, 2, \dots, n\} \tag{4.16}$$

and their volumes

$$v(I) = \prod_{i=1}^{n}(b_i - a_i). \tag{4.17}$$

With $I_1 = [a_1, b_1]$, $I_2 = [a_2, b_2]$, $\ldots, I_n = [a_n, b_n]$, we have

$$I = I_1 \times I_2 \times \cdots \times I_n.$$

Notice that n-dimensional intervals have their edges parallel to the coordinate axes of $\mathbb{R}^n$. When no confusion arises, we shall simply say "interval" rather than "n-dimensional interval."

In preparation for the development of Lebesgue measure, we state two elementary propositions concerning intervals whose proofs will be omitted.

4.19. THEOREM. *Suppose each edge $I_k = [a_k, b_k]$ of an n-dimensional interval I is partitioned into α_k subintervals. The products of these intervals produce a partition of I into $\beta: = \alpha_1 \cdot \alpha_2 \cdots \alpha_n$ subintervals I_i and*

$$v(I) = \sum_{i=1}^{\beta} v(I_i).$$

4.20. THEOREM. *For each interval I and each $\varepsilon > 0$, there exists an interval J whose interior contains I and*

$$v(J) < v(I) + \varepsilon.$$

4.21. DEFINITION. The **Lebesgue outer measure** of an arbitrary set $E \subset \mathbb{R}^n$, denoted by $\lambda^*(E)$, is defined by

$$\lambda^*(E) = \inf\left\{\sum_{k=1}^{\infty} v(I_k)\right\},$$

where the infimum is taken over all countable collections of closed intervals I_k such that

$$E \subset \bigcup_{k=1}^{\infty} I_k.$$

It may be necessary at times to emphasize the dimension of the Euclidean space in which Lebesgue outer measure is defined. When clarification is needed, we will write $\lambda_n^*(E)$ in place of $\lambda^*(E)$.

Our next result shows that Lebesgue outer measure is an extension of volume.

4.22. THEOREM. *For a closed interval $I \subset \mathbb{R}^n$, $\lambda^*(I) = v(I)$.*

PROOF. The inequality $\lambda^*(I) \leq v(I)$ holds, since $\mathcal{S}$ consisting of I alone can be taken as one of the admissible competitors in Definition 4.21.

To prove the opposite inequality, choose $\varepsilon > 0$ and let $\{I_k\}_{k=1}^{\infty}$ be a sequence of closed intervals such that

$$I \subset \bigcup_{k=1}^{\infty} I_k \quad \text{and} \quad \sum_{k=1}^{\infty} v(I_k) < \lambda^*(I) + \varepsilon. \tag{4.18}$$

For each k, refer to Theorem 4.19 to obtain an interval J_k whose interior contains I_k and

$$v(J_k) \leq v(I_k) + \frac{\varepsilon}{2^k}.$$

We therefore have

$$\sum_{k=1}^{\infty} v(J_k) \leq \sum_{k=1}^{\infty} v(I_k) + \varepsilon.$$

Let $\mathcal{F} = \{\text{interior } (J_k) : k \in \mathbb{N}\}$ and observe that $\mathcal{F}$ is an open cover of the compact set I. Let η be the Lebesgue number for $\mathcal{F}$ (see Exercise 4, Section 3.5). By Theorem 4.19, there is a partition of I into finitely many subintervals, $K_1, K_2, \ldots, K_m$, each with diameter less than η and having the property

$$I = \bigcup_{i=1}^{m} K_i \quad \text{and} \quad v(I) = \sum_{i=1}^{m} v(K_i).$$

Each K_i is contained in the interior of some J_k, say J_{k_i}, although more than one K_i may belong to the same J_{k_i}. Thus, if N_m denotes the smallest number of the J_{k_i}'s that contain the K_i's, we have $N_m \leq m$ and

$$v(I) = \sum_{i=1}^{m} v(K_i) \leq \sum_{i=1}^{N_m} v(J_{k_i}) \leq \sum_{k=1}^{\infty} v(J_k) \leq \sum_{k=1}^{\infty} v(I_k) + \varepsilon.$$

From this and (4.18) it follows that

$$v(I) \leq \lambda^*(I) + 2\varepsilon,$$

which yields the desired result, since ε is arbitrary. □

We will now show that Lebesgue outer measure is a Carathéodory outer measure as defined in Definition 4.15. Once we have established this result, we then will be able to apply the important results established in Section 4.2, such as Theorem 4.18, to Lebesgue outer measure.

4.23. THEOREM. *Lebesgue outer measure, λ^*, defined on $\mathbb{R}^n$ is a Carathéodory outer measure.*

PROOF. We first verify that λ^* is an outer measure. The first three conditions of Definition 4.1 are immediate, so we proceed with the proof of condition (iv). Let $\{A_i\}$ be a countable collection of arbitrary sets in $\mathbb{R}^n$ and let $A = \cup_{i=1}^{\infty} A_i$. We may as well assume that $\lambda^*(A_i) < \infty$ for $i = 1, 2, \ldots$, for otherwise, the conclusion is obvious. Choose $\varepsilon > 0$. For each i, the definition of Lebesgue outer measure implies that there exists a countable family of closed intervals, $\{I_j^{(i)}\}_{j=1}^{\infty}$, such that

$$A_i \subset \bigcup_{j=1}^{\infty} I_j^{(i)} \tag{4.19}$$

and

$$\sum_{j=1}^{\infty} v(I_j^{(i)}) < \lambda^*(A_i) + \frac{\varepsilon}{2^i}. \tag{4.20}$$

Now $A \subset \bigcup_{i,j=1}^{\infty} I_j^{(i)}$ and therefore

$$\begin{aligned}\lambda^*(A) &\leq \sum_{i,j=1}^{\infty} v(I_j^{(i)}) = \sum_{i=1}^{\infty}\sum_{j=1}^{\infty} v(I_j^{(i)}) \\ &\leq \sum_{i=1}^{\infty}(\lambda^*(A_i) + \frac{\varepsilon}{2^i}) = \sum_{i=1}^{\infty} \lambda^*(A_i) + \varepsilon.\end{aligned}$$

Since $\varepsilon > 0$ is arbitrary, the countable subadditivity of λ^* is established.

Finally, we verify (4.15) of Definition 4.15. Let A and B be arbitrary sets with $\boldsymbol{d}(A, B) > 0$. From what has just been proved, we know that $\lambda^*(A \cup B) \leq \lambda^*(A) + \lambda^*(B)$. To prove the opposite inequality, choose $\varepsilon > 0$ and, from the definition of Lebesgue outer measure, select closed intervals $\{I_k\}$ whose union contains $A \cup B$ such that

$$\sum_{k=1}^{\infty} v(I_k) \leq \lambda^*(A \cup B) + \varepsilon.$$

By subdividing each interval I_k into smaller intervals if necessary, we may assume that the diameter of each I_k is less than $\boldsymbol{d}(A, B)$. Thus, the family $\{I_k\}$ consists of two subfamilies, $\{I_k'\}$ and $\{I_k''\}$, where the elements of the first have nonempty intersections with A, while the elements of the second have nonempty intersections with B. Consequently,

$$\lambda^*(A) + \lambda^*(B) \leq \sum_{k=1}^{\infty} v(I_k') + \sum_{k=1}^{\infty} v(I_k'') = \sum_{k=1}^{\infty} v(I_k) \leq \lambda^*(A \cup B) + \varepsilon.$$

Since $\varepsilon > 0$ is arbitrary, this shows that

$$\lambda^*(A) + \lambda^*(B) \leq \lambda^*(A \cup B),$$

which completes the proof. □

4.24. Remark. We will henceforth refer to λ^*-measurable sets as Lebesgue measurable sets. Now that we know that Lebesgue outer measure is a Carathéodory outer measure, it follows from Theorem 4.18 that all Borel sets in $\mathbb{R}^n$ are Lebesgue measurable. In particular, each open set and each closed set are Lebesgue measurable. We will denote by λ the set function obtained by restricting λ^* to the family of Lebesgue measurable sets. Thus, whenever E is a Lebesgue measurable set, we have by definition $\lambda(E) = \lambda^*(E)$; λ is called **Lebesgue measure**. Note that the additivity and continuity properties established in Corollary 4.12 apply to Lebesgue measure.

In view of Theorem 4.22 and the continuity properties of Lebesgue measure, it is possible to show that the Lebesgue measure of elementary geometric figures in $\mathbb{R}^n$ agrees with the notion of volume. For example, suppose that J is an open interval in $\mathbb{R}^n$, that is, suppose J is the product of open 1-dimensional intervals. It is easily seen that $\lambda(J)$ equals the product of lengths of these intervals, because J can be written as the union of an increasing sequence $\{I_k\}$ of closed intervals. Then

$$\lambda(J) = \lim_{k\to\infty} \lambda(I_k) = \lim_{k\to\infty} \text{vol}\,(I_k) = \text{vol}\,(J).$$

Next, we give several characterizations of Lebesgue measurable sets. We recall Definition 3.46, in which the concepts of G_δ and F_σ sets are introduced.

4.25. THEOREM. *The following five conditions are equivalent for Lebesgue outer measure, λ^*, on $\mathbb{R}^n$:*
(i) *$E \subset \mathbb{R}^n$ is λ^*-measurable.*
(ii) *For each $\varepsilon > 0$, there is an open set $U \supset E$ such that $\lambda^*(U \setminus E) < \varepsilon$.*
(iii) *There is a G_δ set $U \supset E$ such that $\lambda^*(U \setminus E) = 0$.*
(iv) *For each $\varepsilon > 0$, there is a closed set $F \subset E$ such that $\lambda^*(E \setminus F) < \varepsilon$.*
(v) *There is an F_σ set $F \subset E$ such that $\lambda^*(E \setminus F) = 0$.*

PROOF. (i) $\Rightarrow$ (ii). We first assume that $\lambda(E) < \infty$. For arbitrary $\varepsilon > 0$, the definition of Lebesgue outer measure implies the existence of closed n-dimensional intervals I_k whose union contains E such that

$$\sum_{k=1}^{\infty} v(I_k) < \lambda^*(E) + \frac{\varepsilon}{2}.$$

Now, for each k, let I_k' be an open interval containing I_k such that $v(I_k') < v(I_k) + \varepsilon/2^{k+1}$. Then, defining $U = \cup_{k=1}^{\infty} I_k'$, we have that U is open, and from (4.3), that

$$\lambda(U) \le \sum_{k=1}^{\infty} v(I_k') < \sum_{k=1}^{\infty} v(I_k) + \varepsilon/2 < \lambda^*(E) + \varepsilon.$$

Thus, $\lambda(U) < \lambda^*(E) + \varepsilon$, and since E is a Lebesgue measurable set of finite measure, we may appeal to Corollary 4.12 (i) to conclude that

$$\lambda(U \setminus E) = \lambda^*(U \setminus E) = \lambda^*(U) - \lambda^*(E) < \varepsilon.$$

If $\lambda(E) = \infty$, for each positive integer i let E_i denote $E \cap B(i)$, where $B(i)$ is the open ball of radius i centered at the origin. Then E_i is a Lebesgue measurable set of finite measure, and thus we may apply the previous step to find an open set $U_i \supset E_i$ such that $\lambda(U_i \setminus E_i) < \varepsilon/2^i$. Let $U = \cup_{i=1}^{\infty} U_i$

and observe that $U \setminus E \subset \cup_{i=1}^{\infty}(U_i \setminus E_i)$. Now use the subadditivity of λ to conclude that

$$\lambda(U \setminus E) \leq \sum_{i=1}^{\infty} \lambda(U_i \setminus E_i) < \sum_{i=1}^{\infty} \frac{\varepsilon}{2^i} = \varepsilon,$$

which establishes the implication (i) $\Rightarrow$ (ii).

(ii) $\Rightarrow$ (iii). For each positive integer i, let U_i denote an open set with the property that $U_i \supset E$ and $\lambda^*(U_i \setminus E) < 1/i$. If we define $U = \cap_{i=1}^{\infty} U_i$, then

$$\lambda^*(U \setminus E) = \lambda^*[\bigcap_{i=1}^{\infty}(U_i \setminus E)] \leq \lim_{i \to \infty} \frac{1}{i} = 0.$$

(iii) $\Rightarrow$ (i). This is obvious, since both U and $(U \setminus E)$ are Lebesgue measurable sets with $E = U \setminus (U \setminus E)$.

(i) $\Rightarrow$ (iv). Assume that E is a measurable set and thus that $\tilde{E}$ is measurable. We know that (ii) is equivalent to (i), and thus for every $\varepsilon > 0$, there is an open set $U \supset \tilde{E}$ such that $\lambda(U \setminus \tilde{E}) < \varepsilon$. Note that

$$E \setminus \tilde{U} = E \cap U = U \setminus \tilde{E}.$$

Since $\tilde{U}$ is closed, $\tilde{U} \subset E$, and $\lambda(E \setminus \tilde{U}) < \varepsilon$, we see that (iv) holds with $F = \tilde{U}$.

The proofs of (iv) $\Rightarrow$ (v) and (v) $\Rightarrow$ (i) are analogous to those of (ii) $\Rightarrow$ (iii) and (iii) $\Rightarrow$ (i), respectively. □

4.26. Remark. The above proof is direct and uses only the definition of Lebesgue measure to establish the various regularity properties. However, another proof, which is not as long but is perhaps less transparent, proceeds as follows. Using only the definition of Lebesgue measure, it can be shown that for every set $A \subset \mathbb{R}^n$, there is a G_δ set $G \supset A$ such that $\lambda(G) = \lambda^*(A)$ (see Exercise 9, Section 4.3). Since λ^* is a Carathéodory outer measure (Theorem 4.23), its measurable sets contain the Borel sets (Theorem 4.18). Consequently, λ^* is a Borel regular outer measurem and thus we may appeal to Corollary 4.56 below to conclude that assertions (ii) and (iv) of Theorem 4.25 hold for every Lebesgue measurable set. The remaining properties follow easily from these two.

Exercises for Section 4.3

1. With λ_t defined by

$$\lambda_t(E) = \lambda(|t|\, E),$$

prove that $\lambda_t(N) = 0$ whenever $\lambda(N) = 0$.

2. Let $I, I_1, I_2, \ldots, I_k$ be intervals in $\mathbb{R}$ such that $I \subset \cup_{i=1}^{k} I_i$. Prove that

$$v(I) \leq \sum_{i=1}^{k} v(I_i),$$

where $v(I)$ denotes the length of the interval I.

3. Complete the proofs of (iv) $\Rightarrow$ (v) and (v) $\Rightarrow$ (i) in Theorem 4.25.
4. Let $E \subset \mathbb{R}$, and for each real number t, let $E + t = \{x + t : x \in E\}$. Prove that $\lambda^*(E) = \lambda^*(E + t)$. From this show that if E is Lebesgue measurable, then so is $E + t$.
5. Prove that Lebesgue measure on $\mathbb{R}^n$ is independent of the choice of coordinate system. That is, prove that Lebesgue outer measure is invariant under rigid motions in $\mathbb{R}^n$.
6. Let P denote an arbitrary $(n-1)$-dimensional hyperplane in $\mathbb{R}^n$. Prove that $\lambda(P) = 0$.
7. In this problem, we want to show that every Lebesgue measurable subset of $\mathbb{R}$ must be "densely populated" in some interval. Thus, let $E \subset \mathbb{R}$ be a Lebesgue measurable set, $\lambda(E) > 0$. For each $\varepsilon > 0$, show that there exists an interval I such that $\dfrac{\lambda(E \cap I)}{\lambda(I)} > 1 - \varepsilon$.
8. Suppose $E \subset \mathbb{R}^n$, $\lambda^*(E) < \infty$, is an arbitrary set with the property that there exists an F_σ-set $F \subset E$ with $\lambda(F) = \lambda^*(E)$. Prove that E is a Lebesgue measurable set.
9. Prove that every set $A \subset \mathbb{R}^n$ is contained within a G_δ-set G with the property $\lambda(G) = \lambda^*(A)$.
10. Let $\{E_k\}$ be a sequence of Lebesgue measurable sets contained in a compact set $K \subset \mathbb{R}^n$. Assume for some $\varepsilon > 0$ that $\lambda(E_k) > \varepsilon$ for all k. Prove that there is some point that belongs to infinitely many E_k's.
11. Let $T\colon \mathbb{R}^n \to \mathbb{R}^n$ be a Lipschitz map. Prove that if $\lambda(E) = 0$, then $\lambda(T(E)) = 0$.

4.4. The Cantor Set

The Cantor set construction discussed in this section provides a method of generating a wide variety of important, and often unexpected, examples in real analysis. One of our main interests here is to show how the Cantor set exhibits the disparities in measuring the "size" of a set by the methods discussed so far, namely, by cardinality, topological density, or Lebesgue measure.

The Cantor set is a subset of the interval $[0,1]$. We will describe it by constructing its complement in $[0,1]$. The construction will proceed in stages. At the first step, let $I_{1,1}$ denote the open interval $(\frac{1}{3}, \frac{2}{3})$. Thus, $I_{1,1}$ is the open middle third of the interval $I = [0,1]$. The second step involves performing the first step on each of the two remaining intervals of $I - I_{1,1}$. That is, we produce two open intervals, $I_{2,1}$ and $I_{2,2}$, each being the open middle third of one of the two intervals that constitute $I - I_{1,1}$. At the ith step we produce 2^{i-1} open intervals, $I_{i,1}, I_{i,2}, \dots, I_{i,2^{i-1}}$, each of length $(\frac{1}{3})^i$. The $(i+1)$th step consists in producing middle thirds of each of the intervals of

$$I - \bigcup_{j=1}^{i} \bigcup_{k=1}^{2^{j-1}} I_{j,k}.$$

With C denoting the Cantor set, we define its complement by

$$I \setminus C = \bigcup_{j=1}^{\infty} \bigcup_{k=1}^{2^{j-1}} I_{j,k}.$$

Note that C is a closed set and that its Lebesgue measure is 0, since

$$\begin{aligned}\lambda(I \setminus C) &= \frac{1}{3} + 2\left(\frac{1}{3^2}\right) + 2^2\left(\frac{1}{3^3}\right) + \cdots \\ &= \sum_{k=0}^{\infty} \frac{1}{3}\left(\frac{2}{3}\right)^k \\ &= 1.\end{aligned}$$

Note that since C is closed, $C = \overline{C}$, and so $\lambda(\overline{C}) = \lambda(C) = 0$. Therefore $\overline{C}$ does not contain any open set, since otherwise, we would have $\lambda(\overline{C}) > 0$. This implies that C is nowhere dense.

Thus, the Cantor set is small in the sense of both measure and topology. We now will determine its cardinality.

Every number $x \in [0,1]$ has a ternary expansion of the form

$$x = \sum_{i=1}^{\infty} \frac{x_i}{3^i},$$

where each x_i is 0, 1, or 2 and we write $x = 0.x_1x_2\ldots$. This expansion is unique except when

$$x = \frac{a}{3^n},$$

where a and n are positive integers with $0 < a < 3^n$ and where 3 does not divide a. In this case, x has the form

$$x = \sum_{i=1}^{n} \frac{x_i}{3^i},$$

where x_i is either 1 or 2. If $x_n = 2$, we will use this expression to represent x. However, if $x_n = 1$, we will use the following representation for x:

$$x = \frac{x_1}{3} + \frac{x_2}{3^2} + \cdots + \frac{x_{n-1}}{3^{n-1}} + \frac{0}{3^n} + \sum_{i=n+1}^{\infty} \frac{2}{3^i}.$$

Thus, with this convention, each number $x \in [0,1]$ has a unique ternary expansion.

Let $x \in I$ and consider its ternary expansion $x = 0.x_1x_2\ldots$, bearing in mind the convention we have adopted above. Observe that $x \notin I_{1,1}$ if and only if $x_1 \neq 1$. Also, if $x_1 \neq 1$, then $x \notin I_{2,1} \cup I_{2,2}$ if and only if $x_2 \neq 1$. Continuing in this way, we see that $x \in C$ if and only if $x_i \neq 1$ for each positive integer i. Thus, there is a one-to-one correspondence between elements of C and all sequences $\{x_i\}$ in which each x_i is either 0 or 2. The cardinality of the latter is $2^{\aleph_0}$, which, in view of Theorem 2.30, is c.

The Cantor construction is very general, and its variations lead to many interesting constructions. For example, if $0 < \alpha < 1$, it is possible to produce a Cantor-type set C_α in $[0,1]$ whose Lebesgue measure is $1-\alpha$. The method of construction is the same as above, except that at the ith step, each of the intervals removed has length $\alpha 3^{-i}$. We leave it as an exercise to show that C_α is nowhere dense and has cardinality c.

Exercises for Section 4.4

1. Prove that the Cantor-type set C_α described at the end of Section 4.4 is nowhere dense, has cardinality c, and has Lebesgue measure $1-\alpha$.
2. Construct an open set $U \subset [0,1]$ such that U is dense in $[0,1]$, $\lambda(U) < 1$, and $\lambda(U \cap (a,b)) > 0$ for every interval $(a,b) \subset [0,1]$.
3. Consider the Cantor-type set $C(\gamma)$ constructed in Section 4.8. Show that this set has the same properties as the standard Cantor set; namely, it has measure zero, it is nowhere dense, and it has cardinality c.
4. Prove that the family of Borel subsets of $\mathbb{R}$ has cardinality c. From this deduce the existence of a Lebesgue measurable set that is not a Borel set.
5. Let E be the set of numbers in $[0,1]$ whose ternary expansions have only finitely many 1's. Prove that $\lambda(E) = 0$.

4.5. Existence of Nonmeasurable Sets

The existence of a subset of $\mathbb{R}$ that is not Lebesgue measurable is intertwined with the fundamentals of set theory. Vitali showed that if the axiom of choice is accepted, then it is possible to establish the existence of nonmeasurable sets. However, in 1970, Solovay proved that using the usual axioms of set theory, but excluding the axiom of choice, it is impossible to prove the existence of a nonmeasurable set.

4.27. THEOREM. *There exists a set $E \subset \mathbb{R}$ that is not Lebesgue measurable.*

PROOF. We define a relation on elements of the real line by saying that x and y are equivalent (written $x \sim y$) if $x - y$ is a rational number. It is easily verified that $\sim$ is an equivalence relation as defined in Definition 1.1. Therefore, the real numbers are decomposed into disjoint equivalence classes. Denote the equivalence class that contains x by E_x. Note that if x is rational, then E_x contains all rational numbers. Note also that each equivalence class is countable, and therefore, since $\mathbb{R}$ is uncountable, there must be an uncountable number of equivalence classes. We now appeal to the axiom of choice, Proposition 1.4, to assert the existence of a set S such that for each equivalence class E, $S \cap E$ consists of precisely one point. If x and y are arbitrary elements of S, then $x - y$ is an irrational number, for otherwise, they would belong to the same equivalence class, contrary to the definition of S. Thus, the set of differences, defined by

$$D_S \colon = \{x - y : x, y \in S\},$$

is a subset of the irrational numbers and therefore cannot contain any interval. Since the Lebesgue outer measure of a set is invariant under translation and $\mathbb{R}$ is the union of the translates of S by every rational number, it follows that $\lambda^*(S) \neq 0$. Thus, if S were a measurable set, we would have $\lambda(S) > 0$. If $\lambda(S) < \infty$, then Lemma 4.28 is contradicted, since D_S cannot contain an interval. If $\lambda(S) = \infty$, then there exists a closed interval I such that $0 < \lambda(S \cap I) < \infty$, and $S \cap I$ is measurable. But this contradicts Lemma 4.28, since $D_{S\cap I} \subset D_S$ cannot contain an interval. □

4.28. LEMMA. *If $S \subset \mathbb{R}$ is a Lebesgue measurable set of positive and finite measure, then the set of differences $D_S := \{x - y : x, y \in S\}$ contains an interval.*

PROOF. For each $\varepsilon > 0$, there is an open set $U \supset S$ with $\lambda(U) < (1 + \varepsilon)\lambda(S)$. Now U is the union of a countable number of disjoint open intervals,

$$U = \bigcup_{k=1}^{\infty} I_k.$$

Therefore,

$$S = \bigcup_{k=1}^{\infty} S \cap I_k \quad \text{and} \quad \lambda(S) = \sum_{k=1}^{\infty} \lambda(S \cap I_k).$$

Since $\lambda(U) = \sum_{k=1}^{\infty} \lambda(I_k) < (1 + \varepsilon)\lambda(S) = (1 + \varepsilon) \sum_{k=1}^{\infty} \lambda(S \cap I_k)$, it follows that $\lambda(I_{k_0}) < (1 + \varepsilon)\lambda(S \cap I_{k_0})$ for some k_0. With the choice of $\varepsilon = \frac{1}{3}$, we have

$$\lambda(S \cap I_{k_0}) > \frac{3}{4}\lambda(I_{k_0}). \tag{4.21}$$

Now select any number t with $0 < |t| < \frac{1}{2}\lambda(I_{k_0})$ and consider the translate of the set $S \cap I_{k_0}$ by t, denoted by $(S \cap I_{k_0}) + t$. Then $(S \cap I_{k_0}) \cup ((S \cap I_{k_0}) + t)$ is contained within an interval of length less than $\frac{3}{2}\lambda(I_{k_0})$. Using the fact that the Lebesgue measure of a set remains unchanged under a translation, we conclude that the sets $S \cap I_{k_0}$ and $(S \cap I_{k_0}) + t$ must intersect, for otherwise, we would contradict (4.21). This means that for each t with $|t| < \frac{1}{2}\lambda(I_{k_0})$, there are points $x, y \in S \cap I_{k_0}$ such that $x - y = t$. That is, the set

$$D_S \supset \{x - y : x, y \in S \cap I_{k_0}\}$$

contains an open interval centered at the origin of length $\lambda(I_{k_0})$. □

Exercises for Section 4.5

1. Referring to the proof of Theorem 4.27, prove that every subset of $\mathbb{R}$ with positive outer Lebesgue measure contains a nonmeasurable subset.
2. Let $\mathcal{N}$ denote the nonmeasurable set constructed in this section. Show that if E is a measurable subset of $\mathcal{N}$, then $\lambda(E) = 0$.

4.6. Lebesgue–Stieltjes Measure

Lebesgue–Stieltjes measure on $\mathbb{R}$ is another important outer measure that is often encountered in applications. A Lebesgue–Stieltjes measure is generated by a nondecreasing function, f, and its definition differs from Lebesgue measure in that the length of an interval appearing in the definition of Lebesgue measure is replaced by the oscillation of f over that interval. We will show that it is a Carathéodory outer measure.

Lebesgue measure is defined using the primitive concept of volume in $\mathbb{R}^n$. In $\mathbb{R}$, the length of a closed interval is used. If f is a nondecreasing function defined on $\mathbb{R}$, then the "length" of a half-open interval $(a, b]$, denoted by $\alpha_f((a, b])$, can be defined by

$$\alpha_f((a,b]) = f(b) - f(a). \tag{4.22}$$

Based on this notion of length, a measure analogous to Lebesgue measure can be generated. This establishes an important connection between measures on $\mathbb{R}$ and monotone functions. To make this connection precise, it is necessary to use half-open intervals in (4.22) rather than closed intervals. It is also possible to develop this procedure in $\mathbb{R}^n$, but it becomes more complicated, cf. [Sa].

4.29. Definition. The Lebesgue–Stieltjes outer measure of an arbitrary set $E \subset \mathbb{R}$ is defined by

$$\lambda_f^*(E) = \inf\left\{ \sum_{h_k \in \mathcal{F}} \alpha_f(h_k) \right\}, \tag{4.23}$$

where the infimum is taken over all countable collections $\mathcal{F}$ of half-open intervals h_k of the form $(a_k, b_k]$ such that

$$E \subset \bigcup_{h_k \in \mathcal{F}} h_k.$$

Later in this section, we will show that there is an identification between Lebesgue–Stieltjes measures and nondecreasing right-continuous functions. This explains why we use half-open intervals of the form $(a, b]$. We could have chosen intervals of the form $[a, b)$, and then we would show that the corresponding Lebesgue–Stieltjes measure could be identified with a left-continuous function.

4.30. Remark. Also, observe that the length of each interval $(a_k, b_k]$ that appears in (4.23) can be assumed to be arbitrarily small, because

$$\alpha_f((a,b]) = f(b) - f(a) = \sum_{k=1}^{N} [f(a_k) - f(a_{k-1})] = \sum_{k=1}^{N} \alpha_f((a_{k-1}, a_k])$$

whenever $a = a_0 < a_1 < \cdots < a_N = b$.

4.31. Theorem. *If $f\colon \mathbb{R} \to \mathbb{R}$ is a nondecreasing function, then λ_f^* is a Carathéodory outer measure on $\mathbb{R}$.*

PROOF. Referring to Definitions 4.1 and 4.15, we need only show that λ_f^* is monotone and countably subadditive, and that it satisfies property (4.15). Verification of the remaining properties is elementary.

For the proof of monotonicity, let $A_1 \subset A_2$ be arbitrary sets in $\mathbb{R}$ and assume, without loss of generality, that $\lambda_f^*(A_2) < \infty$. Choose $\varepsilon > 0$ and consider a countable family of half-open intervals $h_k = (a_k, b_k]$ such that

$$A_2 \subset \bigcup_{k=1}^{\infty} h_k \text{ and } \sum_{k=1}^{\infty} \alpha_f(h_k) \leq \lambda_f^*(A_2) + \varepsilon.$$

Then, since $A_1 \subset \cup_{k=1}^{\infty} h_k$,

$$\lambda_f^*(A_1) \leq \sum_{k=1}^{\infty} \alpha_f(h_k) \leq \lambda_f^*(A_2) + \varepsilon,$$

which establishes the desired inequality, since ε is arbitrary.

The proof of countable subadditivity is virtually identical to the proof of the corresponding result for Lebesgue measure given in Theorem 4.23 and thus will not be repeated here.

Similarly, the proof of property (4.9) of Definition 4.15 runs parallel to the one given in the proof of Theorem 4.23 for Lebesgue measure. Indeed, by Remark 4.30, we may assume that the length of each $(a_k, b_k]$ is less than $\boldsymbol{d}(A, B)$. □

Now that we know that λ_f^* is a Carathéodory outer measure, it follows that the family of λ_f^*-measurable sets contains the Borel sets. As in the case of Lebesgue measure, we denote by λ_f the measure obtained by restricting λ_f^* to its family of measurable sets.

In the case of Lebesgue measure, we proved that $\lambda(I) = \text{vol}\,(I)$ for all intervals $I \subset \mathbb{R}^n$. A natural question is whether the analogous property holds for λ_f.

4.32. THEOREM. *If $f : \mathbb{R} \to \mathbb{R}$ is nondecreasing and right-continuous, then*

$$\lambda_f((a, b]) = f(b) - f(a).$$

PROOF. The proof is similar to that of Theorem 4.22, and as in that situation, it suffices to show that

$$\lambda_f((a, b]) \geq f(b) - f(a).$$

Let $\varepsilon > 0$ and select a cover of $(a, b]$ by a countable family of half-open intervals $(a_i, b_i]$ such that

$$\sum_{i=1}^{\infty} f(b_i) - f(a_i) < \lambda_f((a, b]) + \varepsilon. \tag{4.24}$$

Since f is right-continuous, it follows that for each i,

$$\lim_{t \to 0^+} \alpha_f((a_i, b_i + t]) = \alpha_f((a_i, b_i]).$$

Consequently, we may replace each $(a_i, b_i]$ with $(a_i, b_i']$, where $b_i' > b_i$ and $f(b_i') - f(a_i) < f(b_i) - f(a_i) + \varepsilon/2^i$, thus causing no essential change to (4.24), and thus allowing

$$(a, b] \subset \bigcup_{i=1}^{\infty} (a_i, b_i').$$

Let $a' \in (a, b)$. Then

$$[a', b] \subset \bigcup_{i=1}^{\infty} (a_i, b_i'). \tag{4.25}$$

Let η be the Lebesgue number of this open cover of the compact set $[a', b]$ (see Exercise 3.4). Partition $[a', b]$ into a finite number, say m, of intervals, each of length less than η. We then have

$$[a', b] = \bigcup_{k=1}^{m} [t_{k-1}, t_k],$$

where $t_0 = a'$ and $t_m = b$ and each $[t_{k-1}, t_k]$ is contained in some element of the open cover in (4.25), say $(a_{i_k}, b_{i_k}']$. Furthermore, we can relabel the elements of our partition so that each $[t_{k-1}, t_k]$ is contained in precisely one $(a_{i_k}, b_{i_k}']$. Then

$$\begin{aligned} f(b) - f(a') &= \sum_{k=1}^{m} f(t_k) - f(t_{k-1}) \\ &\leq \sum_{k=1}^{m} f(b_{i_k}') - f(a_{i_k}) \\ &\leq \sum_{k=1}^{\infty} f(b_i') - f(a_i) \\ &\leq \lambda_f((a, b]) + 2\varepsilon. \end{aligned}$$

Since ε is arbitrary, we have

$$f(b) - f(a') \leq \lambda_f((a, b]).$$

Furthermore, the right continuity of f implies

$$\lim_{a' \to a^+} f(a') = f(a)$$

and hence

$$f(b) - f(a) \leq \lambda_f((a, b]),$$

as desired. □

We have just seen that a nondecreasing function f gives rise to a Borel outer measure on $\mathbb{R}$. The converse is readily seen to hold, for if μ is a finite Borel outer measure on $\mathbb{R}$ (see Definition 4.14), let

$$f(x) = \mu((-\infty, x]).$$

Then f is nondecreasing and right-continuous (see Exercise 4.1), and

$$\mu((a,b]) = f(b) - f(a) \quad \text{whenever} \quad a < b.$$

(Incidentally, this now shows why half-open intervals are used in the development.) With f defined in this way, note from our previous result, Theorem 4.32, that the corresponding Lebesgue–Stieltjes measure, λ_f, satisfies

$$\lambda_f((a,b]) = f(b) - f(a),$$

thus proving that μ and λ_f agree on all half-open intervals. Since every open set in $\mathbb{R}$ is a countable union of disjoint half-open intervals, it follows that μ and λ_f agree on all open sets. Consequently, it seems plausible that these measures should agree on all Borel sets. In fact, this is true, because both μ and λ_f^* are outer measures with the approximation property described in Theorem 4.52 below. Consequently, we have the following result.

4.33. THEOREM. *Suppose μ is a finite Borel outer measure on $\mathbb{R}$ and let*

$$f(x) = \mu((-\infty, x]).$$

Then the Lebesgue–Stieltjes measure, λ_f, agrees with μ on all Borel sets.

Exercises for Section 4.6

1. Suppose μ is a finite Borel measure defined on $\mathbb{R}$.
   Let $f(x) = \mu((-\infty, x])$. Prove that f is right continuous.
2. Let $f\colon \mathbb{R} \to \mathbb{R}$ be a nondecreasing function and let λ_f be the Lebesgue–Stieltjes measure generated by f. Prove that $\lambda_f(\{x_0\}) = 0$ if and only if f is left-continuous at x_0.
3. Let f be a nondecreasing function defined on $\mathbb{R}$. Define a Lebesgue–Stieltjes-type measure as follows: For $A \subset \mathbb{R}$ an arbitrary set,

$$\Lambda_f^*(A) = \inf\left\{ \sum_{h_k \in \mathcal{F}} [f(b_k) - f(a_k)] \right\}, \tag{4.26}$$

where the infimum is taken over all countable collections $\mathcal{F}$ of closed intervals of the form $h_k := [a_k, b_k]$ such that

$$E \subset \bigcup_{h_k \in \mathcal{F}} h_k.$$

In other words, the definition of $\Lambda_f^*(A)$ is the same as $\lambda_f^*(A)$ except that closed intervals $[a_k, b_k]$ are used instead of half-open intervals $(a_k, b_k]$.

As in the case of Lebesgue–Stieltjes measure it can be easily seen that Λ_f^* is a Carathéodory measure. (You need not prove this.)

   (a) Prove that $\Lambda_f^*(A) \le \lambda_f^*(A)$ for all sets $A \subset \mathbb{R}^n$.
   (b) Prove that $\Lambda_f^*(B) = \lambda_f^*(B)$ for all Borel sets B if f is left-continuous.

4.7. Hausdorff Measure

As a final illustration of a Carathéodory measure, we introduce s-dimensional Hausdorff (outer) measure in $\mathbb{R}^n$, where s is any nonnegative real number. It will be shown that the only significant values of s are those for which $0 \leq s \leq n$ and that for s in this range, Hausdorff measure provides meaningful measurements of small sets. For example, sets of Lebesgue measure zero may have positive Hausdorff measure.

4.34. Definitions. Hausdorff measure is defined in terms of an auxiliary set function that we introduce first. Let $0 \leq s < \infty$, $0 < \varepsilon \leq \infty$, and let $A \subset \mathbb{R}^n$. Define

$$(4.27)\quad H^s_\varepsilon(A) = \inf\left\{\sum_{i=1}^{\infty} \alpha(s) 2^{-s} (\operatorname{diam} E_i)^s : A \subset \bigcup_{i=1}^{\infty} E_i,\ \operatorname{diam} E_i < \varepsilon\right\},$$

where $\alpha(s)$ is a normalization constant defined by

$$\alpha(s) = \frac{\pi^{\frac{s}{2}}}{\Gamma(\frac{s}{2}+1)},$$

with

$$\Gamma(t) = \int_0^\infty e^{-x} x^{t-1}\, dx, \quad 0 < t < \infty.$$

It follows from the definition that if $\varepsilon_1 < \varepsilon_2$, then $H^s_{\varepsilon_1}(E) \geq H^s_{\varepsilon_2}(E)$. This allows the following, which is the definition of **s-dimensional Hausdorff measure**:

$$H^s(A) = \lim_{\varepsilon \to 0} H^s_\varepsilon(A) = \sup_{\varepsilon > 0} H^s_\varepsilon(A).$$

When s is a positive integer, it turns out that $\alpha(s)$ is the Lebesgue measure of the unit ball in $\mathbb{R}^s$. This makes it possible to prove that H^s assigns to elementary sets the value one would expect. For example, consider $n = 3$. In this case $\alpha(3) = \frac{\pi^{3/2}}{\Gamma(\frac{3}{2}+1)} = \frac{\pi^{3/2}}{\Gamma(\frac{5}{2})} = \frac{\pi^{3/2}}{\frac{3}{4}\pi^{1/2}} = \frac{4}{3}\pi$. Note that $\alpha(3)2^{-3}(\operatorname{diam} B(x,r))^3 = \frac{4}{3}\pi r^3 = \lambda(B(x,r))$. In fact, it can be shown that $H^n(B(x,r)) = \lambda(B(x,r))$ for every ball $B(x,r)$ (see Exercise 1, Section 4.7). In Definition 4.27 we have fixed $n > 0$ and we have defined, for all $0 \leq s < \infty$, the s-dimensional Hausdorff measures H^s. However, for every $A \subset \mathbb{R}^n$ and $s > n$ we have $H^s(A) = 0$. That is, $H^s \equiv 0$ on $\mathbb{R}^n$ for all $s > n$ (see Exercise 3, Section 4.7).

Before deriving the basic properties of H^s, a few observations are in order.

4.35. Remark.

(i) Hausdorff measure could be defined in any metric space, since the essential part of the definition depends only on the notion of diameter of a set.

(ii) The sets E_i in the definition of $H^s_\varepsilon(A)$ are arbitrary subsets of $\mathbb{R}^n$. However, they could be taken to be closed sets, since $\operatorname{diam} E_i = \operatorname{diam} \overline{E_i}$.

(iii) The reason for the restriction of coverings by sets of small diameter is to produce an accurate measurement of sets that are geometrically complicated. For example, consider the set $A = \{(x, \sin(1/x)) : 0 < x \leq 1\}$ in $\mathbb{R}^2$. We will see in Section 7.8 that $H^1(A)$ is the length of the set A, so that in this case, $H^1(A) = \infty$ (it is an instructive exercise to prove this directly from the definition). If no restriction on the diameter of the covering sets were imposed, the measure of A would be finite.

(iv) Often Hausdorff measure is defined without the inclusion of the constant $\alpha(s)2^{-s}$. Then the resulting measure differs from our definition by a constant factor, which is not important unless one is interested in the precise value of the Hausdorff measure.

We now proceed to derive some of the basic properties of Hausdorff measure.

4.36. THEOREM. *For each nonnegative number s, H^s is a Carathéodory outer measure.*

PROOF. We must show that the four conditions of Definition 4.1 are satisfied as well as condition (4.15). The first three conditions of Definition 4.1 are immediate, and so we proceed to show that H^s is countably subadditive. For this, suppose $\{A_i\}$ is a sequence of sets in $\mathbb{R}^n$ and select sets $\{E_{i,j}\}$ such that

$$A_i \subset \bigcup_{j=1}^{\infty} E_{i,j}, \quad \text{diam } E_{i,j} \leq \varepsilon, \quad \sum_{j=1}^{\infty} \alpha(s)2^{-s}(\text{diam } E_{i,j})^s < H^s_{\varepsilon}(A_i) + \frac{\varepsilon}{2^i}.$$

Then, as i and j range through the positive integers, the sets $\{E_{i,j}\}$ produce a countable covering of A, and therefore,

$$\begin{aligned} H^s_{\varepsilon}\left(\bigcup_{i=1}^{\infty} A_i\right) &\leq \sum_{i=1}^{\infty}\sum_{j=1}^{\infty} \alpha(s)2^{-s}(\text{ diam } E_{i,j})^s \\ &= \sum_{i=1}^{\infty}\left[H^s_{\varepsilon}(A_i) + \frac{\varepsilon}{2^i}\right]. \end{aligned}$$

Now $H^s_{\varepsilon}(A_i) \leq H^s(A_i)$ for each i, so that

$$H^s_{\varepsilon}\left(\bigcup_{i=1}^{\infty} A_i\right) \leq \sum_{i=1}^{\infty} H^s(A_i) + \varepsilon.$$

Now taking limits as $\varepsilon \to 0$, we obtain

$$H^s\left(\bigcup_{i=1}^{\infty} A_i\right) \leq \sum_{i=1}^{\infty} H^s(A_i),$$

which establishes countable subadditivity.

Now we will show that condition (4.15) is satisfied. Choose $A, B \subset \mathbb{R}^n$ with $\boldsymbol{d}(A, B) > 0$ and let ε be any positive number less than $\boldsymbol{d}(A, B)$. Let $\{E_i\}$ be a covering of $A \cup B$ with $\text{diam } E_i \leq \varepsilon$. Thus no set E_i intersects both

A and B. Let $\mathcal{A}$ be the collection of the E_i that intersect A, and $\mathcal{B}$ those that intersect B. Then

$$\begin{aligned}\sum_{i=1}^{\infty} \alpha(s)2^{-s}(\operatorname{diam} E_i)^s &\geq \sum_{E\in\mathcal{A}} \alpha(s)2^{-s}(\operatorname{diam} E_i)^s \\ &\quad + \sum_{E\in\mathcal{B}} \alpha(s)2^{-s}(\operatorname{diam} E_i)^s \\ &\geq H_\varepsilon^s(A) + H_\varepsilon^s(B).\end{aligned}$$

Taking the infimum over all such coverings $\{E_i\}$, we obtain

$$H_\varepsilon^s(A \cup B) \geq H_\varepsilon^s(A) + H_\varepsilon^s(B),$$

where ε is any number less than $\boldsymbol{d}(A, B)$. Finally, taking the limit as $\varepsilon \to 0$, we have

$$H^s(A \cup B) \geq H^s(A) + H^s(B).$$

Since we already established (countable) subadditivity of H^s, property (4.15) is thus established, and the proof is concluded. □

Since H^s is a Carathéodory outer measure, it follows from Theorem 4.18 that all Borel sets are H^s-measurable. We next show that H^s is, in fact, a Borel regular outer measure in the sense of Definition 4.14.

4.37. THEOREM. *For each $A \subset \mathbb{R}^n$, there exists a Borel set $B \supset A$ such that*

$$H^s(B) = H^s(A).$$

PROOF. From the previous comment, we already know that H^s is a Borel outer measure. To show that it is a Borel regular outer measure, recall from (ii) in Remark 4.35 above that the sets $\{E_i\}$ in the definition of Hausdorff measure can be taken as closed sets. Suppose $A \subset \mathbb{R}^n$ with $H^s(A) < \infty$, thus implying that $H_\varepsilon^s(A) < \infty$ for all $\varepsilon > 0$. Let $\{\varepsilon_j\}$ be a sequence of positive numbers such that $\varepsilon_j \to 0$, and for each positive integer j, choose closed sets $\{E_{i,j}\}$ such that $\operatorname{diam} E_{i,j} \leq \varepsilon_j$, $A \subset \cup_{i=1}^{\infty} E_{i,j}$, and

$$\sum_{i=1}^{\infty} \alpha(s)2^{-s}(\operatorname{diam} E_{i,j})^s \leq H_{\varepsilon_j}^s(A) + \varepsilon_j.$$

Set

$$A_j = \bigcup_{i=1}^{\infty} E_{i,j} \quad \text{and} \quad B = \bigcap_{j=1}^{\infty} A_j.$$

Then B is a Borel set, and since $A \subset A_j$ for each j, we have $A \subset B$. Furthermore, since

$$B \subset \bigcup_{i=1}^{\infty} E_{i,j}$$

for each j, we have

$$H_{\varepsilon_j}^s(B) \leq \sum_{i=1}^{\infty} \alpha(s)2^{-s}(\operatorname{diam} E_{i,j})^s \leq H_{\varepsilon_j}^s(A) + \varepsilon_j.$$

Since $\varepsilon_j \to 0$ as $j \to \infty$, we obtain $H^s(B) \le H^s(A)$. But $A \subset B$, so that we have $H^s(A) = H^s(B)$. □

4.38. REMARK. The preceding result can be improved. In fact, there is a G_δ set G containing A such that $H^s(G) = H^s(A)$; see Exercise 8, Section 4.7.

4.39. THEOREM. *Suppose* $A \subset \mathbb{R}^n$ *and* $0 \le s < t < \infty$. *Then*

(i) *If* $H^s(A) < \infty$ *then* $H^t(A) = 0$.
(ii) *If* $H^t(A) > 0$ *then* $H^s(A) = \infty$.

PROOF. We need only prove (i), because (ii) is simply a restatement of (i). We state (ii) only to emphasize its importance.

For the proof of (i), choose $\varepsilon > 0$ and a covering of A by sets $\{E_i\}$ with diam $E_i < \varepsilon$ such that

$$\sum_{i=1}^{\infty} \alpha(s)2^{-s}(\operatorname{diam} E_i)^s \le H^s_\varepsilon(A) + 1 \le H^s(A) + 1.$$

Then

$$\begin{aligned} H^t_\varepsilon(A) &\le \sum_{i=1}^{\infty} \alpha(t)2^{-t}(\operatorname{diam} E_i)^t \\ &= \frac{\alpha(t)}{\alpha(s)}2^{s-t}\sum_{i=1}^{\infty} \alpha(s)2^{-s}(\operatorname{diam} E_i)^s(\operatorname{diam} E_i)^{t-s} \\ &\le \frac{\alpha(t)}{\alpha(s)}2^{s-t}\varepsilon^{t-s}[H^s(A) + 1]. \end{aligned}$$

Now let $\varepsilon \to 0$ to obtain $H^t(A) = 0$. □

4.40. DEFINITION. The **Hausdorff dimension** of an arbitrary set $A \subset \mathbb{R}^n$ is the number $0 \le \delta_A \le n$ such that

$$\delta_A = \inf\{t : H^t(A) = 0\} = \sup\{s : H^s(A) = \infty\}.$$

In other words, the Hausdorff dimension δ_A is the unique number such that

$$\begin{aligned} s < \delta_A &\text{ implies } H^s(A) = \infty, \\ t > \delta_A &\text{ implies } H^t(A) = 0. \end{aligned}$$

The existence and uniqueness of δ_A follows directly from Theorem 4.39.

4.41. REMARK. If $s = \delta_A$, then one of the following three possibilities has to occur: $H^s(A) = 0$, $H^s(A) = \infty$, $0 < H^s(A) < \infty$. On the other hand, if $0 < H^s(A) < \infty$, then it follows from Theorem 4.39 that $\delta_A = s$.

The notion of Hausdorff dimension is not very intuitive. Indeed, the Hausdorff dimension of a set need not be an integer. Moreover, if the dimension of a set is an integer k, the set need not resemble a "k-dimensional surface" in any usual sense. See Falconer [27] or Federer [28] for examples of pathological Cantor-like sets with integer Hausdorff dimension. However, we can at least be reassured by the fact that the Hausdorff dimension of an open set $U \subset \mathbb{R}^n$ is n. To verify this, it is sufficient to assume that U is bounded and to prove that

$$0 < H^n(U) < \infty. \tag{4.28}$$

Exercise 2, Section 4.7, deals with the proof of this. Also, it is clear that every countable set has Hausdorff dimension zero; however, there are uncountable sets with dimension zero (see Exercise 7, Section 4.7).

Exercises for Section 4.7

1. If $A \subset \mathbb{R}$ is an arbitrary set, show that $H^1(A) = \lambda^*(A)$.

   4.42. Remark. In this problem, you will see the importance of the constant that appears in the definition of Hausdorff measure. The constant $\alpha(s)$ that appears in the definition of H^s-measure is equal to 2 when $s = 1$. That is, $\alpha(1) = 2$, and therefore, the definition of $H^1(A)$ can be written as

   $$H^1(A) = \lim_{\varepsilon \to 0} H^1_\varepsilon(A),$$

   where

   $$H^1_\varepsilon(A) = \inf\left\{ \sum_{i=1}^{\infty} \operatorname{diam} E_i : A \subset \bigcup_{i=1}^{\infty} E_i \subset \mathbb{R},\ \operatorname{diam} E_i < \varepsilon \right\}.$$

   This result is also true in $\mathbb{R}^n$ but is more difficult to prove. The isodiametric inequality $\lambda^*(A) \le \alpha(n)\left(\frac{\operatorname{diam} A}{2}\right)^n$ (whose proof is omitted in this book) can be used to prove that $H^n(A) = \lambda^*(A)$ for all $A \subset \mathbb{R}^n$.

2. For $A \subset \mathbb{R}^n$, use the isodiametric inequality introduced in Exercise 4.1 to show that

   $$\lambda^*(A) \le H^n(A) \le \alpha(n)\left(\frac{\sqrt{n}}{2}\right)^n \lambda^*(A).$$

3. Show that $H^s \equiv 0$ on $\mathbb{R}^n$ for all $s > n$.
4. Let $C \subset \mathbb{R}^2$ denote the circle of radius 1 and regard C as a topological space with the topology induced from $\mathbb{R}^2$. Define an outer measure $\mathcal{H}$ on C by

   $$\mathcal{H}(A) := \frac{1}{2\pi} H^1(A) \quad \text{for any set} A \subset C.$$

   Later in the book, we will prove that $H^1(C) = 2\pi$. Thus, you may assume that. Show that
   (i) $\mathcal{H}(C) = 1$.
   (ii) Prove that $\mathcal{H}$ is a Borel regular outer measure.
   (iii) Prove that $\mathcal{H}$ is rotationally invariant; that is, prove that for every set $A \subset C$, one has $\mathcal{H}(A) = \mathcal{H}(A')$, where A' is obtained by rotating A through an arbitrary angle.
   (iv) Prove that $\mathcal{H}$ is the only outer measure defined on C that satisfies the previous three conditions.

5. Another Hausdorff-type measure that is frequently used is **Hausdorff spherical measure,** H^s_S. It is defined in the same way as Hausdorff measure (Definition 4.34) except that the sets E_i are replaced by n-balls. Clearly, $H^s(E) \leq H^s_S(E)$ for every set $E \subset \mathbb{R}^n$. Prove that $H^s_S(E) \leq 2^s H^s(E)$ for every set $E \subset \mathbb{R}^n$.
6. Prove that a countable set $A \subset \mathbb{R}^n$ has Hausdorff dimension 0. The following problem shows that the converse is not true.
7. Let $S = \{a_i\}$ be any sequence of real numbers in $(0, 1/2)$. We will now construct a Cantor set $C(S)$ similar in construction to that of $C(\lambda)$ except that the length of the intervals $I_{k,j}$ at the kth stage will not be a constant multiple of those in the preceding stage. Instead, we proceed as follows: Define $I_{0,1} = [0,1]$ and then define the intervals $I_{1,1}, I_{1,2}$ to have length a_1. Proceeding inductively, the intervals $I_{k,i}$ at the kth stage will have length $a_k l(I_{k-1,i})$. Consequently, at the kth stage, we obtain 2^k intervals $I_{k,j}$ each of length

$$s_k = a_1 a_2 \cdots a_k.$$

It can be easily verified that the resulting Cantor set $C(S)$ has cardinality c and is nowhere dense.

The focus of this problem is to determine the Hausdorff dimension of $C(S)$. For this purpose, consider the following function defined on $(0, \infty)$:

$$h(r) := \begin{cases} \frac{r^s}{\log(1/r)} & \text{when } 0 \leq s < 1, \\ r^s \log(1/r) & \text{where } 0 < s \leq 1. \end{cases} \tag{4.29}$$

Note that h is increasing and $\lim_{r \to 0} h(r) = 0$. Corresponding to this function, we will construct a Cantor set $C(S_h)$ that will have interesting properties. We will select inductively numbers $a_1, a_2, \dots$ such that

$$h(s_k) = 2^{-k}. \tag{4.30}$$

That is, a_1 is chosen so that $h(a_1) = 1/2$, i.e., $a_1 = h^{-1}(1/2)$. Now that a_1 has been chosen, let a_2 be the number such that $h(a_1 a_2) = 1/2^2$. In this way, we can choose a sequence $S_h := \{a_1, a_2, \dots\}$ such that (4.30) is satisfied. Now consider the following Hausdorff-type measure:

$$H^h_\varepsilon(A) := \inf \left\{ \sum_{i=1}^{\infty} h(\text{diam } E_i) : A \subset \bigcup_{i=1}^{\infty} E_i \subset \mathbb{R}^n,\ h(\text{diam } E_i) < \varepsilon \right\},$$

and

$$H^h(A) := \lim_{\varepsilon \to 0} H^h_\varepsilon(A).$$

With the Cantor set $C(S_h)$ that was constructed above, it follows that

$$\frac{1}{4} \leq H^h(C(S_h)) \leq 1. \tag{4.31}$$

The proof of this proceeds in precisely the same way as in Section 4.8.

(a) With $s = 0$, our function h in (4.29) becomes $h(r) = 1/\log(1/r)$, and we obtain a corresponding Cantor set $C(S_h)$. With the help of (4.31) prove that the Hausdorff dimension of $C(S_h)$ is zero, thus showing that the converse of Problem 1 is not true.

(b) Now take $s = 1$, and then our function h in (4.29) becomes $h(r) = r\log(1/r)$, and again we obtain a corresponding Cantor set $C(S_h)$. Prove that the Hausdorff dimension of $C(S_h)$ is 1, which shows that there are sets other than intervals in $\mathbb{R}$ that have dimension 1.

8. Prove that for each set $A \subset \mathbb{R}^n$, there exists a G_δ set $B \supset A$ such that

$$H^s(B) = H^s(A). \tag{4.32}$$

4.43. REMARK. This result shows that all three of our primary measures, namely Lebesgue measure, Lebesgue–Stieltjes measure, and Hausdorff measure, share the same important regularity property (4.32).

9. If $A \subset \mathbb{R}^n$ is an arbitrary set and $0 \le t \le n$, prove that if $H^t_\varepsilon(A) = 0$ for some $0 < \varepsilon \le \infty$, then $H^t(A) = 0$.

10. Let $f : \mathbb{R}^n \to \mathbb{R}^m$ be Lipschitz (see Definition 3.65), $E \subset \mathbb{R}^n$, $0 \le s < \infty$. Prove that then

$$\mathcal{H}^s(f(A)) \le C_f^s \mathcal{H}^s(A).$$

4.8. Hausdorff Dimension of Cantor Sets

In this section the Hausdorff dimension of Cantor sets will be determined. Note that for H^1 defined in $\mathbb{R}$, the constant $\alpha(s)2^{-s}$ in (4.34) equals 1.

4.44. DEFINITION. [General Cantor set] Let $0 < \gamma < 1/2$ and set $I_{0,1} = [0,1]$. Let $I_{1,1}$ and $I_{1,2}$ denote the intervals $[0,\gamma]$ and $[1-\gamma, 1]$ respectively. They result by deleting the open middle interval of length $1-2\gamma$. At the next stage, delete the open middle interval of length $\gamma(1-2\gamma)$ of each of the intervals $I_{1,1}$ and $I_{1,2}$. There remain 2^2 closed intervals each of length γ^2. Continuing this process, at the kth stage there are 2^k closed intervals each of length γ^k. Denote these intervals by $I_{k,1}, \ldots, I_{k,2^k}$. We define the generalized Cantor set as

$$C(\gamma) = \bigcap_{k=0}^{\infty} \bigcup_{j=1}^{2^k} I_{k,j}.$$

Note that $C(1/3)$ is the Cantor set discussed in Section 4.4.

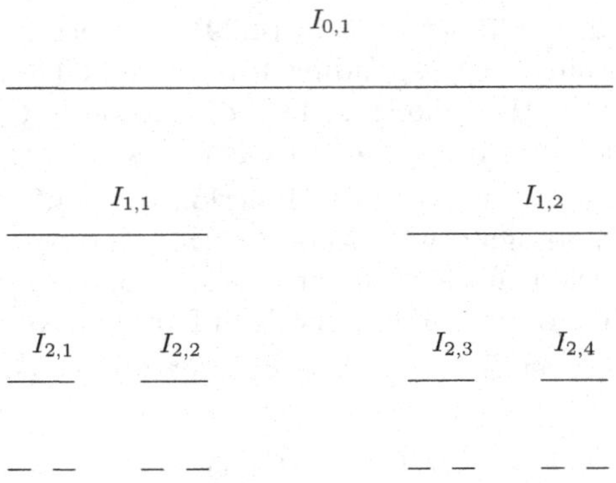

$I_{0,1}$

$I_{1,1}$ $I_{1,2}$

$I_{2,1}$ $I_{2,2}$ $I_{2,3}$ $I_{2,4}$

Since $C(\gamma) \subset \bigcup_{j=1}^{2^k} I_{k,j}$ for each k, it follows that

$$H^s_{\gamma^k}(C(\gamma)) \leq \sum_{j=1}^{2^k} l(I_{k,j})^s = 2^k \gamma^{ks} = (2\gamma^s)^k,$$

where

$$l(I_{k,j}) \text{ denotes the length of } I_{k,j}.$$

If s is chosen so that $2\gamma^s = 1$ (if $s = \log 2 / \log(1/\gamma)$), we have

$$H^s(C(\gamma)) = \lim_{k\to\infty} H^s_{\gamma^k}(C(\gamma)) \leq 1. \tag{4.33}$$

It is important to observe that our choice of s implies that the sum of the sth powers of the lengths of the intervals at any stage is equal to 1; that is,

$$\sum_{j=1}^{2^k} l(I_{k,j})^s = 1. \tag{4.34}$$

Next, we show that $H^s(C(\gamma)) \geq 1/4$, which along with (4.33), implies that the Hausdorff dimension of $C(\gamma)$ equals $\log(2)/\log(1/\gamma)$. We will establish this by showing that if

$$C(\gamma) \subset \bigcup_{i=1}^{\infty} J_i$$

is an open covering $C(\gamma)$ by intervals J_i, then

$$\sum_{i=1}^{\infty} l(J_i)^s \geq \frac{1}{4}. \tag{4.35}$$

Since this is an open cover of the compact set $C(\gamma)$, we can employ the Lebesgue number of this covering to conclude that each interval $I_{k,j}$ of the kth stage is contained in some J_i, provided k is sufficiently large. We will show for every open interval I and fixed ℓ that

$$\sum_{I_{\ell,i}\subset I} l(I_{\ell,i})^s \leq 4l(I)^s. \tag{4.36}$$

This will establish (4.35), since

$$4\sum_i l(J_i)^s \geq \sum_i \sum_{I_{k,j}\subset J_i} l(I_{k,j})^s \quad \text{by (4.36)}$$

$$\geq \sum_{j=1}^{\infty} l(I_{k,j})^s = 1, \quad \text{because } \bigcup_{i=1}^{2^k} I_{k,i} \subset \bigcup_{i=1}^{\infty} J_i \text{ and by(4.34).}$$

To verify (4.36), assume that I contains some interval $I_{\ell,i}$ from the ℓth stage, and let k denote the smallest integer for which I contains some interval $I_{k,j}$ from the kth stage. Then $k \leq \ell$. By considering the construction of our set $C(\gamma)$, it follows that no more than four intervals from the kth stage can intersect I, for otherwise, I would contain some $I_{k-1,i}$. Call the intervals I_{k,k_m}, $m = 1, 2, 3, 4$. Thus,

$$4l(I)^s \geq \sum_{m=1}^{4} l(I_{k,k_m})^s = \sum_{m=1}^{4} \sum_{I_{\ell,i}\subset I_{k,k_m}} l(I_{\ell,i})^s$$

$$\geq \sum_{I_{\ell,i}\subset I} l(I_{\ell,i})^s,$$

which establishes (4.36). This proves that the dimension of $C(\gamma)$ is equal to $\log 2/\log(1/\gamma)$.

4.45. Remark. It can be shown that (4.35) can be improved to read

$$\sum_{i}^{\infty} l(J_i)^s \geq 1, \tag{4.37}$$

which implies the precise result $H^s(C(\gamma)) = 1$ if

$$s = \frac{\log 2}{\log(1/\gamma)}.$$

4.46. Remark. The Cantor sets $C(\gamma)$ are prototypical examples of sets that possess self-similar properties. A set is **self-similar** if it can be decomposed into parts that are geometrically similar to the whole set. For example, the sets $C(\gamma) \cap [0, \gamma]$ and $C(\gamma) \cap [1 - \gamma, 1]$ when magnified by the factor $1/\gamma$ yield a translate of $C(\gamma)$. Self-similarity is the characteristic property of **fractals**.

4.9. Measures on Abstract Spaces

Given an arbitrary set X and a σ-algebra, $\mathcal{M}$, of subsets of X, a nonnegative countably additive set function defined on $\mathcal{M}$ is called a measure. In this section we extract the properties of outer measures when restricted to their measurable sets.

Before proceeding, recall the development of the first three sections of this chapter. We began with the concept of an outer measure on an arbitrary set X and proved that the family of measurable sets forms a σ-algebra. Furthermore, we showed that the outer measure is countably additive on measurable sets. In order to ensure that there are situations in which the family of measurable sets is large, we investigated Carathéodory outer measures on a metric space and established that their measurable sets always contain the Borel sets. We then introduced Lebesgue measure as the primary example of a Carathéodory outer measure. In this development, we begin to see that countable additivity plays a central and indispensable role, and thus, we now call upon a common practice in mathematics of placing a crucial concept in an abstract setting in order to isolate it from the clutter and distractions of extraneous ideas. We begin with the following definition:

4.47. DEFINITION. Let X be a set and $\mathcal{M}$ a σ-algebra of subsets of X. A **measure** on $\mathcal{M}$ is a function $\mu\colon \mathcal{M} \to [0,\infty]$ satisfying the properties

(i) $\mu(\emptyset) = 0$;

(ii) if $\{E_i\}$ is a sequence of disjoint sets in $\mathcal{M}$, then

$$\mu\Big(\bigcup_{i=1}^{\infty} E_i\Big) = \sum_{i=1}^{\infty} \mu(E_i).$$

Thus, a measure is a countably additive set function defined on $\mathcal{M}$. Sometimes the notion of **finite additivity** is useful. It states that

(ii)′ If $E_1, E_2, \ldots, E_k$ is any finite family of disjoint sets in $\mathcal{M}$, then

$$\mu\Big(\bigcup_{i=1}^{k} E_i\Big) = \sum_{i=1}^{k} \mu(E_i).$$

If μ satisfies (i) and (ii′) but not necessarily (ii), then μ is called a **finitely additive measure**. The triple $(X, \mathcal{M}, \mu)$ is called a **measure space**, and the sets that constitute $\mathcal{M}$ are called **measurable sets**. To be precise, these sets should be referred to as $\mathcal{M}$-measurable, to indicate their dependence on $\mathcal{M}$. However, in most situations, it will be clear from the context which σ-algebra is intended, and thus the more involved notation will not be required. If $\mathcal{M}$ constitutes the family of Borel sets in a metric space X, then μ is called a **Borel measure**. A measure μ is said to be **finite** if $\mu(X) < \infty$, and **σ-finite** if X can be written as $X = \cup_{i=1}^{\infty} E_i$, where $\mu(E_i) < \infty$ for each i. A measure μ with the property that all subsets of sets of μ-measure zero are measurable is said to be **complete**, and $(X, \mathcal{M}, \mu)$ is called a **complete measure space**. A Borel measure on a topological space X that is finite on compact sets is called a **Radon measure.** Thus, Lebesgue measure on $\mathbb{R}^n$ is a Radon measure, but s-dimensional Hausdorff measure, $0 \le s < n$, is not (see Theorem 4.63 for regularity properties of Borel measures).

We emphasize that the notation $\mu(E)$ implies that E is an element of $\mathcal{M}$, since μ is defined only on $\mathcal{M}$. Thus, when we write $\mu(E_i)$ as in the definition above, it should be understood that the sets E_i are necessarily elements of $\mathcal{M}$.

4.48. EXAMPLES. Here are some examples of measures.

(i) $(\mathbb{R}^n, \mathcal{M}, \lambda)$, where λ is Lebesgue measure and $\mathcal{M}$ is the family of Lebesgue measurable sets.

(ii) $(X, \mathcal{M}, \varphi)$, where φ is an outer measure on an abstract set X and $\mathcal{M}$ is the family of φ-measurable sets.

(iii) $(X, \mathcal{M}, \delta_{x_0})$, where X is an arbitrary set and δ_{x_0} is an outer measure defined by

$$\delta_{x_0}(E) = \begin{cases} 1 & \text{if } x_0 \in E, \\ 0 & \text{if } x_0 \notin E. \end{cases}$$

The point $x_0 \in X$ is selected arbitrarily. It can easily be shown that all subsets of X are δ_{x_0}-measurable, and therefore $\mathcal{M}$ is taken as the family of all subsets of X.

(iv) $(\mathbb{R}, \mathcal{M}, \mu)$, where $\mathcal{M}$ is the family of all Lebesgue measurable sets, and $x_0 \in \mathbb{R}$ and μ is defined by

$$\mu(E) = \lambda(E \setminus \{x_0\}) + \delta_{x_0}(E)$$

whenever $E \in \mathcal{M}$.

(v) $(X, \mathcal{M}, \mu)$, where $\mathcal{M}$ is the family of all subsets of an arbitrary space X and where $\mu(E)$ is defined as the number (possibly infinite) of points in $E \in \mathcal{M}$.

The proof of Corollary 4.12 used only the properties of an outer measure that an abstract measure possesses, and therefore most of the following do not require a proof.

4.49. THEOREM. *Let $(X, \mathcal{M}, \mu)$ be a measure space and suppose $\{E_i\}$ is a sequence of sets in $\mathcal{M}$.*

(i) *(Monotonicity) If $E_1 \subset E_2$, then $\mu(E_1) \leq \mu(E_2)$.*

(ii) *(Subtractivity) If $E_1 \subset E_2$ and $\mu(E_1) < \infty$, then $\mu(E_2 - E_1) = \mu(E_2) - \mu(E_1)$.*

(iii) *(Countable subadditivity)*

$$\mu\Big(\bigcup_{i=1}^{\infty} E_i\Big) \leq \sum_{i=1}^{\infty} \mu(E_i).$$

(iv) *(Continuity from the left) If $\{E_i\}$ is an increasing sequence of sets, that is, if $E_i \subset E_{i+1}$ for each i, then*

$$\mu\Big(\bigcup_{i=1}^{\infty} E_i\Big) = \mu(\lim_{i\to\infty} E_i) = \lim_{i\to\infty} \mu(E_i).$$

(v) (*Continuity from the right*) *If* $\{E_i\}$ *is a decreasing sequence of sets, that is, if* $E_i \supset E_{i+1}$ *for each* i, *and if* $\mu(E_{i_0}) < \infty$ *for some* i_0, *then*

$$\mu(\bigcap_{i=1}^{\infty} E_i) = \mu(\lim_{i\to\infty} E_i) = \lim_{i\to\infty} \mu(E_i).$$

(vi)

$$\mu(\liminf_{i\to\infty} E_i) \leq \liminf_{i\to\infty} \mu(E_i).$$

(vii) *If*

$$\mu(\bigcup_{i=i_0}^{\infty} E_i) < \infty$$

for some positive integer i_0, *then*

$$\mu(\limsup_{i\to\infty} E_i) \geq \limsup_{i\to\infty} \mu(E_i).$$

PROOF. Only (i) and (iii) have not been established in Corollary 4.12. For (i), observe that if $E_1 \subset E_2$, then $\mu(E_2) = \mu(E_1) + \mu(E_2 - E_1) \geq \mu(E_1)$.

(iii) Refer to Lemma 4.7 to obtain a sequence of disjoint measurable sets $\{A_i\}$ such that $A_i \subset E_i$ and

$$\bigcup_{i=1}^{\infty} E_i = \bigcup_{i=1}^{\infty} A_i.$$

Then,

$$\mu(\bigcup_{i=1}^{\infty} E_i) = \mu(\bigcup_{i=1}^{\infty} A_i) = \sum_{i=1}^{\infty} \mu(A_i) \leq \sum_{i=1}^{\infty} \mu(E_i).$$ □

One property that is characteristic of an outer measure φ but is not enjoyed by abstract measures in general is the following: if $\varphi(E) = 0$, then E is φ-measurable, and consequently, so is every subset of E. Not all measures are complete, but this is not a crucial defect, since every measure can easily be completed by enlarging its domain of definition to include all subsets of sets of measure zero.

4.50. THEOREM. *Suppose* $(X, \mathcal{M}, \mu)$ *is a measure space. Define* $\overline{\mathcal{M}} = \{A \cup N : A \in \mathcal{M},\ N \subset B$ *for some* $B \in \mathcal{M}$ *such that* $\mu(B) = 0\}$ *and define* $\bar{\mu}$ *on* $\overline{\mathcal{M}}$ *by* $\bar{\mu}(A \cup N) = \mu(A)$. *Then* $\overline{\mathcal{M}}$ *is a* σ-*algebra,* $\bar{\mu}$ *is a complete measure on* $\overline{\mathcal{M}}$, *and* $(X, \overline{\mathcal{M}}, \bar{\mu})$ *is a complete measure space. Moreover,* $\bar{\mu}$ *is the only complete measure on* $\overline{\mathcal{M}}$ *that is an extension of* μ.

PROOF. It is easy to verify that $\overline{\mathcal{M}}$ is closed under countable unions, since this is true for sets of measure zero. To show that $\overline{\mathcal{M}}$ is closed under complementation, note that with sets A, N, and B as in the definition of $\overline{\mathcal{M}}$, it may be assumed that $A \cap N = \emptyset$, because $A \cup N = A \cup (N \setminus A)$ and $N \setminus A$ is a subset of a measurable set of measure zero, namely $B \setminus A$. It can be readily verified that

$$A \cup N = (A \cup B) \cap ((\tilde{B} \cup N) \cup (A \cap B))$$

and therefore

$$\begin{aligned}(A \cup N)^{\sim} &= (A \cup B)^{\sim} \cup ((\tilde{B} \cup N) \cup (A \cap B))^{\sim} \\ &= (A \cup B)^{\sim} \cup ((B \cap \tilde{N}) \cap (A \cap B)^{\sim}) \\ &= (A \cup B)^{\sim} \cup ((B \setminus N) \setminus A \cap B).\end{aligned}$$

Since $(A \cup B)^{\sim} \in \mathcal{M}$ and $(B \setminus N) \setminus A \cap B$ is a subset of a set of measure zero, it follows that $\overline{\mathcal{M}}$ is closed under complementation. Consequently, $\overline{\mathcal{M}}$ is a σ-algebra.

To show that the definition of $\bar{\mu}$ is unambiguous, suppose $A_1 \cup N_1 = A_2 \cup N_2$, where $N_i \subset B_i$, $i = 1, 2$. Then $A_1 \subset A_2 \cup N_2$ and

$$\bar{\mu}(A_1 \cup N_1) = \mu(A_1) \leq \mu(A_2) + \mu(B_2) = \mu(A_2) = \bar{\mu}(A_2 \cup N_2).$$

Similarly, we have the opposite inequality. It is easily verified that $\bar{\mu}$ is complete, since $\bar{\mu}(N) = \bar{\mu}(\emptyset \cup N) = \mu(\emptyset) = 0$. Uniqueness is left as Exercise 2, Section 4.9. □

Exercises for Section 4.9

1. Let $\{\mu_k\}$ be a sequence of measures on a measure space such that $\mu_{k+1}(E) \geq \mu_k(E)$ for each measurable set E. With μ defined as $\mu(E) = \lim_{k\to\infty} \mu_k(E)$, prove that μ is a measure.
2. Prove that the measure $\bar{\mu}$ introduced in Theorem 4.50 is a unique extension of μ.
3. Let μ be finite Borel measure on $\mathbb{R}^2$. For fixed $r > 0$, let $C_x = \{y : |y - x| = r\}$ and define $f\colon \mathbb{R}^2 \to \mathbb{R}$ by $f(x) = \mu[C_x]$. Prove that f is continuous at x_0 if and only if $\mu[C_{x_0}] = 0$.
4. Let μ be finite Borel measure on $\mathbb{R}^2$. For fixed $r > 0$, define $f\colon \mathbb{R}^2 \to \mathbb{R}$ by $f(x) = \mu[B(x, r)]$. Prove that f is continuous at x_0 if and only if $\mu[C_{x_0}] = 0$.
5. This problem is set within the context of Theorem 4.50 of the text. With μ given as in Theorem 4.50, define an outer measure μ^* on all subsets of X in the following way: For an arbitrary set $A \subset X$ let

$$\mu^*(A) := \inf \left\{ \sum_{i=1}^{\infty} \mu(E_i) \right\}$$

where the infimum is taken over all countable collections $\{E_i\}$ such that

$$A \subset \bigcup_{i=1}^{\infty} E_i, \qquad E_i \in \mathcal{M}.$$

Prove that $\overline{\mathcal{M}} = \mathcal{M}^*$ where $\mathcal{M}^*$ denotes the σ-algebra of μ^*-measurable sets and that $\bar{\mu} = \mu^*$ on $\overline{\mathcal{M}}$.

6. In an abstract measure space $(X, \mathcal{M}, \mu)$, if $\{A_i\}$ is a countable disjoint family of sets in $\mathcal{M}$, we know that

$$\mu(\bigcup_{i=1}^{\infty} A_i) = \sum_{i=1}^{\infty} \mu(A_i).$$

Prove that converse is essentially true. That is, under the assumption that $\mu(X) < \infty$, prove that if $\{A_i\}$ is a countable family of sets in $\mathcal{M}$ with the property that

$$\mu(\bigcup_{i=1}^{\infty} A_i) = \sum_{i=1}^{\infty} \mu(A_i),$$

then $\mu(A_i \cap A_j) = 0$ whenever $i \neq j$.

7. Recall that an **algebra** in a space X is a nonempty collection of subsets of X that is closed under the operations of finite unions and complements. Also recall that a **measure on an algebra,** $\mathcal{A}$, is a function $\mu\colon \mathcal{A} \to [0, \infty]$ satisfying the properties
   (i) $\mu(\emptyset) = 0$,
   (ii) if $\{A_i\}$ is a disjoint sequence of sets in $\mathcal{A}$ whose union is also in $\mathcal{A}$, then

$$\mu\left(\bigcup_{i=1}^{\infty} A_i\right) = \sum_{i=1}^{\infty} \mu(A_i).$$

Finally, recall that a measure μ on an algebra $\mathcal{A}$ generates a set function μ^* defined on all subsets of X in the following way: for each $E \subset X$, let

$$\mu^*(E) := \inf\left\{\sum_{i=1}^{\infty} \mu(A_i)\right\}, \tag{4.38}$$

where the infimum is taken over countable collections $\{A_i\}$ such that

$$E \subset \bigcup_{i=1}^{\infty} A_i, \quad A_i \in \mathcal{A}.$$

Assuming that $\mu(X) < \infty$, prove that μ^* is a regular outer measure.

8. Give an example of two σ-algebras in a set X whose union is not an algebra.
9. Prove that if the union of two σ-algebras is an algebra, then it is necessarily a σ-algebra.
10. Let φ be an outer measure on a set X and let $\mathcal{M}$ denote the σ-algebra of φ-measurable sets. Let μ denote the measure defined by $\mu(E) = \varphi(E)$ whenever $E \in \mathcal{M}$; that is, μ is the restriction of φ to $\mathcal{M}$. Since, in particular, $\mathcal{M}$ is an algebra, we know that μ generates an outer measure μ^*. Prove:

(a) $\mu^*(E) \geq \varphi(E)$ whenever $E \in \mathcal{M}$.
(b) $\mu^*(A) = \varphi(A)$ for $A \subset X$ if and only if there exists $E \in \mathcal{M}$ such that $E \supset A$ and $\varphi(E) = \varphi(A)$.
(c) $\mu^*(A) = \varphi(A)$ for all $A \subset X$ if φ is regular.

4.10. Regular Outer Measures

In any context, the ability to approximate a complex entity by a simpler one is very important. The following result is one of many such approximations that occur in measure theory; it states that for outer measures with rather general properties, it is possible to approximate Borel sets by both open and closed sets. Note the strong parallel to similar results for Lebesgue measure and Hausdorff measure; see Theorems 4.25 and 4.37 along with Exercise 9, Section 4.3.

4.51. THEOREM. *If φ is a regular outer measure on X, then*
(i) *If $A_1 \subset A_2 \subset \ldots$ is an increasing sequence of arbitrary sets, then*

$$\varphi\left(\bigcup_{i=1}^{\infty} A_i\right) = \lim_{i\to\infty} \varphi(A_i).$$

(ii) *If $A \cup B$ is φ-measurable, $\varphi(A) < \infty$, $\varphi(B) < \infty$, and $\varphi(A \cup B) = \varphi(A) + \varphi(B)$, then both A and B are φ-measurable.*

PROOF. (i): Choose φ-measurable sets $C_i \supset A_i$ with $\varphi(C_i) = \varphi(A_i)$. The φ-measurable sets

$$B_i := \bigcap_{j=i}^{\infty} C_j$$

form an ascending sequence that satisfies the conditions $A_i \subset B_i \subset C_i$ as well as

$$\varphi\left(\bigcup_{i=1}^{\infty} A_i\right) \leq \varphi\left(\bigcup_{i=1}^{\infty} B_i\right) = \lim_{i\to\infty} \varphi(B_i) \leq \lim_{i\to\infty} \varphi(C_i) = \lim_{i\to\infty} \varphi(A_i).$$

Hence, it follows that

$$\varphi\left(\bigcup_{i=1}^{\infty} A_i\right) \leq \lim_{i\to\infty} \varphi(A_i).$$

The opposite inequality is immediate, since

$$\varphi\left(\bigcup_{i=1}^{\infty} A_i\right) \geq \varphi(A_k) \text{ for each} k \in \mathbb{N}.$$

(ii): Choose a φ-measurable set $C' \supset A$ such that $\varphi(C') = \varphi(A)$. Then, with $C := C' \cap (A \cup B)$, we have a φ-measurable set C with $A \subset C \subset A \cup B$ and $\varphi(C) = \varphi(A)$. Note that

$$\varphi(B \cap C) = 0, \tag{4.39}$$

because the φ-measurability of C implies

$$\varphi(B) = \varphi(B \cap C) + \varphi(B \setminus C)$$

and

$$\begin{aligned}\varphi(C)+\varphi(B) &= \varphi(A)+\varphi(B)\\ &= \varphi(A\cup B)\\ &= \varphi((A\cup B)\cap C)+\varphi((A\cup B)\setminus C)\\ &= \varphi(C)+\varphi(B\setminus C)\\ &= \varphi(C)+\varphi(B)-\varphi(B\cap C).\end{aligned}$$

This implies that $\varphi(B\cap C)=0$, because $\varphi(B)+\varphi(C)<\infty$. Since $C\subset A\cup B$, we have
$C\setminus A\subset B$, which leads to $(C\setminus A)\subset B\cap C$. Then (4.39) implies $\varphi(C\setminus A)=0$, which yields the φ-measurability of A, since $A=C\setminus(C\setminus A)$. Finally, B is also φ-measurable, since the roles of A and B are interchangeable. □

4.52. THEOREM. *Suppose φ is an outer measure on a metric space X whose measurable sets contain the Borel sets; that is, φ is a* **Borel outer measure**. *Then for each Borel set $B\subset X$ with $\varphi(B)<\infty$ and each $\varepsilon>0$, there exists a closed set $F\subset B$ such that*

$$\varphi(B\setminus F)<\varepsilon.$$

Furthermore, suppose

$$B\subset\bigcup_{i=1}^{\infty}V_i,$$

where each V_i is an open set with $\varphi(V_i)<\infty$. Then for each $\varepsilon>0$, there is an open set $W\supset B$ such that

$$\varphi(W\setminus B)<\varepsilon.$$

PROOF. For the proof of the first part, select a Borel set B with $\varphi(B)<\infty$ and define a set function μ by

$$\mu(A)=\varphi(A\cap B) \tag{4.40}$$

whenever $A\subset X$. It is easy to verify that μ is an outer measure on X whose measurable sets include all φ-measurable sets (see Exercise 3, Section 4.2) and thus all open sets. The outer measure μ is introduced merely to allow us to work with an outer measure for which $\mu(X)<\infty$.

Let $\mathcal{D}$ be the family of all μ-measurable sets $A\subset X$ with the following property: for each $\varepsilon>0$, there is a closed set $F\subset A$ such that $\mu(A\setminus F)<\varepsilon$. The first part of the theorem will be established by proving that $\mathcal{D}$ contains all Borel sets. Obviously, $\mathcal{D}$ contains all closed sets. It also contains all open sets. Indeed, if U is an open set, then the closed sets

$$F_i=\{x:\boldsymbol{d}(x,\tilde{U})\geq 1/i\}$$

have the property that $F_1\subset F_2\subset\ldots$ and

$$U=\bigcup_{i=1}^{\infty}F_i$$

and therefore that

$$\bigcap_{i=1}^{\infty}(U \setminus F_i) = \emptyset.$$

Therefore, since $\mu(X) < \infty$, Corollary 4.12 (iv) yields

$$\lim_{i\to\infty} \mu(U \setminus F_i) = 0,$$

which shows that $\mathcal{D}$ contains all open sets U.

Since $\mathcal{D}$ contains all open and closed sets, according to Theorem 4.17, we need only show that $\mathcal{D}$ is closed under countable unions and countable intersections to conclude that it also contains all Borel sets. For this purpose, suppose $\{A_i\}$ is a sequence of sets in $\mathcal{D}$ and for given $\varepsilon > 0$, choose closed sets $C_i \subset A_i$ with $\mu(A_i \setminus C_i) < \varepsilon/2^i$. Since

$$\bigcap_{i=1}^{\infty} A_i \setminus \bigcap_{i=1}^{\infty} C_i \subset \bigcup_{i=1}^{\infty}(A_i \setminus C_i)$$

and

$$\bigcup_{i=1}^{\infty} A_i \setminus \bigcup_{i=1}^{\infty} C_i \subset \bigcup_{i=1}^{\infty}(A_i \setminus C_i),$$

it follows that

$$\mu\Big[\bigcap_{i=1}^{\infty} A_i \setminus \bigcap_{i=1}^{\infty} C_i\Big] \le \mu\Big[\bigcup_{i=1}^{\infty}(A_i \setminus C_i)\Big] < \sum_{i=1}^{\infty}\frac{\varepsilon}{2^i} = \varepsilon \tag{4.41}$$

and

$$\lim_{k\to\infty}\mu\Big[\bigcup_{i=1}^{\infty} A_i \setminus \bigcup_{i=1}^{k} C_i\Big] = \mu\Big[\bigcup_{i=1}^{\infty} A_i \setminus \bigcup_{i=1}^{\infty} C_i\Big] \le \mu\Big[\bigcup_{i=1}^{\infty}(A_i \setminus C_i)\Big] < \varepsilon. \tag{4.42}$$

Consequently, there exists a positive integer k such that

$$\mu\Big[\bigcup_{i=1}^{\infty} A_i \setminus \bigcup_{i=1}^{k} C_i\Big] < \varepsilon. \tag{4.43}$$

We have used the fact that $\cup_{i=1}^{\infty} A_i$ and $\cap_{i=1}^{\infty} A_i$ are μ-measurable, and in (4.42), we again have used (iv) of Corollary 4.12. Since the sets $\cap_{i=1}^{\infty} C_i$ and $\cup_{i=1}^{k} C_i$ are closed subsets of $\cap_{i=1}^{\infty} A_i$ and $\cup_{i=1}^{\infty} A_i$ respectively, it follows from (4.41) and (4.43) that $\mathcal{D}$ is closed under the operations of countable unions and intersections.

To prove the second part of the theorem, consider the Borel sets $V_i \setminus B$ and use the first part to find closed sets $C_i \subset (V_i \setminus B)$ such that

$$\varphi[(V_i \setminus C_i) \setminus B] = \varphi[(V_i \setminus B) \setminus C_i] < \frac{\varepsilon}{2^i}.$$

For the desired set W in the statement of the theorem, let $W = \cup_{i=1}^{\infty}(V_i \setminus C_i)$ and observe that

$$\varphi(W \setminus B) \le \sum_{i=1}^{\infty}\varphi[(V_i \setminus C_i) \setminus B] < \sum_{i=1}^{\infty}\frac{\varepsilon}{2^i} = \varepsilon.$$

Moreover, since $B \cap V_i \subset V_i \setminus C_i$, we have

$$B = \bigcup_{i=1}^{\infty} (B \cap V_i) \subset \bigcup_{i=1}^{\infty} (V_i \setminus C_i) = W. \qquad \square$$

4.53. COROLLARY. *If two finite Borel outer measures agree on all open (or closed) sets, then they agree on all Borel sets. In particular, in $\mathbb{R}$, if they agree on all half-open intervals, then they agree on all Borel sets.*

4.54. REMARK. The preceding theorem applies directly to every Carathéodory outer measure, since its measurable sets contain the Borel sets. In particular, the result applies to both Lebesgue–Stieltjes measure λ_f^* and Lebesgue measure and thereby furnishes an alternative proof of Theorem 4.25.

4.55. REMARK. In order to underscore the importance of Theorem 4.52, let us return to Theorem 4.33. There we are given a Borel outer measure μ with $\mu(\mathbb{R}) < \infty$. Then define a function f by

$$f(x) := \mu((-\infty, x])$$

and observe that f is nondecreasing and right-continuous. Consequently, f produces a Lebesgue–Stieltjes measure λ_f^* with the property that

$$\lambda_f^*((a, b]) = f(b) - f(a)$$

for each half-open interval. However, it is clear from the definition of f that μ also enjoys the same property:

$$\mu((a, b]) = f(b) - f(a).$$

Thus, μ and λ_f^* agree on all half-open intervals and therefore they agree on all open sets, since every open set is the disjoint union of half-open intervals. Hence, from Corollary 4.53, they agree on all Borel sets. This allows us to conclude that there is a unique correspondence between nondecreasing right-continuous functions and finite Borel measures on $\mathbb{R}$.

It is natural to ask whether the previous theorem remains true if B is assumed to be only φ-measurable rather than being a Borel set. In general the answer is no, but it is true if φ is assumed to be a Borel regular outer measure. To see this, observe that if φ is a Borel regular outer measure and A is a φ-measurable set with $\varphi(A) < \infty$, then there exist Borel sets B_1 and B_2 such that

$$B_2 \subset A \subset B_1 \quad \text{and} \quad \varphi(B_1 \setminus B_2) = 0. \tag{4.44}$$

PROOF. For this, first choose a Borel set $B_1 \supset A$ with $\varphi(B_1) = \varphi(A)$. Then choose a Borel set $D \supset B_1 \setminus A$ such that $\varphi(D) = \varphi(B_1 \setminus A)$. Note that since A and B_1 are φ-measurable, we have $\varphi(B_1 \setminus A) = \varphi(B_1) - \varphi(A) = 0$. Now take $B_2 = B_1 \setminus D$. Thus, we have the following corollary. $\square$

4.56. COROLLARY. *In the previous theorem, if φ is assumed to be a* **Borel regular outer measure**, *then the conclusions remain valid if the phrase "for each Borel set B" is replaced by "for each φ-measurable set B."*

Although not all Carathéodory outer measures are Borel regular, the following theorems show that they do agree with Borel regular outer measures on the Borel sets.

4.57. THEOREM. *Let φ be a Carathéodory outer measure. For each set $A \subset X$, define*

$$\psi(A) = \inf\{\varphi(B) : B \supset A,\ B \text{ a Borel set}\}. \tag{4.45}$$

Then ψ is a Borel regular outer measure on X that agrees with φ on all Borel sets.

PROOF. We leave it as an easy exercise (Exercise 4, Section 4.2) to show that ψ is an outer measure on X. To show that all Borel sets are ψ-measurable, suppose $D \subset X$ is a Borel set. Then, by Definition 4.3, we must show that

$$\psi(A) \geq \psi(A \cap D) + \psi(A \setminus D) \tag{4.46}$$

whenever $A \subset X$. For this we may as well assume $\psi(A) < \infty$. For $\varepsilon > 0$, choose a Borel set $B \supset A$ such that $\varphi(B) < \psi(A) + \varepsilon$. Then, since φ is a Borel outer measure (Theorem 4.18), we have

$$\begin{aligned} \varepsilon + \psi(A) \geq \varphi(B) &\geq \varphi(B \cap D) + \varphi(B \setminus D) \\ &\geq \psi(A \cap D) + \psi(A \setminus D), \end{aligned}$$

which establishes (4.46), since ε is arbitrary. Also, if B is a Borel set, we claim that $\psi(B) = \varphi(B)$. Half the claim is obvious, because $\psi(B) \leq \varphi(B)$ by definition. As for the opposite inequality, choose a sequence of Borel sets $D_i \subset X$ with $D_i \supset B$ and $\lim_{i\to\infty} \varphi(D_i) = \psi(B)$. Then, with $D = \liminf_{i\to\infty} D_i$, we have by Corollary 4.12 (v),

$$\varphi(B) \leq \varphi(D) \leq \liminf_{i\to\infty} \varphi(D_i) = \psi(B),$$

which establishes the claim. Finally, since φ and ψ agree on Borel sets, we have for arbitrary $A \subset X$,

$$\begin{aligned} \psi(A) &= \inf\{\varphi(B) : B \supset A,\ B \text{ a Borel set}\} \\ &= \inf\{\psi(B) : B \supset A,\ B \text{ a Borel set}\}. \end{aligned}$$

For each positive integer i, let $B_i \supset A$ be a Borel set with $\psi(B_i) < \psi(A) + 1/i$. Then

$$B = \bigcap_{i=1}^{\infty} B_i \supset A$$

is a Borel set with $\psi(B) = \psi(A)$, which shows that ψ is Borel regular. $\square$

4.11. Outer Measures Generated by Measures

Thus far we have seen that with every outer measure there is an associated measure. This measure is defined by restricting the outer measure to its measurable sets. In this section, we consider the situation in reverse. It is shown that a measure defined on an abstract space generates an outer measure and that if this measure is σ-finite, the extension is unique. An important consequence of this development is that every finite Borel measure is necessarily regular.

We begin by describing a process by which a measure generates an outer measure. This method is reminiscent of the one used to define Lebesgue–Stieltjes measure. Actually, this method does not require the measure to be defined on a σ-algebra, but only on an algebra of sets. We make this precise in the following definition.

4.58. DEFINITIONS. An **algebra** in a space X is defined as a nonempty collection of subsets of X that is closed under the operations of finite unions and complements. Thus, the only difference between an algebra and a σ-algebra is that the latter is closed under countable unions. By a **measure on an algebra,** $\mathcal{A}$, we mean a function $\mu\colon \mathcal{A} \to [0, \infty]$ satisfying the properties

(i) $\mu(\emptyset) = 0$,

(ii) if $\{A_i\}$ is a disjoint sequence of sets in $\mathcal{A}$ whose union is also in $\mathcal{A}$, then

$$\mu\left(\bigcup_{i=1}^{\infty} A_i\right) = \sum_{i=1}^{\infty} \mu(A_i).$$

Consequently, a measure on an algebra $\mathcal{A}$ is a measure (in the sense of Definition 4.47) if and only if $\mathcal{A}$ is a σ-algebra. A measure on $\mathcal{A}$ is called **σ-finite** if X can be written

$$X = \bigcup_{i=1}^{\infty} A_i$$

with $A_i \in \mathcal{A}$ and $\mu(A_i) < \infty$.

A measure μ on an algebra $\mathcal{A}$ generates a set function μ^* defined on all subsets of X in the following way: for each $E \subset X$, let

$$\mu^*(E) := \inf\left\{\sum_{i=1}^{\infty} \mu(A_i)\right\}, \tag{4.47}$$

where the infimum is taken over countable collections $\{A_i\}$ such that

$$E \subset \bigcup_{i=1}^{\infty} A_i, \quad A_i \in \mathcal{A}.$$

Note that this definition is in the same spirit as that used to define Lebesgue measure or more generally, Lebesgue–Stieltjes measure.

4.59. THEOREM. *Let μ be a measure on an algebra $\mathcal{A}$ and let μ^* be the corresponding set function generated by μ. Then*

(i) *μ^* is an outer measure.*
(ii) *μ^* is an extension of μ; that is, $\mu^*(A) = \mu(A)$ whenever $A \in \mathcal{A}$.*
(iii) *Each $A \in \mathcal{A}$ is μ^*-measurable.*

PROOF. The proof of (i) is similar to showing that λ^* is an outer measure (see the proof of Theorem 4.23) and is left as an exercise.

(ii) From the definition, $\mu^*(A) \leq \mu(A)$ whenever $A \in \mathcal{A}$. For the opposite inequality, consider $A \in \mathcal{A}$ and let $\{A_i\}$ be any sequence of sets in $\mathcal{A}$ with

$$A \subset \bigcup_{i=1}^{\infty} A_i.$$

Set

$$B_i = A \cap A_i \setminus (A_{i-1} \cup A_{i-2} \cup \cdots \cup A_1).$$

These sets are disjoint. Furthermore, $B_i \in \mathcal{A}$, $B_i \subset A_i$, and $A = \cup_{i=1}^{\infty} B_i$. Hence, by the countable additivity of μ,

$$\mu(A) = \sum_{i=1}^{\infty} \mu(B_i) \leq \sum_{i=1}^{\infty} \mu(A_i).$$

Since by definition, the infimum of the right-side of this expression tends to $\mu^*(A)$, this shows that $\mu(A) \leq \mu^*(A)$.

(iii) For $A \in \mathcal{A}$, we must show that

$$\mu^*(E) \geq \mu^*(E \cap A) + \mu^*(E \setminus A)$$

whenever $E \subset X$. For this we may assume that $\mu^*(E) < \infty$. Given $\varepsilon > 0$, there is a sequence of sets $\{A_i\}$ in $\mathcal{A}$ such that

$$E \subset \bigcup_{i=1}^{\infty} A_i \quad \text{and} \quad \sum_{i=1}^{\infty} \mu(A_i) < \mu^*(E) + \varepsilon.$$

Since μ is additive on $\mathcal{A}$, we have

$$\mu(A_i) = \mu(A_i \cap A) + \mu(A_i \setminus A).$$

In view of the inclusions

$$E \cap A \subset \bigcup_{i=1}^{\infty} (A_i \cap A) \quad \text{and} \quad E \setminus A \subset \bigcup_{i=1}^{\infty} (A_i \setminus A),$$

we have

$$\begin{aligned} \mu^*(E) + \varepsilon &> \sum_{i=1}^{\infty} \mu(A_i \cap A) + \sum_{i=1}^{\infty} \mu(A_i \cap \tilde{A}) \\ &> \mu^*(E \cap A) + \mu^*(E \setminus A). \end{aligned}$$

Since ε is arbitrary, the desired result follows. □

4.60. EXAMPLE. Let us see how the previous result can be used to produce Lebesgue–Stieltjes measure. Let $\mathcal{A}$ be the algebra formed by including $\emptyset$, $\mathbb{R}$, all intervals of the form $(-\infty, a]$, $(b, +\infty)$, along with all possible finite disjoint unions of these and intervals of the form $(a, b]$. Suppose that f

is a nondecreasing right-continuous function and define μ on intervals $(a, b]$ in $\mathcal{A}$ by

$$\mu((a,b]) = f(b) - f(a),$$

and then extend μ to all elements of $\mathcal{A}$ by additivity. Then we see that the outer measure μ^* generated by μ using (4.47) agrees with the definition of Lebesgue–Stieltjes measure defined by (4.23). Our previous result states that $\mu^*(A) = \mu(A)$ for all $A \in \mathcal{A}$, which agrees with Theorem 4.32.

4.61. REMARK. In the previous example, the right continuity of f is needed to ensure that μ is in fact a measure on $\mathcal{A}$. For example, if

$$f(x) := \begin{cases} 0 & x \leq 0, \\ 1 & x > 0, \end{cases}$$

then $\mu((0,1]) = 1$. But $(0,1] = \bigcup_{k=1}^{\infty} (\frac{1}{k+1}, \frac{1}{k}]$ and

$$\begin{aligned} \mu\left(\bigcup_{k=1}^{\infty} (\frac{1}{k+1}, \frac{1}{k}]\right) &= \sum_{k=1}^{\infty} \mu((\frac{1}{k+1}, \frac{1}{k}]) \\ &= 0, \end{aligned}$$

which shows that μ is not a measure.

Next is the main result of this section, which in addition to restating the results of Theorem 4.59, ensures that the outer measure generated by μ is unique.

4.62. THEOREM. (Carathéodory–Hahn extension theorem). *Let μ be a measure on an algebra $\mathcal{A}$, let μ^* be the outer measure generated by μ, and let $\mathcal{A}^*$ be the σ-algebra of μ^*-measurable sets.*

(i) *Then $\mathcal{A}^* \supset \mathcal{A}$ and $\mu^* = \mu$ on $\mathcal{A}$.*
(ii) *Let $\mathcal{M}$ be a σ-algebra with $\mathcal{A} \subset \mathcal{M} \subset \mathcal{A}^*$ and suppose ν is a measure on $\mathcal{M}$ that agrees with μ on $\mathcal{A}$. Then $\nu = \mu^*$ on $\mathcal{M}$ if μ is σ-finite.*

PROOF. As noted above, (i) is a restatement of Theorem 4.59.

(ii) Given $E \in \mathcal{M}$, note that $\nu(E) \leq \mu^*(E)$, since if $\{A_i\}$ is a countable collection in $\mathcal{A}$ whose union contains E, then

$$\nu(E) \leq \nu\left(\bigcup_{i=1}^{\infty} A_i\right) \leq \sum_{i=1}^{\infty} \nu(A_i) = \sum_{i=1}^{\infty} \mu(A_i).$$

To prove equality let $A \in \mathcal{A}$ with $\mu(A) < \infty$. Then we have

$$\nu(E) + \nu(A \setminus E) = \nu(A) = \mu^*(A) = \mu^*(E) + \mu^*(A \setminus E). \tag{4.48}$$

Note that $A \setminus E \in \mathcal{M}$, and therefore $\nu(A \setminus E) \leq \mu^*(A \setminus E)$ from what we have just proved. Since all terms in (4.48) are finite, we deduce that

$$\nu(A \cap E) = \mu^*(A \cap E)$$

whenever $A \in \mathcal{A}$ with $\mu(A) < \infty$. Since μ is σ-finite, there exist $A_i \in \mathcal{A}$ such that

$$X = \bigcup_{i=1}^{\infty} A_i$$

with $\mu(A_i) < \infty$ for each i. We may assume that the A_i are disjoint (Lemma 4.7) and therefore

$$\nu(E) = \sum_{i=1}^{\infty} \nu(E \cap A_i) = \sum_{i=1}^{\infty} \mu^*(E \cap A_i) = \mu^*(E).$$

□

Let us consider a special case of this result, namely, the situation in which $\mathcal{A}$ is the family of Borel sets in a metric space X. If μ is a finite measure defined on the Borel sets, the previous result states that the outer measure, μ^*, generated by μ agrees with μ on the Borel sets. Theorem 4.52 asserts that μ^* enjoys certain regularity properties. Since μ and μ^* agree on Borel sets, it follows that μ also enjoys these regularity properties. This implies the remarkable fact that every finite Borel measure is automatically regular. We state this as our next result.

4.63. THEOREM. *Suppose $(X, \mathcal{M}, \mu)$ is a measure space, where X is a metric space and μ is a finite Borel measure (that is, $\mathcal{M}$ denotes the Borel sets of X and $\mu(X) < \infty$). Then for each $\varepsilon > 0$ and each Borel set B, there exist an open set U and a closed set F such that $F \subset B \subset U$, $\mu(B \setminus F) < \varepsilon$, and $\mu(U \setminus B) < \varepsilon$.*

If μ is a measure defined on a σ-algebra $\mathcal{M}$ rather than on an algebra $\mathcal{A}$, there is another method for generating an outer measure. In this situation, we define μ^{**} on an arbitrary set $E \subset X$ by

$$\mu^{**}(E) = \inf\{\mu(B) : B \supset E, B \in \mathcal{M}\}. \tag{4.49}$$

We have the following result.

4.64. THEOREM. *Consider a measure space $(X, \mathcal{M}, \mu)$. The set function μ^{**} defined above is an outer measure on X. Moreover, μ^{**} is a regular outer measure and $\mu(B) = \mu^{**}(B)$ for each $B \in \mathcal{M}$.*

PROOF. The proof proceeds exactly as in Theorem 4.57. One need only replace each reference to a Borel set in that proof with $\mathcal{M}$-measurable set. □

4.65. THEOREM. *Suppose $(X, \mathcal{M}, \mu)$ is a measure space and let μ^* and μ^{**} be the outer measures generated by μ as described in (4.47) and (4.49), respectively. Then for each $E \subset X$ with $\mu(E) < \infty$ there exists $B \in \mathcal{M}$ such that $B \supset E$,*

$$\mu(B) = \mu^*(B) = \mu^*(E) = \mu^{**}(E).$$

PROOF. We will show that for every $E \subset X$ there exists $B \in \mathcal{M}$ such that $B \supset E$ and $\mu(B) = \mu^*(B) = \mu^*(E)$. From the previous result, it will then follow that $\mu^*(E) = \mu^{**}(E)$.

Note that for each $\varepsilon > 0$ and every set E, there exists a sequence $\{A_i\} \in \mathcal{M}$ such that

$$E \subset \bigcup_{i=1}^{\infty} A_i \quad \text{and} \quad \sum_{i=1}^{\infty} \mu(A_i) \leq \mu^*(E) + \varepsilon.$$

Setting $A = \cup A_i$, we have

$$\mu(A) < \mu^*(E) + \varepsilon.$$

For each positive integer k, use this observation with $\varepsilon = 1/k$ to obtain a set $A_k \in \mathcal{M}$ such that $A_k \supset E$ and $\mu(A_k) < \mu^*(E) + 1/k$. Let

$$B = \bigcap_{k=1}^{\infty} A_k.$$

Then $B \in \mathcal{M}$, and since $E \subset B \subset A_k$, we have

$$\mu^*(E) \leq \mu^*(B) \leq \mu(B) \leq \mu(A_k) < \mu^*(E) + 1/k.$$

Since k is arbitrary, it follows that $\mu(B) = \mu^*(B) = \mu^*(E)$. □

CHAPTER 5

Measurable Functions

5.1. Elementary Properties of Measurable Functions

The class of measurable functions will play a critical role in the theory of integration. It is shown that this class remains closed under the usual elementary operations, although special care must be taken in the case of composition of functions. The main results of this chapter are the theorems of Egorov and Lusin. Roughly, they state that pointwise convergence of a sequence of measurable functions is "nearly" uniform convergence and that a measurable function is "nearly" continuous.

Throughout this chapter, we will consider an abstract measure space $(X, \mathcal{M}, \mu)$, where μ is a measure defined on the σ-algebra $\mathcal{M}$. Virtually all the material in this first section depends only on the σ-algebra and not on the measure μ. This is a reflection of the fact that the elementary properties of measurable functions are set-theoretic and are not related to μ. Also, we will consider functions $f\colon X \to \overline{\mathbb{R}}$, where $\overline{\mathbb{R}} = \mathbb{R} \cup \{-\infty\} \cup \{+\infty\}$ is the set of **extended real numbers**. For convenience, we will write ∞ for $+\infty$. Arithmetic operations on $\overline{\mathbb{R}}$ are subject to the following conventions. For $x \in \mathbb{R}$, we define

$$x + (\pm\infty) = (\pm\infty) + x = \pm\infty$$

and

$$(\pm\infty) + (\pm\infty) = \pm\infty, \; (\pm\infty) - (\mp\infty) = \pm\infty,$$

but

$$(\pm\infty) + (\mp\infty), \text{ and } (\pm\infty) - (\pm\infty)$$

are undefined. Also, for the operation of multiplication, we define

$$x(\pm\infty) = (\pm\infty)x = \begin{cases} \pm\infty, & x > 0, \\ 0, & x = 0, \\ \mp\infty, & x < 0, \end{cases}$$

for each $x \in \mathbb{R}$ and let

$$(\pm\infty) \cdot (\pm\infty) = +\infty \qquad \text{and} \qquad (\pm\infty) \cdot (\mp\infty) = -\infty.$$

W.P. Ziemer, *Modern Real Analysis*, Graduate Texts in Mathematics 278,
https://doi.org/10.1007/978-3-319-64629-9_5

The operations

$$\frac{\infty}{-\infty}, \frac{-\infty}{\infty}, \frac{\infty}{\infty} \text{ and } \frac{-\infty}{-\infty}$$

are undefined.

We endow $\overline{\mathbb{R}}$ with a topology called the order topology in the following manner. For each $a \in \mathbb{R}$ let

$$L_a = \overline{\mathbb{R}} \cap \{x : x < a\} = [-\infty, a) \qquad \text{and} \qquad R_a = \overline{\mathbb{R}} \cap \{x : x > a\} = (a, \infty].$$

The collection $\mathcal{S} = \{L_a : a \in \mathbb{R}\} \cup \{R_a : a \in \mathbb{R}\}$ is taken as a subbasis for this topology. A basis for the topology is given by

$$\mathcal{S} \cup \{R_a \cap L_b : a, b \in \mathbb{R}, a < b\}.$$

Observe that the topology on $\mathbb{R}$ induced by the order topology on $\overline{\mathbb{R}}$ is precisely the usual topology on $\mathbb{R}$.

Suppose X and Y are topological spaces. Recall that a mapping $f: X \to Y$ is continuous if and only if $f^{-1}(U)$ is open whenever $U \subset Y$ is open. We define a measurable mapping analogously.

5.1. Definitions. Suppose $(X, \mathcal{M})$ and $(Y, \mathcal{N})$ are measure spaces. A mapping $f: X \to Y$ is called **measurable with respect to $\mathcal{M}$ and $\mathcal{N}$** if

$$f^{-1}(E) \in \mathcal{M} \text{ whenever } E \in \mathcal{N}. \tag{5.1}$$

If there is no danger of confusion, reference to $\mathcal{M}$ and $\mathcal{N}$ will be omitted, and we will simply use the term "measurable mapping."

If Y is a topological space, a restriction is placed on $\mathcal{N}$. In this case it is always assumed that $\mathcal{N}$ is the σ-algebra of Borel sets $\mathcal{B}$. Thus, in this situation, a mapping $(X, \mathcal{M}) \xrightarrow{f} (Y, \mathcal{B})$ is measurable if

$$f^{-1}(E) \in \mathcal{M} \text{ whenever } E \in \mathcal{B}. \tag{5.2}$$

The reason for imposing this condition is to ensure that continuous mappings will be measurable. That is, if both X and Y are topological spaces, $X \xrightarrow{f} Y$ is continuous, and $\mathcal{M}$ contains the Borel sets of X, then f is measurable, since $f^{-1}(E) \in \mathcal{M}$ whenever E is a Borel set; see Exercise 1, Section 5.1. One of the most important situations occurs when Y is taken as $\overline{\mathbb{R}}$ (endowed with the order topology) and $(X, \mathcal{M})$ is a topological space with $\mathcal{M}$ the collection of Borel sets. Then f is called a **Borel measurable function**. Another important example of this occurs when $X = \mathbb{R}^n$, $\mathcal{M}$ is the class of Lebesgue measurable sets and $Y = \overline{\mathbb{R}}$. Here, it is required that $f^{-1}(E)$ be Lebesgue measurable whenever $E \subset \overline{\mathbb{R}}$ is Borel, in which case f is called a **Lebesgue measurable function**. The definitions imply that E is a measurable set if and only if χ_E is a measurable function.

If the mapping $(X, \mathcal{M}) \xrightarrow{f} (Y, \mathcal{B})$ is measurable, where $\mathcal{B}$ is the σ-algebra of Borel sets, then we can make the following observation, which will be useful in the development. Define

$$\Sigma = \{E : E \subset Y \text{ and } f^{-1}(E) \in \mathcal{M}\}. \tag{5.3}$$

Note that Σ is closed under countable unions. It is also closed under complementation, since

$$f^{-1}(E^{\sim}) = [f^{-1}(E)]^{\sim} \in \mathcal{M} \tag{5.4}$$

for $E \in \Sigma$, and thus Σ is a σ-algebra.

In view of (5.3) and (5.4), note that a continuous mapping is a Borel measurable function (Exercise 1, Section 5.1).

If $f\colon X \to \overline{\mathbb{R}}$, it will be convenient to characterize measurability in terms of the sets $X \cap \{x : f(x) > a\}$ for $a \in \mathbb{R}$. To simplify notation, we simply write $\{f > a\}$ to denote these sets. The sets $\{f > a\}$ are called the **superlevel sets of** f. The behavior of a function f is to a large extent reflected in the properties of its superlevel sets. For example, if f is a continuous function on a metric space X, then $\{f > a\}$ is an open set for each real number a. If the function is nicer, then we should expect better behavior of the superlevel sets. Indeed, if f is an infinitely differentiable function defined on $\mathbb{R}^n$ with nonvanishing gradient, then not only is each $\{f > a\}$ an open set, but an application of the implicit function theorem shows that its boundary is a smooth manifold of dimension $n-1$ as well.

We begin by showing that the definition of an $\overline{\mathbb{R}}$-valued measurable function could just as well be stated in terms of its level sets.

5.2. THEOREM. *Let $f\colon X \to \overline{\mathbb{R}}$, where $(X, \mathcal{M})$ is a measure space. The following conditions are equivalent:*

(i) *f is measurable.*

(ii) *$\{f > a\} \in \mathcal{M}$ for each $a \in \mathbb{R}$.*

(iii) *$\{f \geq a\} \in \mathcal{M}$ for each $a \in \mathbb{R}$.*

(iv) *$\{f < a\} \in \mathcal{M}$ for each $a \in \mathbb{R}$.*

(v) *$\{f \leq a\} \in \mathcal{M}$ for each $a \in \mathbb{R}$.*

PROOF. (i) implies (ii) by definition, since $\{f > a\} = f^{-1}((a, \infty])$ and $(a, \infty]$ is open in the order topology. In view of $\{f \geq a\} = \cap_{k=1}^{\infty}\{f > a - 1/k\}$, (ii) implies (iii). The set $\{f < a\}$ is the complement of $\{f \geq a\}$, thus establishing the next implication. Similarly to the proof of the first implication, we have $\{f \leq a\} = \cap_{k=1}^{\infty}\{f < a + 1/k\}$, which shows that (iv) implies (v). For the proof that (v) implies (i), in view of (5.3) and (5.4) with $Y = \overline{\mathbb{R}}$, it is sufficient to show that $f^{-1}(U) \in \mathcal{M}$ whenever $U \subset \overline{\mathbb{R}}$ is open. Since f^{-1} preserves unions and intersections and U can be written as a countable union of elements of the basis, we need only consider $f^{-1}(J)$, where J assumes the form $J_1 = [-\infty, a)$, $J_2 = (a, b)$, and $J_3 = (b, \infty]$ for $a, b \in \mathbb{R}$. By assumption, $\{f \leq b\} \in \mathcal{M}$ and therefore $f^{-1}(J_3) = \{x : f(x) \leq b\}^{\sim} \in \mathcal{M}$. Also,

$$J_1 = \bigcup_{k=1}^{\infty} [-\infty, a_k],$$

where $a_k < a$ and $a_k \to a$ as $k \to \infty$. Hence, $f^{-1}(J_1) = \bigcup_{k=1}^{\infty} f^{-1}([-\infty, a_k]) = \bigcup_{k=1}^{\infty} \{x : f(x) \leq a_k\} \in \mathcal{M}$. Finally, $f^{-1}(J_2) \in \mathcal{M}$, since $J_2 = J_1 \cap J_3$. □

5.3. THEOREM. *A function $f\colon X \to \overline{\mathbb{R}}$ is measurable if and only if*

(i) $f^{-1}\{-\infty\} \in \mathcal{M}$ *and* $f^{-1}\{\infty\} \in \mathcal{M}$ *and*

(ii) $f^{-1}(a,b) \in \mathcal{M}$ *for all open intervals* $(a,b) \subset \mathbb{R}$.

PROOF. If f is measurable, then (i) and (ii) are satisfied, since $\{\infty\}$, $\{-\infty\}$ and (a,b) are Borel subsets of $\overline{\mathbb{R}}$.

In order to prove that f is measurable, we need to show that $f^{-1}(E) \in \mathcal{M}$ whenever E is a Borel subset of $\overline{\mathbb{R}}$. From (i), and since $E \subset \overline{\mathbb{R}}$ is Borel if and only if $E \cap \mathbb{R}$ is Borel, we have only to show that $f^{-1}(E) \in \mathcal{M}$ whenever $E \subset \mathbb{R}$ is a Borel set. Since f^{-1} preserves unions of sets and since every open set in $\mathbb{R}$ is the disjoint union of open intervals, we see from (ii) that $f^{-1}(U) \in \mathcal{M}$ whenever $U \subset \mathbb{R}$ is an open set. If we define Σ as in (5.3) with $Y = \mathbb{R}$, we see that Σ is a σ-algebra that contains the open sets of $\mathbb{R}$ and therefore it contains all Borel sets. □

We now proceed to show that measurability is preserved under elementary arithmetic operations on measurable functions. For this, the following will be useful.

5.4. LEMMA. *If f and g are measurable functions, then the following sets are measurable:*

(i) $X \cap \{x : f(x) > g(x)\}$,

(ii) $X \cap \{x : f(x) \geq g(x)\}$,

(iii) $X \cap \{x : f(x) = g(x)\}$.

PROOF. If $f(x) > g(x)$, then there is a rational number r such that $f(x) > r > g(x)$. Therefore, it follows that

$$\{f > g\} = \bigcup_{r \in \mathbb{Q}} (\{f > r\} \cap \{g < r\}),$$

and (i) easily follows. The set (ii) is the complement of the set (i) with f and g interchanged, and it is therefore measurable. The set (iii) is the intersection of two measurable sets of type (ii), and so it too is measurable. □

Since all functions under discussion are extended real-valued functions, we must take some care in defining the sum and product of such functions. If f and g are measurable functions, then $f + g$ is undefined at points where it would be of the form $\infty - \infty$. This difficulty is overcome if we define

$$(f+g)(x) \colon = \begin{cases} f(x) + g(x), & x \in X - B, \\ \alpha, & x \in B, \end{cases} \tag{5.5}$$

where $\alpha \in \overline{\mathbb{R}}$ is chosen arbitrarily and where

$$B\colon = (f^{-1}\{\infty\} \bigcap g^{-1}\{-\infty\}) \bigcup (f^{-1}\{-\infty\} \bigcap g^{-1}\{\infty\}). \tag{5.6}$$

With this definition we have the following.

5.5. THEOREM. *If $f, g\colon X \to \overline{\mathbb{R}}$ are measurable functions, then $f+g$ and fg are measurable.*

PROOF. We will treat the case that f and g have values in $\mathbb{R}$. The proof is similar in the general case and is left as an exercise (see Exercise 2, Section 5.1).

To prove that the sum is measurable, define $F\colon X \to \mathbb{R} \times \mathbb{R}$ by

$$F(x) = (f(x), g(x))$$

and $G\colon \mathbb{R} \times \mathbb{R} \to \mathbb{R}$ by

$$G(x, y) = x + y.$$

Then $G \circ F(x) = f(x) + g(x)$, so it suffices to show that $G \circ F$ is measurable. Referring to Theorem 5.3, we need only show that $(G \circ F)^{-1}(J) \in \mathcal{M}$ whenever $J \subset \mathbb{R}$ is an open interval. Now $U\colon = G^{-1}(J)$ is an open set in $\mathbb{R}^2$, since G is continuous. Furthermore, U is the union of a countable family, $\mathcal{F}$, of 2-dimensional intervals I of the form $I = I_1 \times I_2$, where I_1 and I_2 are open intervals in $\mathbb{R}$. Since

$$F^{-1}(I) = f^{-1}(I_1) \cap g^{-1}(I_2),$$

we have

$$F^{-1}(U) = F^{-1}\left(\bigcup_{I \in \mathcal{F}} I\right) = \bigcup_{I \in \mathcal{F}} F^{-1}(I),$$

which is a measurable set. Thus $G \circ F$ is measurable, since

$$(G \circ F)^{-1}(J) = F^{-1}(U).$$

The product is measurable by essentially the same proof. □

5.6. REMARK. In the situation of abstract measure spaces, if

$$(X, \mathcal{M}) \xrightarrow{f} (Y, \mathcal{N}) \xrightarrow{g} (Z, \mathcal{P})$$

are measurable functions, the definitions immediately imply that the composition $g \circ f$ is measurable. Because of this, one might be tempted to conclude that the composition of Lebesgue measurable functions is again Lebesgue measurable. Let's look at this closely. Suppose f and g are Lebesgue measurable functions:

$$\mathbb{R} \xrightarrow{f} \mathbb{R} \xrightarrow{g} \mathbb{R}.$$

Thus, here we have $X = Y = Z = \mathbb{R}$. Since $Z = \mathbb{R}$, our convention requires that we take $\mathcal{P}$ to be the Borel sets. Moreover, since f is assumed to be Lebesgue measurable, the definition requires $\mathcal{M}$ to be the σ-algebra of Lebesgue measurable sets. If $g \circ f$ were to be Lebesgue measurable, it would be necessary that $f^{-1}(g^{-1}(E))$ be Lebesgue measurable whenever E is a Borel set in $\mathbb{R}$. The definitions imply that it would be necessary for $g^{-1}(E)$ to be a Borel set set whenever E is a Borel set in $\mathbb{R}$. The following example shows that this is not generally true.

5.7. EXAMPLE. (The Cantor–Lebesgue function) Our example is based on the construction of the Cantor ternary set. Recall (p. 94) that the Cantor set C can be expressed as

$$C = \bigcap_{j=1}^{\infty} C_j,$$

where C_j is the union of the 2^j closed intervals that remain after the jth step of the construction. Each of these intervals has length 3^{-j}. Thus, the set

$$D_j = [0,1] - C_j$$

consists of the $2^j - 1$ open intervals that are deleted at the jth step. Let these intervals be denoted by $I_{j,k}$, $k = 1, 2, \ldots, 2^j - 1$, and order them in the obvious way from left to right. Now define a continuous function f_j on [0, 1] by

$$\begin{aligned} f_j(0) &= 0, \\ f_j(1) &= 1, \\ f_j(x) &= \frac{k}{2^j} \quad \text{for } x \in I_{j,k}, \end{aligned}$$

and define f_j linearly on each interval of C_j. The function f_j is continuous and nondecreasing, and it satisfies

$$|f_j(x) - f_{j+1}(x)| < \frac{1}{2^j} \quad \text{for } x \in [0,1].$$

Since

$$|f_j - f_{j+m}| < \sum_{i=j}^{j+m-1} \frac{1}{2^i} < \frac{1}{2^{j-1}},$$

it follows that the sequence $\{f_j\}$ is uniformly Cauchy in the space of continuous functions and thus converges uniformly to a continuous function f, called the **Cantor–Lebesgue function.**

Note that f is nondecreasing and is constant on each interval in the complement of the Cantor set. Furthermore, $f\colon [0,1] \to [0,1]$ is onto. In fact, it is easy to see that $f(C) = [0,1]$, because $f(C)$ is compact and $f([0,1] - C)$ is countable.

We use the Cantor–Lebesgue function to show that the composition of Lebesgue measurable functions need not be Lebesgue measurable. Let $h(x) = f(x) + x$ and observe that h is strictly increasing, since f is nondecreasing. Thus, h is a homeomorphism from $[0,1]$ onto $[0,2]$. Furthermore, it is clear that h carries the complement of the Cantor set onto an open set of measure 1.

Therefore, h maps the Cantor set onto a set P of measure 1. Now let N be a non-Lebesgue measurable subset of P; see Exercise 1, Section 4.5. Then, with $A = h^{-1}(N)$, we have $A \subset C$, and therefore A is Lebesgue measurable, since $\lambda(A) = 0$. Thus we have that h carries a measurable set onto a nonmeasurable set.

Note that h^{-1} is measurable, since it is continuous. Let $F := h^{-1}$. Observe that A is not a Borel set, for if it were, then $F^{-1}(A)$ would be a Borel set. But $F^{-1}(A) = h(A) = N$ and N is not a Borel set. Now χ_A is a Lebesgue measurable function, since A is a Lebesgue measurable set. Let $g := \chi_A$. Observe that $g^{-1}(1) = A$, and thus g is an example of a Lebesgue measurable function that does not preserve Borel sets. Also,

$$g \circ F = \chi_A \circ h^{-1} = \chi_N,$$

which shows that this composition of Lebesgue measurable functions is not Lebesgue measurable. To summarize the properties of the Cantor–Lebesgue function, we have the following corollary.

5.8. COROLLARY. *The Cantor–Lebesgue function f and its associate $h(x) := f(x) + x$ described above have the following properties:*

(i) *$f(C) = [0, 1]$; that is, f maps a set of measure 0 onto a set of positive measure.*
(ii) *h maps a Lebesgue measurable set onto a nonmeasurable set.*
(iii) *The composition of Lebesgue measurable functions need not be Lebesgue measurable.*

Although the example above shows that Lebesgue measurable functions are not generally closed under composition, a positive result can be obtained if the outer function in the composition is assumed to be Borel measurable. The proof of the following theorem is a direct consequence of the definitions.

5.9. THEOREM. *Suppose $f\colon X \to \overline{\mathbb{R}}$ is measurable and $g\colon \overline{\mathbb{R}} \to \overline{\mathbb{R}}$ is Borel measurable. Then $g \circ f$ is measurable. In particular, if $X = \mathbb{R}^n$ and f is Lebesgue measurable, then $g \circ f$ is Lebesgue measurable.*

The function g is required to have $\overline{\mathbb{R}}$ as its domain of definition because f is an extended real-valued function; however, every Borel measurable function g defined on $\mathbb{R}$ can be extended to $\overline{\mathbb{R}}$ by assigning arbitrary values to ∞ and $-\infty$.

As a consequence of this result, we have the following corollary, which complements Theorem 5.5.

5.10. COROLLARY. *Let $f\colon X \to \overline{\mathbb{R}}$ be a measurable function.*

(i) *Let $\varphi(x) = |f(x)|^p$, $0 < p < \infty$, and let φ assume arbitrary extended values on the sets $f^{-1}(\infty)$ and $f^{-1}(-\infty)$. Then φ is measurable.*
(ii) *Let $\varphi(x) = \dfrac{1}{f(x)}$, and let φ assume arbitrary extended values on the sets $f^{-1}(0), f^{-1}(\infty)$ and $f^{-1}(-\infty)$. Then φ is measurable.*

In particular, if $X = \mathbb{R}^n$ *and* f *is Lebesgue measurable, then* φ *is Lebesgue measurable in* (i) *and* (ii).

PROOF. For (i), define $g(t) = |t|^p$ for $t \in \mathbb{R}$ and assign arbitrary values to $g(\infty)$ and $g(-\infty)$. Now apply the previous theorem.

For (ii), proceed in a similar way by defining $g(t) = \dfrac{1}{t}$ when $t \neq 0, \infty, -\infty$ and assigning arbitrary values to $g(0)$, $g(\infty)$, and $g(-\infty)$. □

For much of the development thus far, the measure μ in $(X, \mathcal{M}, \mu)$ has played no role. We have used only the fact that $\mathcal{M}$ is a σ-algebra. Later it will be necessary to deal with functions that are not necessarily defined on all of X but only on the complement of some set of μ-measure 0. That is, we will deal with functions that are defined only μ-almost everywhere. A measurable set N is called a **μ-null set** if $\mu(N) = 0$. A property that holds for all $x \in X$ except for those x in some μ-null set is said to hold **μ-almost everywhere**. The term "μ-almost everywhere" is often written in abbreviated form, "μ-a.e.." If it is clear from context that the measure μ is under consideration, we will simply use the terms "null set" and "almost everywhere."

The next result shows that a measurable function on a complete measure space remains measurable if it is altered on an arbitrary set of measure 0.

5.11. THEOREM. *Let* $(X, \mathcal{M}, \mu)$ *be a complete measure space and let* f, g *be extended real-valued functions defined on* X. *If* f *is measurable and* $f = g$ *almost everywhere, then* g *is measurable.*

PROOF. Let $N = \{x : f(x) = g(x)\}$. Then $\mu(\widetilde{N}) = 0$, and thus $\widetilde{N}$ as well as all subsets of $\widetilde{N}$ are measurable. For $a \in \mathbb{R}$, we have

$$\begin{aligned}\{g > a\} &= (\{g > a\} \cap N) \cup (\{g > a\} \cap \widetilde{N}) \\ &= (\{f > a\} \cap N) \cup (\{g > a\} \cap \widetilde{N}) \in \mathcal{M}.\end{aligned}$$ □

5.12. REMARK. If μ is a complete measure, this result allows us to attach the meaning of measurability to a function f that is defined merely almost everywhere. Indeed, if N is the null set on which f is not defined, we modify the definition of measurability by saying that f is measurable if $\{f > a\} \cap \widetilde{N}$ is measurable for each $a \in \mathbb{R}$. This is tantamount to saying that $\overline{f}$ is measurable, where $\overline{f}$ is an extension of f obtained by assigning arbitrary values to f on N. This is easily seen because

$$\{\overline{f} > a\} = (\{\overline{f} > a\} \cap N) \cup (\{f > a\} \cap \widetilde{N});$$

the first set on the right is of measure zero, because μ is complete, and therefore measurable. Furthermore, for functions f, g that are finite-valued at μ-almost every point, we may define $f + g$ as $(f + g)(x) = f(x) + g(x)$ for all $x \in X$ at which both f and g are defined and do not assume infinite values of opposite sign. Then, if both f and g are measurable, $f + g$ is measurable. A similar discussion holds for the product fg.

It therefore becomes apparent that functions that coincide almost everywhere may be considered equivalent. In fact, if we define $f \sim g$ to mean that $f = g$ almost everywhere, then $\sim$ defines an equivalence relation as discussed in Definition 2.9, and thus a function may be regarded as an equivalence class of functions.

It should be kept in mind that this entire discussion pertains only to the situation in which the measure space $(X, \mathcal{M}, \mu)$ is complete. In particular, it applies in the context of Lebesgue measure on $\mathbb{R}^n$, the most important example of a measure space.

We conclude this section by returning to the context of an outer measure φ defined on an arbitrary space X as in Definition 4.1. If $f\colon X \to \overline{\mathbb{R}}$, then according to Theorem 5.2, f is φ-measurable if $\{f \leq a\}$ is a φ-measurable set for each $a \in \mathbb{R}$. That is, with $E_a = \{f \leq a\}$, the φ-measurability of f is equivalent to

$$\varphi(A) = \varphi(A \cap E_a) + \varphi(A - E_a) \tag{5.7}$$

for an arbitrary set $A \subset X$ and each $a \in \mathbb{R}$. The next result is often useful in applications and gives a characterization of φ-measurability that appears to be weaker than (5.7).

5.13. THEOREM. *Suppose φ is an outer measure on a space X. Then an extended real-valued function f on X is φ-measurable if and only if*

$$\varphi(A) \geq \varphi(A \cap \{f \leq a\}) + \varphi(A \cap \{f \geq b\}) \tag{5.8}$$

whenever $A \subset X$ and $a < b$ are real numbers.

PROOF. If f is φ-measurable, then (5.8) holds, since it is implied by (5.7).

To prove the converse, it suffices to show that for every real number r, (5.8) implies that

$$E = \{x : f(x) \leq r\}$$

is φ-measurable. Let $A \subset X$ be an arbitrary set with $\varphi(A) < \infty$ and define

$$B_i = A \cap \left\{x : r + \frac{1}{i+1} \leq f(x) \leq r + \frac{1}{i}\right\}$$

for each positive integer i. First,[1] we will show that

$$\infty > \varphi(A) \geq \varphi\left(\bigcup_{k=1}^{\infty} B_{2k}\right) = \sum_{k=1}^{\infty} \varphi(B_{2k}). \tag{5.9}$$

The proof is by induction, so assume that (5.9) is valid as k runs from 1 to $j-1$. That is, assume

$$\varphi\left(\bigcup_{k=1}^{j-1} B_{2k}\right) = \sum_{k=1}^{j-1} \varphi(B_{2k}). \tag{5.10}$$

---

[1]Note the similarity between the technique used in the following argument and the proof of Theorem 4.16, from (4.11) to the end of that proof

Let

$$A_j = \bigcup_{k=1}^{j-1} B_{2k}.$$

Then, using (5.8), the induction hypothesis, and the fact that

$$(B_{2j} \cup A_j) \cap \left\{ f \leq r + \frac{1}{2j} \right\} = B_{2j} \tag{5.11}$$

and

$$(B_{2j} \cup A_j) \cap \left\{ f \geq r + \frac{1}{2j-1} \right\} = A_j, \tag{5.12}$$

we obtain

$$\begin{aligned}
\varphi\left(\bigcup_{k=1}^{j} B_{2k}\right) &= \varphi(B_{2j} \cup A_j) \\
&\geq \varphi\left[(B_{2j} \cup A_j) \cap \left\{ f \leq r + \frac{1}{2j} \right\}\right] \\
&\quad + \varphi\left[(B_{2j} \cup A_j) \cap \left\{ f \geq r + \frac{1}{2j-1} \right\}\right] && \text{by (5.8)} \\
&= \varphi(B_{2j}) + \varphi(A_j) && \text{by (5.11) and (5.12)} \\
&= \varphi(B_{2j}) + \sum_{k=1}^{j-1} \varphi(B_{2k}) && \text{by the induction hypothesis (5.10)} \\
&= \sum_{k=1}^{j} \varphi(B_{2k}).
\end{aligned}$$

Thus, (5.9) is valid as k runs from 1 to j for every positive integer j. In other words, we obtain

$$\infty > \varphi(A) \geq \varphi\left(\bigcup_{k=1}^{\infty} B_{2k}\right) \geq \varphi\left(\bigcup_{k=1}^{j} B_{2k}\right) = \sum_{k=1}^{j} \varphi(B_{2k}),$$

for every positive integer j. This implies

$$\infty > \varphi(A) \geq \sum_{k=1}^{\infty} \varphi(B_{2k}).$$

Virtually the same argument can be used to obtain

$$\infty > \varphi(A) \geq \sum_{k=1}^{\infty} \varphi(B_{2k-1}),$$

thus implying

$$\infty > 2\varphi(A) \geq \sum_{k=1}^{\infty} \varphi(B_k).$$

Now the tail end of this convergent series can be made arbitrarily small; that is, for each $\varepsilon > 0$ there exists a positive integer m such that

$$\varepsilon > \sum_{i=m}^{\infty} \varphi(B_i) = \varphi\left(\bigcup_{i=m}^{\infty} B_i\right)$$
$$\geq \varphi\left(A \cap \left\{r < f < r + \frac{1}{m}\right\}\right).$$

For ease of notation, define an outer measure $\psi(S) = \varphi(S \cap A)$ whenever $S \subset X$. With this notation, we have shown that

$$\varepsilon > \psi\left(\left\{r < f < r + \frac{1}{m}\right\}\right)$$
$$= \psi\left(\{r < f\} \cap \left\{f < r + \frac{1}{m}\right\}\right)$$
$$\geq \psi(\{f > r\}) - \psi\left(\left\{f \geq r + \frac{1}{m}\right\}\right).$$

The last inequality is implied by the subadditivity of ψ. Therefore,

$$\varphi(A \cap E) + \varphi(A - E) = \psi(E) + \psi(\widetilde{E})$$
$$= \psi(E) + \psi(\{f > r\})$$
$$\leq \psi(E) + \psi\left(\left\{f \geq r + \frac{1}{m}\right\}\right) + \varepsilon$$
$$= \varphi(A \cap E) + \varphi\left(A \cap \left\{f \geq r + \frac{1}{m}\right\}\right) + \varepsilon$$
$$\leq \varphi(A) + \varepsilon. \qquad \text{by (5.8)}$$

Since ε is arbitrary, this proves that E is φ-measurable. □

Exercises for Section 5.1

1. Let $(X, \mathcal{M}) \xrightarrow{f} (Y, \mathcal{B})$ be a continuous mapping, where X and Y are topological spaces, $\mathcal{M}$ is a σ-algebra that contains the Borel sets in X, and $\mathcal{B}$ is the family of Borel sets in Y. Prove that f is measurable.
2. Complete the proof of Theorem 5.5 when f and g have values in $\overline{\mathbb{R}}$.
3. Prove that a function defined on $\mathbb{R}^n$ that is continuous everywhere except for a set of Lebesgue measure zero is a Lebesgue measurable function. In particular, conclude that a nondecreasing function defined on $[0, 1]$ is Lebesgue measurable.

5.2. Limits of Measurable Functions

In order to be useful in applications, it is necessary for measurability to be preserved by virtually all types of limit operations on sequences of measurable functions. In this section, it is shown that measurability is preserved under the operations of upper and lower limits of sequences of functions as well as upper and lower envelopes. It is also shown that on a finite measure space, pointwise a.e. convergence of a sequence of measurable functions implies uniform convergence on the complements of sets of arbitrarily small measure (Egorov's theorem). Finally, the relationship between convergence in measure and pointwise a.e. convergence is investigated.

Throughout this section, it will be assumed that all functions are $\overline{\mathbb{R}}$-valued, unless otherwise stated.

5.14. DEFINITION. Let $(X, \mathcal{M}, \mu)$ be a measure space, and let $\{f_i\}$ be a sequence of measurable functions defined on X. The **upper** and **lower envelopes** of $\{f_i\}$ are defined respectively as

$$\sup_i f_i(x) = \sup\{f_i(x) : i = 1, 2, \ldots\}$$

and

$$\inf_i f_i(x) = \inf\{f_i(x) : i = 1, 2, \ldots\}.$$

Also, the **upper** and **lower limits** of $\{f_i\}$ are defined as

$$\limsup_{i\to\infty} f_i(x) = \inf_{j\geq 1} \left(\sup_{i\geq j} f_i(x)\right)$$

and

$$\liminf_{i\to\infty} f_i(x) = \sup_{j\geq 1} \left(\inf_{i\geq j} f_i(x)\right).$$

5.15. THEOREM. *Let $\{f_i\}$ be a sequence of measurable functions defined on the measure space $(X, \mathcal{M}, \mu)$. Then $\sup_i f_i$, $\inf_i f_i$, $\limsup_{i\to\infty} f_i$, and $\liminf_{i\to\infty} f_i$ are all measurable functions.*

PROOF. For each $a \in \mathbb{R}$ the identity

$$X \cap \{x : \sup_i f_i(x) > a\} = \bigcup_{i=1}^{\infty} \left(X \cap \{f_i(x) > a\}\right)$$

implies that $\sup_i f_i$ is measurable. The measurability of the lower envelope follows from

$$\inf_i f_i(x) = -\sup_i \big(-f_i(x)\big).$$

Now that it has been shown that the upper and lower envelopes are measurable, it is immediate that the upper and lower limits of $\{f_i\}$ are also measurable. □

We begin by investigating what information can be deduced from the pointwise almost everywhere convergence of a sequence of measurable functions on a finite measure space.

5.16. DEFINITION. A sequence of measurable functions, $\{f_i\}$, with the property that

$$\lim_{i\to\infty} f_i(x) = f(x)$$

for μ-almost every $x \in X$ is said to **converge pointwise almost everywhere** (or more briefly, **converge pointwise a.e.**) to f.

We have the following:

5.17. COROLLARY. *Let $X = \mathbb{R}^n$. If $\{f_i\}$ is a sequence of Lebesgue measurable functions that converge pointwise almost everywhere to f, then f is measurable.*

The following is one of the main results of this section.

5.18. THEOREM (Egorov). *Let $(X, \mathcal{M}, \mu)$ be a finite measure space and suppose $\{f_i\}$ and f are measurable functions that are finite almost everywhere on X. Also, suppose that $\{f_i\}$ converges pointwise a.e. to f. Then for each $\varepsilon > 0$ there exists a set $A \in \mathcal{M}$ such that $\mu(\widetilde{A}) < \varepsilon$ and $\{f_i\} \to f$ uniformly on A.*

First, we will prove the following theorem.

5.19. THEOREM (Egorov). *Assume the hypotheses of the previous theorem. Then for each pair of numbers ε, $\delta > 0$, there exist a set $A \in \mathcal{M}$ and an integer i_0 such that $\mu(\widetilde{A}) < \varepsilon$ and*

$$|f_i(x) - f(x)| < \delta$$

whenever $x \in A$ and $i \geq i_0$.

PROOF. Choose ε, $\delta > 0$. Let E denote the set on which the functions f_i, $i = 1, 2, \ldots$, and f are defined and finite. Also, let F be the set on which $\{f_i\}$ converges pointwise to f. With $A_0 \colon = E \cap F$, we have by hypothesis, $\mu(\widetilde{A}_0) = 0$. For each positive integer i, let

$$A_i = A_0 \cap \{x : |f_j(x) - f(x)| < \delta \text{ for all } j \geq i\}.$$

Then, $A_1 \subset A_2 \subset \ldots$ and $\cup_{i=1}^{\infty} A_i = A_0$, and consequently, $\widetilde{A}_1 \supset \widetilde{A}_2 \supset \ldots$ with $\cap_{i=1}^{\infty} \widetilde{A}_i = \widetilde{A}_0$. Since $\mu(\widetilde{A}_1) \leq \mu(X) < \infty$, it follows from Theorem 4.49 (v) that

$$\lim_{i\to\infty} \mu(\widetilde{A}_i) = \mu(\widetilde{A}_0) = 0.$$

The result follows by choosing i_0 such that $\mu(\widetilde{A}_{i_0}) < \varepsilon$ and $A = A_{i_0}$. □

PROOF OF EGOROV'S THEOREM. Choose $\varepsilon > 0$. By the previous lemma, for each positive integer i, there exist a positive integer j_i and a measurable set A_i such that

$$\mu(\widetilde{A}_i) < \frac{\varepsilon}{2^i} \quad \text{and} \quad |f_j(x) - f(x)| < \frac{1}{i}$$

for all $x \in A_i$ and all $j \geq j_i$. With A defined as $A = \cap_{i=1}^{\infty} A_i$, we have

$$\widetilde{A} = \bigcup_{i=1}^{\infty} \widetilde{A}_i$$

and

$$\mu(\widetilde{A}) \leq \sum_{i=1}^{\infty} \mu(\widetilde{A}_i) < \sum_{i=1}^{\infty} \frac{\varepsilon}{2^i} = \varepsilon.$$

Furthermore, if $j \geq j_i$, then

$$\sup_{x \in A} |f_j(x) - f(x)| \leq \sup_{x \in A_i} |f_j(x) - f(x)| \leq \frac{1}{i}$$

for every positive integer i. This implies that $\{f_i\} \to f$ uniformly on A. □

5.20. COROLLARY. *In the previous theorem, assume in addition that X is a metric space and that μ is a Borel measure with $\mu(X) < \infty$. Then A can be taken as a closed set.*

PROOF. The previous theorem provides a set A such that $A \in \mathcal{M}$, $\{f_i\}$ converges uniformly to f and $\mu(\tilde{A}) < \varepsilon/2$. Since μ is a finite Borel measure, we see from Theorem 4.63 (p. 125) that there exists a closed set $F \subset A$ with $\mu(A \setminus F) < \varepsilon/2$. Hence, $\mu(\tilde{F}) < \varepsilon$ and $\{f_i\} \to f$ uniformly on F. □

5.21. DEFINITION. Because of its importance, we attach a name to the type of convergence exhibited in the conclusion of Egorov's theorem. Suppose that $\{f_i\}$ and f are measurable functions that are finite almost everywhere. We say that $\{f_i\}$ converges to f **almost uniformly** if for every $\varepsilon > 0$, there exists a set $A \in \mathcal{M}$ such that $\mu(\widetilde{A}) < \varepsilon$ and $\{f_i\}$ converges to f uniformly on A. Thus, Egorov's theorem states that pointwise a.e. convergence on a finite measure space implies almost uniform convergence. The converse is also true and is left as Exercise 5, Section 5.2.

5.22. REMARK. The hypothesis that $\mu(X) < \infty$ is essential in Egorov's theorem. Consider the case of Lebesgue measure on $\mathbb{R}$ and define a sequence of functions by

$$f_i = \chi_{[i,\infty)},$$

for each positive integer i. Then, $\lim_{i\to\infty} f_i(x) = 0$ for each $x \in \mathbb{R}$, but $\{f_i\}$ does not converge uniformly to 0 on any set A whose complement has finite Lebesgue measure. Indeed, for every such set, it would follow that $\widetilde{A}$ does not contain any $[i, \infty)$; that is, for each i, there would exist $x \in [i, \infty) \cap A$ with $f_i(x) = 1$, thus showing that $\{f_i\}$ does not converge uniformly to 0 on A.

5.23. DEFINITION. A sequence of measurable functions $\{f_i\}$ defined relative to the measure space $(X, \mathcal{M}, \mu)$ is said to **converge in measure** to a measurable function f if for every $\varepsilon > 0$, we have

$$\lim_{i\to\infty} \mu\big(X \cap \{x : |f_i(x) - f(x)| \geq \varepsilon\}\big) = 0.$$

We already encountered a result (Theorem 5.19) that essentially shows that pointwise a.e. convergence on a finite measure space implies convergence in measure. Formally, it is as follows.

5.24. THEOREM. *Let $(X, \mathcal{M}, \mu)$ be a finite measure space, and suppose $\{f_i\}$ and f are measurable functions that are finite a.e. on X. If $\{f_i\}$ converges to f a.e. on X, then $\{f_i\}$ converges to f in measure.*

PROOF. Choose positive numbers ε and δ. According to Theorem 5.19, there exist a set $A \in \mathcal{M}$ and an integer i_0 such that $\mu(\widetilde{A}) < \varepsilon$ and

$$|f_i(x) - f(x)| < \delta$$

whenever $x \in A$ and $i \geq i_0$. Thus,

$$X \cap \{x : |f_i(x) - f(x)| \geq \delta\} \subset \widetilde{A}$$

if $i \geq i_0$. Since $\mu(\widetilde{A}) < \varepsilon$ and $\varepsilon > 0$ is arbitrary, the result follows. □

5.25. REMARK. It is easy to see that the converse is not true. Let $X = [0, 1]$ with μ taken as Lebesgue measure. Consider a sequence of partitions of $[0, 1]$, $\mathcal{P}_i$, each consisting of closed, nonoverlapping intervals of length $1/2^i$. Let $\mathcal{F}$ denote the family of all intervals that form the partitions $\mathcal{P}_i$, $i = 1, 2, \ldots$. Linearly order $\mathcal{F}$ by defining $I \leq I'$ if both I and I' are elements of the same partition $\mathcal{P}_i$ and if I is to the left of I'. Otherwise, define $I \leq I'$ if the length of I is no greater than that of I'. Now put the elements of $\mathcal{F}$ into a one-to-one order-preserving correspondence with the positive integers. With the elements of $\mathcal{F}$ labeled I_k, $k = 1, 2, \ldots$, define a sequence of functions $\{f_k\}$ by $f_k = \chi_{I_k}$. Then it is easy to see that $\{f_k\} \to 0$ in measure but that $\{f_k(x)\}$ does not converge to 0 for any $x \in [0, 1]$.

Although the sequence $\{f_k\}$ converges nowhere to 0, it does have a subsequence that converges to 0 a.e., namely the subsequence

$$f_1, f_2, f_4, \ldots, f_{2^{k-1}}, \ldots .$$

In fact, this sequence converges to 0 at all points except $x = 0$. This illustrates the following general result.

5.26. THEOREM. *Let $(X, \mathcal{M}, \mu)$ be a measure space and let $\{f_i\}$ and f be measurable functions such that $f_i \to f$ in measure. Then there exists a subsequence $\{f_{i_j}\}$ such that*

$$\lim_{j\to\infty} f_{i_j}(x) = f(x)$$

for μ-a.e. $x \in X$.

Proof. Let i_1 be a positive integer such that

$$\mu(X \cap \{x : |f_{i_1}(x) - f(x)| \geq 1\}) < \frac{1}{2}.$$

Assuming that $i_1, i_2, \ldots, i_k$, have been chosen, let $i_{k+1} > i_k$ be such that

$$\mu\left(X \cap \left\{x : |f_{i_{k+1}}(x) - f(x)| \geq \frac{1}{k+1}\right\}\right) \leq \frac{1}{2^{k+1}}.$$

Let

$$A_j = \bigcup_{k=j}^{\infty} \left\{x : |f_{i_k}(x) - f(x)| \geq \frac{1}{k}\right\}$$

and observe that the sequence A_j is descending. Since

$$\mu(A_1) < \sum_{k=1}^{\infty} \frac{1}{2^k} < \infty,$$

with $B = \cap_{j=1}^{\infty} A_j$, it follows that

$$\mu(B) = \lim_{j\to\infty} \mu(A_j) \leq \lim_{j\to\infty} \sum_{k=j}^{\infty} \frac{1}{2^k} = \lim_{j\to\infty} \frac{1}{2^{j-1}} = 0.$$

Now select $x \in \widetilde{B}$. Then there exists an integer $j = j_x$ such that

$$x \in \widetilde{A_{j_x}} = \bigcap_{k=j_x}^{\infty} \left(X \cap \left\{y : |f_{i_k}(y) - f(y)| < \frac{1}{k}\right\}\right).$$

If $\varepsilon > 0$, choose k_0 such that $k_0 \geq j_x$ and $\frac{1}{k_0} \leq \varepsilon$. Then for $k \geq k_0$, we have

$$|f_{i_k}(x) - f(x)| < \frac{1}{k} \leq \varepsilon,$$

which implies that $f_{i_k}(x) \to f(x)$ for all $x \in \widetilde{B}$. □

There is another mode of convergence, **fundamental in measure**, which is discussed in Exercise 6, Section 5.2.

Exercises for Section 5.2

1. Let $\mathcal{F}$ be a family of continuous functions on a metric space (X, ρ). Let f denote the upper envelope of the family $\mathcal{F}$, that is,

$$f(x) = \sup\{g(x) : g \in \mathcal{F}\}.$$

   Prove that for each real number a, the set $\{x : f(x) > a\}$ is open.
2. Let $f(x, y)$ be a function defined on $\mathbb{R}^2$ that is continuous in each variable separately. Prove that f is Lebesgue measurable. Hint: Approximate f in the variable x by piecewise-linear continuous functions f_n such that $f_n \to f$ pointwise.
3. Let $(X, \mathcal{M}, \mu)$ be a finite measure space. Suppose that $\{f_i\}_{i=1}^{\infty}$ and f are measurable functions. Prove that $f_i \to f$ in measure if and only if each subsequence of f_i has a subsequence that converges to f μ-a.e.

4. Show that the supremum of an uncountable family of measurable $\overline{\mathbb{R}}$-valued functions can fail to be measurable.
5. Suppose $(X, \mathcal{M}, \mu)$ is a finite measure space. Prove that almost uniform convergence implies convergence almost everywhere.
6. A sequence $\{f_i\}$ of a.e. finite-valued measurable functions on a measure space $(X, \mathcal{M}, \mu)$ is **fundamental in measure** if for every $\varepsilon > 0$,

$$\mu(\{x : |f_i(x) - f_j(x)| \geq \varepsilon\}) \to 0$$

as i and $j \to \infty$. Prove that if $\{f_i\}$ is fundamental in measure, then there is a measurable function f to which the sequence $\{f_i\}$ converges in measure. Hint: Choose integers $i_{j+1} > i_j$ such that $\mu\{|f_{i_j} - f_{i_{j+1}}| > 2^{-j}\} < 2^{-j}$. The sequence $\{f_{i_j}\}$ converges a.e. to a function f. Then it follows that

$$\{|f_i - f| \geq \varepsilon\} \subset \{|f_i - f_{i_j}| \geq \varepsilon/2\} \cup \{|f_{i_j} - f| \geq \varepsilon/2\}.$$

By hypothesis, the measure of the first term on the right is arbitrarily small if i and i_j are large, and the measure of the second term tends to 0, since almost uniform convergence implies convergence in measure.

5.3. Approximation of Measurable Functions

In Section 3.2 certain fundamental approximation properties of Carathéodory outer measures were established. In particular, it was shown that each Borel set B of finite measure contains a closed set whose measure is arbitrarily close to that of B. The structure of a Borel set can be very complicated, yet this result states that a complicated set can be approximated by one with an elementary topological property. In this section we pursue an analogous situation by showing that each measurable function on a metric space of finite measure is almost continuous (Lusin's theorem). That is, every measurable function is continuous on sets whose complements have arbitrarily small measure. This result is in the same spirit as Egorov's theorem, which states that pointwise a.e. convergence implies almost uniform convergence on a finite measure space.

The characteristic function of a measurable set is the most elementary example of a measurable function. The next level of complexity involves linear combinations of such functions. A **simple function** on X is one that assumes only a finite number of values; thus the range of a simple function, f, is a finite subset of $\mathbb{R}$. If $\operatorname{rng} f = \{a_1, a_2, \ldots, a_k\}$, and $A_i = f^{-1}\{a_i\}$, then f can be written as

$$f = \sum_{i=1}^{k} a_i \chi_{A_i}.$$

If $X = \mathbb{R}^n$, a **step function** is of the form $f = \sum_{k=1}^{N} a_k \chi_{R_k}$, where each R_k is an interval and the a_k are real numbers.

We begin by proving that every measurable function is the pointwise a.e. limit of measurable simple functions.

5.27. THEOREM. *Let $f\colon X \to \overline{\mathbb{R}}$ be an arbitrary (possibly nonmeasurable) function. Then the following hold:*

(i) *There exists a sequence of simple functions, $\{f_i\}$, such that $f_i(x) \to f(x)$ for each $x \in X$.*

(ii) *If f is nonnegative, the sequence can be chosen such that $f_i \uparrow f$.*

(iii) *If f is bounded, the sequence can be chosen such that $f_i \to f$ uniformly on X.*

(iv) *If f is measurable, the f_i can be chosen to be measurable.*

PROOF. Assume first that $f \geq 0$. For each positive integer i, partition $[0, i)$ into $i \cdot 2^i$ half-open intervals of the form $\left[\frac{k-1}{2^i}, \frac{k}{2^i}\right)$, $k = 1, 2, \ldots, i \cdot 2^i$. Label these intervals $H_{i,k}$ and let

$$A_{i,k} = f^{-1}(H_{i,k}) \quad \text{and} \quad A_i = f^{-1}\big([i, \infty]\big).$$

These sets are mutually disjoint and form a partition of X. The approximating simple function f_i on X is defined as

$$f_i(x) = \begin{cases} \frac{k-1}{2^i}, & x \in A_{i,k}, \\ i \quad , & x \in A_i. \end{cases}$$

If f is measurable, then the sets $A_{i,k}$ and A_i are measurable, and thus so are the functions f_i. Moreover, it is easy to see that

$$f_1 \leq f_2 \leq \cdots \leq f.$$

If $f(x) < \infty$, then for every $i > f(x)$ we have

$$|f_i(x) - f(x)| < \frac{1}{2^i},$$

and hence $f_i(x) \to f(x)$. If $f(x) = \infty$, then $f_i(x) = i \to f(x)$. In any case we obtain

$$\lim_{i \to \infty} f_i(x) = f(x) \text{ for } x \in X.$$

Suppose f is bounded by some number, say M; that is, suppose $f(x) \leq M$ for all $x \in X$. Then $A_i = \emptyset$ for all $i > M$ and therefore $|f_i(x) - f(x)| < 1/2^i$ for all $x \in X$, thus showing that $f_i \to f$ uniformly if f is bounded. This establishes the theorem in the case $f \geq 0$.

In general, let $f^+(x) = \max(f(x), 0)$ and $f^-(x) = -\min(f(x), 0)$ denote the positive and negative parts of f. Then f^+ and f^- are nonnegative and $f = f^+ - f^-$. Now apply the previous results to f^+ and f^- to obtain the final form of the theorem. □

The proof of the following corollary is left to the reader (see Exercise 2, Section 5.3).

5.28. COROLLARY. *Let $f : \mathbb{R}^n \to \mathbb{R}$ be a Lebesgue measurable function. Then there exists a sequence of step functions f_i such that $f_i(x) \to f(x)$ for almost every $x \in \mathbb{R}^n$.*

Since it is possible for a measurable function to be discontinuous at every point of its domain, it seems unlikely that an arbitrary measurable function would have any regularity properties. However, the next result gives some information in the positive direction. It states, roughly, that every measurable function f is continuous on a closed set F whose complement has arbitrarily small measure. It is important to note that the result asserts the the function is continuous on F with respect to the relative topology on F. It should not be interpreted to say that f is continuous at every point of F relative to the topology on X.

5.29. THEOREM (Lusin's theorem). *Suppose $(X, \mathcal{M}, \mu)$ is a measure space where X is a metric space and μ is a finite Borel measure. Let $f\colon X \to \overline{\mathbb{R}}$ be a measurable function that is finite almost everywhere. Then for every $\varepsilon > 0$ there is a closed set $F \subset X$ with $\mu(\widetilde{F}) < \varepsilon$ such that f is continuous on F in the relative topology.*

PROOF. Choose $\varepsilon > 0$. For each fixed positive integer i, write $\mathbb{R}$ as the disjoint union of half-open intervals $H_{i,j}$, $j = 1, 2, \ldots$, whose lengths are $1/i$. Consider the disjoint measurable sets

$$A_{i,j} = f^{-1}(H_{i,j})$$

and refer to Theorem 4.63 to obtain disjoint closed sets $F_{i,j} \subset A_{i,j}$ such that $\mu(A_{i,j} - F_{i,j}) < \varepsilon/2^{i+j}$, $j = 1, 2, \ldots$. Let

$$E_k = X - \bigcup_{j=1}^{k} F_{i,j}$$

for $k = 1, 2, \ldots, \infty$. (Keep in mind that i is fixed, so it is not necessary to indicate that E_k depends on i.) Then $E_1 \supset E_2 \supset \ldots$, $\cap_{k=1}^{\infty} E_k = E_\infty$, and

$$\mu(E_\infty) = \mu\left(X - \bigcup_{j=1}^{\infty} F_{i,j}\right) = \sum_{j=1}^{\infty} \mu(A_{i,j} - F_{i,j}) < \frac{\varepsilon}{2^i}.$$

Since $\mu(X) < \infty$, it follows that

$$\lim_{k\to\infty} \mu(E_k) = \mu(E_\infty) < \frac{\varepsilon}{2^i}.$$

Hence, there exists a positive integer $J = J(i)$ such that

$$\mu(E_J) = \mu\left(X - \bigcup_{j=1}^{J} F_{i,j}\right) < \frac{\varepsilon}{2^i}.$$

For each $H_{i,j}$, select an arbitrary point $y_{i,j} \in H_{i,j}$ and let $B_i = \cup_{j=1}^{J} F_{i,j}$. Then define a continuous function g_i on the closed set B_i by

$$g_i(x) = y_{i,j} \quad \text{whenever} \quad x \in F_{i,j},\ j = 1, 2, \ldots, J.$$

The functions g_i are continuous (relative to B_i) because the closed sets $F_{i,j}$ are disjoint. Note that $|f(x) - g_i(x)| < 1/i$ for $x \in B_i$. Therefore, on the closed set

$$F = \bigcap_{i=1}^{\infty} B_i \quad \text{with} \quad \mu(X - F) \leq \sum_{i=1}^{\infty} \mu(X - B_i) < \varepsilon,$$

it follows that the continuous functions g_i converge uniformly to f, thus proving that f is continuous on F. $\square$

Using Corollary 4.56 we can rewrite Lusin's theorem as follows:

5.30. COROLLARY. *Let φ be a Borel regular outer measure on a metric space X. Let $\mathcal{M}$ be the σ-algebra of φ-measurable sets. Consider the measure space $(X, \mathcal{M}, \varphi)$ and let $A \in \mathcal{M}$, $\varphi(A) < \infty$. If $f : A \to \overline{\mathbb{R}}$ is a measurable function that is finite almost everywhere, then for every $\epsilon > 0$ there exists a closed set $F \subset A$ with $\varphi(A \setminus F) < \epsilon$ such that f is continuous on F in the relative topology.*

In particular, since λ^* is a Borel regular outer measure, we conclude that if $f : A \to \mathbb{R}$, $A \subset \mathbb{R}^n$ Lebesgue measurable, is a Lebesgue measurable function with $\lambda(A) < \infty$, then for every $\epsilon > 0$ there exists a closed set $F \subset A$ with $\lambda(A \setminus F) < \epsilon$ such that f is continuous on F in the relative topology.

We close this chapter with a table that reflects the interaction among the various types of convergence that we have encountered so far. A convergence type listed in the first column implies one in the first row if the corresponding entry of the matrix is indicated by $\Uparrow$ (along with the appropriate hypothesis).

	Fundamental in measure	Convergence in measure	Almost uniform convergence	Pointwise a.e. convergence
Fundamental in measure	$\Uparrow$	$\Uparrow$	$\Uparrow$ For a subsequence if $\mu(X) < \infty$	$\Uparrow$ For a subsequence
Convergence in measure	$\Uparrow$	$\Uparrow$	$\Uparrow$ For a subsequence if $\mu(X) < \infty$	$\Uparrow$ For a subsequence
Almost uniform convergence	$\Uparrow$	$\Uparrow$	$\Uparrow$	$\Uparrow$
Pointwise a.e. convergence	$\Uparrow$ if $\mu(X) < \infty$	$\Uparrow$ if $\mu(X) < \infty$	$\Uparrow$ if $\mu(X) < \infty$	$\Uparrow$

Exercises for Section 5.3

1. Let $E \subset \mathbb{R}^n$ be a Lebesgue measurable set with $\lambda(E) < \infty$ and let χ_E be the characteristic function of E. Prove that there is a sequence of step functions $\{\psi_k\}_{k=1}^{\infty}$ that converges pointwise to χ_E almost everywhere. Hint: Show that if $\lambda(E) < \infty$, then there exists a finite union of closed intervals Q_j such that $F = \cup_{j=1}^{N} Q_j$ and $\lambda(E\Delta F) \leq \epsilon$. Recall that $E\Delta F = (E\backslash F) \cup (F\backslash E)$.
2. Use the previous exercise to prove Corollary 5.28.
3. A union of n-dimensional (closed) intervals in $\mathbb{R}^n$ is said to be almost disjoint if the interiors of the intervals are disjoint. Show that every open subset U of $\mathbb{R}^n$, $n \geq 1$, can be written as a countable union of almost disjoint intervals.
4. Let $(X, \mathcal{M}, \mu)$ be a σ-finite measure space and suppose that f, f_k, $k = 1, 2, \ldots$, are measurable functions that are finite almost everywhere and
$$\lim_{k\to\infty} f_k(x) = f(x)$$
for μ almost all $x \in X$. Prove that there are measurable sets $E_0, E_1, E_2, \ldots$, such that $\nu(E_0) = 0$,
$$X = \bigcup_{i=0}^{\infty} E_i,$$
and $\{f_k\} \to f$ uniformly on each $E_i, i > 0$.
5. Use Corollary 4.56 to prove Corollary 5.30.
6. Suppose $f \colon [0,1] \to \mathbb{R}$ is Lebesgue measurable. Show that for every $\varepsilon > 0$, there is a continuous function g on $[0,1]$ such that
$$\lambda([0,1] \cap \{x : f(x) \neq g(x)\}) < \varepsilon.$$

CHAPTER 6

Integration

6.1. Definitions and Elementary Properties

Based on the ideas of H. Lebesgue, a far-reaching generalization of Riemann integration has been developed. In this section we define and deduce the elementary properties of integration with respect to an abstract measure.

We first extend the notion of simple function to allow better approximation of unbounded functions. Throughout this section and the next, we will assume the context of a general measure space $(X, \mathcal{M}, \mu)$.

6.1. DEFINITION. A function $f : X \to \overline{\mathbb{R}}$ is called **countably simple** if it assumes only a **countable** number of values, including possibly $\pm\infty$. Given a measure space $(X, \mathcal{M}, \mu)$, the **integral** of a nonnegative measurable countably simple function $f\colon X \to \overline{\mathbb{R}}$ is defined to be

$$\int_X f\,d\mu = \sum_{i=1}^{\infty} a_i \mu(f^{-1}\{a_i\}),$$

where the range of f is $\{a_1, a_2, \ldots\}$, and by convention, $0 \cdot \infty = \infty \cdot 0 = 0$. Note that the integral may equal ∞.

6.2. DEFINITIONS. For an arbitrary function $f : X \to \overline{\mathbb{R}}$, we define

$$\begin{aligned} f^+(x) &:= \quad f(x) \text{ if } f(x) \geq 0, \\ f^-(x) &:= -f(x) \text{ if } f(x) \geq 0. \end{aligned}$$

Thus, $f = f^+ - f^-$ and $|f| = f^+ + f^-$.

If f is a measurable countably simple function and at least one of $\int_X f^+\,d\mu$ and $\int_X f^-\,d\mu$ is finite, we define

$$\int_X f\,d\mu \colon = \int_X f^+\,d\mu - \int_X f^-\,d\mu.$$

If $f : X \to \overline{\mathbb{R}}$ (not necessarily measurable), we define the **upper integral** of f by

$$\bar{\int_X} f\,d\mu := \inf\left\{\int_X g\,d\mu : g \text{ is measurable, countably simple, and } g \geq f\ \mu\text{-a.e.}\right\},$$

W.P. Ziemer, *Modern Real Analysis*, Graduate Texts in Mathematics 278,
https://doi.org/10.1007/978-3-319-64629-9_6

and the **lower integral** of f by

$$\underline{\int_X} f\,d\mu := \sup\left\{\int_X g\,d\mu : g \text{ is measurable, countably simple, and } g \leq f\ \mu\text{-a.e.}\right\}.$$

The integral (with respect to the measure μ) of a measurable function $f : X \to \overline{\mathbb{R}}$ is said to **exist** if

$$\overline{\int_X} f\,d\mu = \underline{\int_X} f\,d\mu,$$

in which case we write

$$\int_X f d\mu$$

for the common value. If this value is finite, f is said to be **integrable**.

6.3. REMARK. Observe that our definition requires f to be measurable if it is to be integrable. See Exercise 2, Section 6.2, which shows that measurability is necessary for a function to be integrable, provided the measure μ is complete.

6.4. REMARK. If f is a countably simple function such that $\int_X f^-\,d\mu$ is finite, then the definitions immediately imply that the integral of f exists and that

$$\int_X f\,d\mu = \sum_{i=1}^{\infty} a_i\mu(f^{-1}\{a_i\}), \tag{6.1}$$

where the range of f is $\{a_1, a_2, \ldots\}$. Clearly, the integral should not depend on the order in which the terms of (6.1) appear. Consequently, the series converges unconditionally, possibly to $+\infty$ (see Exercise 1, Section 6.1). An analogous statement holds if $\int_X f^+\,d\mu$ is finite.

6.5. REMARK. It is clear from the definitions of upper and lower integrals that if $f = g$ μ-a.e., then $\overline{\int}_X f\,d\mu = \overline{\int}_X g\,d\mu$ and $\underline{\int}_X f\,d\mu = \underline{\int}_X g\,d\mu$. From this observation it follows that if both f and g are measurable, $f = g$ μ-a.e., and f is integrable, then g is integrable and

$$\int_X f\,d\mu = \int_X g\,d\mu.$$

6.6. DEFINITION. If $A \subset X$ (possibly nonmeasurable), we write

$$\overline{\int_A} f\,d\mu : = \overline{\int_X} f\chi_A\,d\mu$$

and use analogous notation for the other integrals.

6.7. THEOREM.

(i) *If f is an integrable function, then f is finite μ-a.e.*

(ii) *If f and g are integrable functions and a, b are constants, then $af + bg$ is integrable and*

$$\int_X (af + bg)\, d\mu = a \int_X f\, d\mu + b \int_X g\, d\mu.$$

(iii) *If f and g are integrable functions and $f \leq g$ μ-a.e., then*

$$\int_X f\, d\mu \leq \int_X g\, d\mu.$$

(iv) *If f is an integrable function and $E \in \mathcal{M}$, then $f\chi_E$ is integrable.*
(v) *A measurable function f is integrable if and only if $|f|$ is integrable.*
(vi) *If f is an integrable function, then*

$$\left| \int_X f\, d\mu \right| \leq \int_X |f|\, d\mu.$$

PROOF. Each of the assertions above is easily seen to hold if the functions are countably simple. We leave these proofs as exercises.

(i) If f is integrable, then there are integrable countably simple functions g and h such that $g \leq f \leq h$ μ-a.e. Thus f is finite μ-a.e.

(ii) Suppose f is integrable and c is a constant. If $c > 0$, then for every integrable countably simple function g,

$$cg \leq cf \ \text{ if and only if } \ g \leq f.$$

Since $\int_X cg\, d\mu = c \int_X g\, d\mu$, it follows that

$$\underline{\int}_X cf\, d\mu = c \underline{\int}_X f\, d\mu$$

and

$$\overline{\int}_X cf\, d\mu = c \overline{\int}_X f\, d\mu.$$

Clearly $-f$ is integrable and $\int_X -f\, d\mu = -\int_X f\, d\mu$. Thus if $c < 0$, then $cf = |c|\,(-f)$ is integrable and

$$\int_X cf\, d\mu = |c| \int_X (-f)\, d\mu = -|c| \int_X f\, d\mu = c \int_X f\, d\mu.$$

Now suppose f, g are integrable and f_1, g_1 are integrable countably simple functions such that $f_1 \leq f$, $g_1 \leq g$ μ-a.e. Then $f_1 + g_1 \leq f + g$ μ-a.e. and

$$\underline{\int}_X (f+g)\, d\mu \geq \int_X (f_1 + g_1)\, d\mu = \int_X f_1\, d\mu + \int_X g_1\, d\mu.$$

Thus

$$\int_X f\, d\mu + \int_X g\, d\mu \leq \underline{\int}_X (f+g)\, d\mu.$$

An analogous argument shows that

$$\overline{\int}_X (f+g)\, d\mu \leq \int_X f\, d\mu + \int_X g\, d\mu,$$

and assertion (ii) follows.

(iii) If f, g are integrable and $f \le g$ μ-a.e., then by (ii), $g - f$ is integrable and $g - f \ge 0$ μ-a.e. Clearly $\underline{\int}_X (g-f)\,d\mu = \overline{\int}_X (g-f)\,d\mu \ge 0$, and hence by (ii) again,

$$\int_X g\,d\mu = \int_X f\,d\mu + \int_X (g-f)\,d\mu \ge \int_X f\,d\mu.$$

(iv) If f is integrable, then given $\varepsilon > 0$ there are integrable countably simple functions g, h such that $g \le f \le h$ μ-a.e. and

$$\int_X (h-g)\,d\mu < \varepsilon.$$

Thus

$$\int_X (h-g)\chi_E\,d\mu \le \varepsilon$$

for $E \in \mathcal{M}$. Thus

$$0 \le \overline{\int_X} f\chi_E\,d\mu - \underline{\int_X} f\chi_E\,d\mu < \varepsilon,$$

and since

$$-\infty < \int_X g\chi_E\,d\mu \le \underline{\int_X} f\chi_E\,d\mu \le \overline{\int_X} f\chi_E\,d\mu \le \int_X h\chi_E\,d\mu < \infty,$$

it follows that $f\chi_E$ is integrable.

(v) If f is integrable, then by (iv), $f^+ := f\chi_{\{x:f(x)>0\}}$ and $f^- := -f\chi_{\{x:f(x)<0\}}$ are integrable, and by (ii), $|f| = f^+ + f^-$ is integrable. If $|f|$ is integrable, then by (iv), $f^+ = |f|\,\chi_{\{x:f(x)>0\}}$ and $f^- = |f|\,\chi_{\{x:f(x)<0\}}$ are integrable, and hence $f = f^+ - f^-$ is integrable.

(vi) If f is integrable, then by (v), $f^\pm$ are integrable and

$$\left|\int_X f\,d\mu\right| = \left|\int_X f^+\,d\mu - \int_X f^-\,d\mu\right| \le \int_X f^+\,d\mu + \int_X f^-\,d\mu = \int_X |f|\,d\mu. \ \square$$

The next result, whose proof is very simple, is remarkably strong in view of the weak hypothesis. In particular, it implies that every bounded, nonnegative, measurable function is μ-integrable. This exhibits a striking difference between the Lebesgue and Riemann integrals (see Theorem 6.19 below).

6.8. Theorem. *If f is μ-measurable and $f \ge 0$ μ-a.e., then the integral of f exists: that is,*

$$\underline{\int_X} f\,d\mu = \overline{\int_X} f\,d\mu.$$

PROOF. If the lower integral is infinite, then the upper and lower integrals are both infinite. Thus we may assume that the lower integral is finite and, in particular, $\mu(\{x : f(x) = \infty\}) = 0$. For $t > 1$ and $k = 0, \pm 1, \pm 2, \dots$ set

$$E_k = \{x : t^k \leq f(x) < t^{k+1}\}$$

and

$$g_t = \sum_{k=-\infty}^{\infty} t^k \chi_{E_k}.$$

Since each set E_k is measurable, it follows that g_t is a measurable countably simple function and $g_t \leq f \leq t g_t$ μ-a.e. Thus

$$\overline{\int_X} f \, d\mu \leq \int_X t g_t \, d\mu = t \int_X g_t \, d\mu \leq t \underline{\int_X} f \, d\mu$$

for each $t > 1$, and therefore on letting $t \to 1^+$,

$$\overline{\int_X} f \, d\mu \leq \underline{\int_X} f \, d\mu,$$

which implies our conclusion, since

$$\underline{\int_X} f \, d\mu \leq \overline{\int_X} f \, d\mu$$

is always true. □

6.9. THEOREM. *If f is a nonnegative measurable function and g is an integrable function, then*

$$\underline{\int_X} (f+g) \, d\mu = \overline{\int_X} (f+g) \, d\mu = \int_X f \, d\mu + \int_X g \, d\mu.$$

PROOF. If f is integrable, the assertion follows from Theorem 6.7, so assume that $\int_X f d\mu = \infty$. Let h be a countably simple function such that $0 \leq h \leq f$, and let k be an integrable countably simple function such that $k \geq |g|$. Then by Exercise 6.1,

$$\underline{\int_X} (f+g) \, d\mu \geq \underline{\int_X} (f - |g|) \, d\mu \geq \int_X (h-k) \, d\mu = \int_X h \, d\mu - \int_X k \, d\mu,$$

from which it follows that $\underline{\int_X}(f+g) \, d\mu = \infty$, and the assertion is proved. □

One of the main applications of this result is the following.

6.10. COROLLARY. *If f is measurable, and if either f^+ or f^- is integrable, then the following integral exists:*

$$\int_X f \, d\mu.$$

PROOF. For example, if f^+ is integrable, take $g := -f^+$ and $f := f^-$ in the previous theorem to conclude that the integrals

$$\int_X (f^- - f^+)\, d\mu = \int_X -f\, d\mu$$

exist and therefore that

$$\int_X f\, d\mu$$

exists. □

6.11. THEOREM. *If f is μ-measurable, g is integrable, and $|f| \le |g|$ μ-a.e., then f is integrable.*

PROOF. This follows immediately from Theorem 6.7 (v) and Theorem 6.8. □

Exercises for Section 6.1

1. A series $\sum_{i=1}^{\infty} c_i$ is said to converge unconditionally if it converges and for every one-to-one mapping σ of $\mathbb{N}$ onto $\mathbb{N}$ the series $\sum_{i=1}^{\infty} c_{\sigma(i)}$ converges to the same limit. Verify the assertion in Remark (6.4). That is, suppose $\mathbb{N}_1$ and $\mathbb{N}_2$ are both infinite subsets of $\mathbb{N}$ such that $\mathbb{N}_1 \cap \mathbb{N}_2 = \emptyset$ and $\mathbb{N}_1 \cup \mathbb{N}_2 = \mathbb{N}$. Suppose $\{a_i : i \in \mathbb{N}\}$ are real numbers such that $\{a_i : i \in \mathbb{N}_1\}$ are all nonpositive and that $\{a_i : i \in \mathbb{N}_2\}$ are all positive numbers. If
$$-\sum_{i\in\mathbb{N}_1} a_i < \infty \quad \text{and} \quad \sum_{i\in\mathbb{N}_2} a_i = \infty,$$
prove that
$$\sum_{\sigma(i)\in\mathbb{N}} a_{\sigma(i)} = \infty$$
for every bijection $\sigma\colon \mathbb{N} \to \mathbb{N}$. Also, show that
$$\sum_{\sigma(i)\in\mathbb{N}} a_{\sigma(i)} < \infty \quad \text{and} \quad \sum_{i=1}^{\infty} |a_i| < \infty$$
if $\sum_{i\in\mathbb{N}_2} a_i < \infty$. Use the assertion to show that if f is a nonnegative countably simple function and g is an integrable countably simple function, then
$$\int_X (f+g)\, d\mu = \int_X f\, d\mu + \int_X g\, d\mu.$$
2. Verify the assertions of Theorem 6.7 for countably simple functions.
3. Suppose f is a nonnegative measurable function. Show that
$$\int_X f\ d\mu = \sup \sum_{k=1}^{N} (\inf_{x\in E_k} f(x))\mu(E_k),$$

where the supremum is taken over all finite measurable partitions of X, i.e., over all finite collections $\{E_k\}_{k=1}^N$ of disjoint measurable subsets of X such that $X = \bigcup_{k=1}^{N} E_k$.

4. Suppose f is a nonnegative integrable function with the property that
$$\int_X f \, d\mu = 0.$$
Show that $f = 0$ μ-a.e.

5. Suppose f is an integrable function with the property that
$$\int_E f \, d\mu = 0$$
whenever E is a μ-measurable set. Show that $f = 0$ μ-a.e.

6. Show that if f is measurable, g is μ-integrable, and $f \geq g$, then f^- is μ-integrable and
$$\underline{\int}_X f \, d\mu = \overline{\int}_X f \, d\mu = \int_X f^+ \, d\mu - \int_X f^- \, d\mu.$$

7. Suppose $(X, \mathcal{M}, \mu)$ is a measure space and $Y \in \mathcal{M}$. Set
$$\mu_Y(E) = \mu(E \cap Y)$$
for each $E \in \mathcal{M}$. Show that μ_Y is a measure on $(X, \mathcal{M})$ and that
$$\int_X g \, d\mu_Y = \int_X g\chi_Y \, d\mu$$
for each nonnegative measurable function g on X.

8. A function $f\colon (a,b) \to \mathbb{R}$ is **convex** if
$$f[(1-t)x + ty] \leq (1-t)f(x) + tf(y)$$
for all $x, y \in (a,b)$ and $t \in [0,1]$. Prove that this is equivalent to
$$\frac{f(y) - f(x)}{y - x} \leq \frac{f(z) - f(y)}{z - y}$$
whenever $a < x < y < z < b$.

6.2. Limit Theorems

The most important results in integration theory are those related to the continuity of the integral operator. That is, if $\{f_i\}$ converges to f in some sense, how are $\int f$ and $\lim_{i\to\infty} \int f_i$ related? There are three fundamental results that address this question: Fatou's lemma, the monotone convergence theorem, and Lebesgue's dominated convergence theorem. These will be discussed along with associated results.

Our first result concerning the behavior of sequences of integrals is Fatou's lemma. Note the similarity between this result and its measure-theoretic counterpart, Theorem 4.49 (vi).

We continue to assume the context of a general measure space $(X, \mathcal{M}, \mu)$.

6.12. LEMMA (Fatou's lemma). *If $\{f_k\}_{k=1}^{\infty}$ is a sequence of nonnegative μ-measurable functions, then*

$$\int_X \liminf_{k\to\infty} f_k \, d\mu \le \liminf_{k\to\infty} \int_X f_k \, d\mu.$$

PROOF. Let g be any measurable countably simple function such that $0 \le g \le \liminf_{k\to\infty} f_k$ μ-a.e. For each $x \in X$, set

$$g_k(x) = \inf\{f_m(x) : m \ge k\}$$

and observe that $g_k \le g_{k+1}$ and

$$\lim_{k\to\infty} g_k = \liminf_{k\to\infty} f_k \ge g \ \mu\text{-a.e.}$$

Write $g = \sum_{j=1}^{\infty} a_j \chi_{A_j}$, where $A_j := g^{-1}(a_j)$. Therefore $A_j \cap A_i = \emptyset$ if $i \ne j$ and $X = \bigcup_{k=1}^{\infty} A_j$. For $0 < t < 1$ set

$$B_{j,k} = A_j \cap \{x : g_k(x) > ta_j\}.$$

Then each $B_{j,k}$ is in $\mathcal{M}$, $B_{j,k} \subset B_{j,k+1}$, and since $\lim_{k\to\infty} g_k \ge g$ μ-a.e., we have

$$\bigcup_{k=1}^{\infty} B_{j,k} = A_j \text{ and } \lim_{k\to\infty} \mu(B_{j,k}) = \mu(A_j)$$

for $j = 1, 2, \ldots$. Noting that

$$\sum_{j=1}^{\infty} ta_j \chi_{B_{j,k}} \le g_k \le f_m$$

for each $m \ge k$, we obtain

$$t\int_X g \, d\mu = \sum_{j=1}^{\infty} ta_j \mu(A_j) = \lim_{k\to\infty} \sum_{j=1}^{\infty} ta_j \mu(B_{j,k}) \le \liminf_{k\to\infty} \int_X f_k \, d\mu.$$

Thus, on letting $t \to 1^-$, we obtain

$$\int_X g \, d\mu \le \liminf_{k\to\infty} \int_X f_k \, d\mu.$$

By taking the supremum of the left-hand side over all countably simple functions g with $g \le \liminf_{k\to\infty} f_k$, we have

$$\underline{\int}_X \liminf_{k\to\infty} f_k \, d\mu \le \liminf_{k\to\infty} \int_X f_k \, d\mu.$$

Since $\liminf_{k\to\infty} f_k$ is a nonnegative measurable function, we can apply Theorem 6.8 to obtain our desired conclusion. □

6.13. THEOREM (Monotone convergence theorem). *If $\{f_k\}_{k=1}^{\infty}$ is a sequence of nonnegative μ-measurable functions such that $f_k \le f_{k+1}$ for $k = 1, 2, \ldots$, then*

$$\lim_{k\to\infty} \int_X f_k \, d\mu = \int_X \lim_{k\to\infty} f_k \, d\mu.$$

PROOF. Set $f = \lim_{k\to\infty} f_k$. Then f is μ-measurable,

$$\int_X f_k \, d\mu \leq \int_X f \, d\mu \quad \text{for} \quad k = 1, 2, \ldots,$$

and

$$\lim_{k\to\infty} \int_X f_k \, d\mu \leq \int_X f \, d\mu.$$

The opposite inequality follows from Fatou's lemma. □

6.14. THEOREM. *If $\{f_k\}_{k=1}^{\infty}$ is a sequence of nonnegative μ-measurable functions, then*

$$\int_X \sum_{k=1}^{\infty} f_k \, d\mu = \sum_{k=1}^{\infty} \int_X f_k \, d\mu.$$

PROOF. With $g_m := \sum_{k=1}^{m} f_k$ we have $g_m \uparrow \sum_{k=1}^{\infty} f_k$, and the conclusion follows easily from the monotone convergence theorem and Theorem 6.7 (ii). □

6.15. THEOREM. *If f is integrable and $\{E_k\}_{k=1}^{\infty}$ is a sequence of disjoint measurable sets such that $X = \bigcup_{k=1}^{\infty} E_k$, then*

$$\int_X f \, d\mu = \sum_{k=1}^{\infty} \int_{E_k} f \, d\mu.$$

PROOF. Assume first that $f \geq 0$, set $f_k = f\chi_{E_k}$, and apply the previous theorem. For arbitrary integrable f use the fact that $f = f^+ - f^-$. □

6.16. COROLLARY. *If $f \geq 0$ is integrable and if ν is a set function defined by*

$$\nu(E) := \int_E f \, d\mu$$

for every measurable set E, then ν is a measure.

6.17. THEOREM (Lebesgue's dominated convergence theorem). *Suppose g is integrable, f is measurable, $\{f_k\}_{k=1}^{\infty}$ is a sequence of μ-measurable functions such that $|f_k| \leq g$ μ-a.e. for $k = 1, 2, \ldots,$ and*

$$\lim_{k\to\infty} f_k(x) = f(x)$$

for μ-a.e. $x \in X$. Then

$$\lim_{k\to\infty} \int_X |f_k - f| \, d\mu = 0.$$

PROOF. Clearly $|f| \leq g$ μ-a.e. and hence f and each f_k are integrable. Set $h_k = 2g - |f_k - f|$. Then $h_k \geq 0$ μ-a.e., and by Fatou's lemma,

$$\begin{aligned} 2\int_X g\,d\mu = \int_X \liminf_{k\to\infty} h_k\,d\mu &\leq \liminf_{k\to\infty}\int_X h_k\,d\mu \\ &= 2\int_X g\,d\mu - \limsup_{k\to\infty}\int_X |f_k - f|\,d\mu. \end{aligned}$$

Thus

$$\limsup_{k\to\infty}\int_X |f_k - f|\,d\mu = 0. \qquad \square$$

Exercises for Section 6.2

1. Let $(X, \mathcal{M}, \mu)$ be an arbitrary measure space. For an arbitrary $X \xrightarrow{f} \overline{\mathbb{R}}$ prove that there is a measurable function with $g \geq f$ μ-a.e. such that
$$\int_X g\,d\mu = \overline{\int_X} f\,d\mu.$$
2. Suppose $(X, \mathcal{M}, \mu)$ is a measure space, $f : X \to \overline{\mathbb{R}}$, and
$$\underline{\int_X} f\ d\mu = \overline{\int_X} f\ d\mu < \infty.$$
Show that there exists an integrable (and measurable) function g such that $f = g$ μ-a.e. Thus if $(X, \mathcal{M}, \mu)$ is complete, f is measurable.
3. Suppose $\{f_k\}$ is a sequence of measurable functions, g is a μ-integrable function, and $f_k \geq g$ μ-a.e. for each k. Show that
$$\int_X \liminf_{k\to\infty} f_k\ d\mu \leq \liminf_{k\to\infty}\int_X f_k\ d\mu.$$
4. Let $(X, \mathcal{M}, \mu)$ be an arbitrary measure space. For arbitrary nonnegative functions $f_i \colon X \to \overline{\mathbb{R}}$, prove that
$$\overline{\int_X} \liminf_{i\to\infty} f_i\,d\mu \leq \liminf_{i\to\infty}\overline{\int_X} f_i\,d\mu.$$
Hint: See Exercise 1, Section 6.2.
5. If $\{f_k\}$ is an increasing sequence of measurable functions, g is μ-integrable, and $f_k \geq g$ μ-a.e. for each k, show that
$$\lim_{k\to\infty}\int_X f_k\ d\mu = \int_X \lim_{k\to\infty} f_k\ d\mu.$$
6. Show that there exists a sequence of bounded Lebesgue measurable functions mapping $\mathbb{R}$ into $\mathbb{R}$ such that
$$\liminf_{i\to\infty}\int_{\mathbb{R}} f_i\,d\lambda < \int_{\mathbb{R}} \liminf_{i\to\infty} f_i\,d\lambda.$$

7. Let f be a bounded function on the unit square Q in $\mathbb{R}^2$. Suppose for each fixed y that f is a measurable function of x. For each $(x, y) \in Q$ let the partial derivative $\dfrac{\partial f}{\partial y}$ exist. Under the assumption that $\dfrac{\partial f}{\partial y}$ is bounded in Q, prove that

$$\frac{d}{dy} \int_0^1 f(x, y)\, d\lambda(x) = \int_0^1 \frac{\partial f}{\partial y}\, d\lambda(x).$$

6.3. Riemann and Lebesgue Integration: A Comparison

The Riemann and Lebesgue integrals are compared, and it is shown that a bounded function is Riemann integrable if and only if it is continuous almost everywhere.

We first recall the definition and some elementary facts concerning Riemann integration. Suppose $[a, b]$ is a closed interval in $\mathbb{R}$. By a **partition** $\mathcal{P}$ of $[a, b]$ we mean a finite set of points $\{x_i\}_{i=0}^m$ such that $a = x_0 < x_1 < \cdots < x_m = b$. Let

$$\|\mathcal{P}\| : = \max\{x_i - x_{i-1} \,:\, 1 \le i \le m\}.$$

For each $i \in \{1, 2, \ldots, m\}$ let x_i^* be an arbitrary point of the interval $[x_{i-1}, x_i]$. A bounded function $f : [a, b] \to \mathbb{R}$ is **Riemann integrable** if

$$\lim_{\|\mathcal{P}\| \to 0} \sum_{i=1}^m f(x_i^*)(x_i - x_{i-1})$$

exists, in which case the value is the Riemann integral of f over $[a, b]$, which we will denote by

$$(R) \int_a^b f(x)\, dx.$$

Given a partition $\mathcal{P} = \{x_i\}_{i=0}^m$ of $[a, b]$ set

$$U(\mathcal{P}) = \sum_{i=1}^m \left[\sup_{x \in [x_{i-1}, x_i]} f(x) \right] (x_i - x_{i-1}),$$

$$L(\mathcal{P}) = \sum_{i=1}^m \left[\inf_{x \in [x_{i-1}, x_i]} f(x) \right] (x_i - x_{i-1}).$$

Then

$$L(\mathcal{P}) \le \sum_{i=1}^m f(x_i^*)(x_i - x_{i-1}) \le U(\mathcal{P})$$

for every choice of the x_i^*. Since the supremum (infimum) of $\sum_{i=1}^m f(x_i^*)$ $(x_i - x_{i-1})$ over all choices of the x_i^* is equal to $U(\mathcal{P})$ $(L(\mathcal{P}))$, we see that a bounded function f is Riemann integrable if and only if

$$\lim_{\|\mathcal{P}\| \to 0} (U(\mathcal{P}) - L(\mathcal{P})) = 0. \tag{6.2}$$

We next examine the effects of using a finer partition. Suppose then that $\mathcal{P} = \{x_i\}_{i=1}^m$ is a partition of $[a, b]$, $z \in [a, b] - \mathcal{P}$, and $Q = \mathcal{P} \cup \{z\}$. Thus

$\mathcal{P} \subset \mathcal{Q}$, and $\mathcal{Q}$ is called a refinement of $\mathcal{P}$. Then $z \in (x_{i-1}, x_i)$ for some $1 \leq i \leq m$ and

$$\sup_{x\in[x_{i-1},z]} f(x) \leq \sup_{x\in[x_{i-1},x_i]} f(x),$$
$$\sup_{x\in[z,x_i]} f(x) \leq \sup_{x\in[x_{i-1},x_i]} f(x).$$

Thus

$$U(\mathcal{Q}) \leq U(\mathcal{P}).$$

An analogous argument shows that $L(\mathcal{P}) \leq L(\mathcal{Q})$. It follows by induction on the number of points in Q that

$$L(\mathcal{P}) \leq L(\mathcal{Q}) \leq U(\mathcal{Q}) \leq U(\mathcal{P})$$

whenever $\mathcal{P} \subset \mathcal{Q}$. Thus, U does not increase and L does not decrease when a refinement of the partition is used.

We will say that a Lebesgue measurable function f on $[a,b]$ is **Lebesgue integrable** if f is integrable with respect to Lebesgue measure λ on $[a,b]$.

6.18. THEOREM. *If $f : [a,b] \to \mathbb{R}$ is a bounded Riemann integrable function, then f is Lebesgue integrable and*

$$(R)\int_a^b f(x)\,dx = \int_{[a,b]} f\,d\lambda.$$

PROOF. Let $\{\mathcal{P}_k\}_{k=1}^{\infty}$ be a sequence of partitions of $[a,b]$ such that $\mathcal{P}_k \subset \mathcal{P}_{k+1}$ and $\|\mathcal{P}_k\| \to 0$ as $k \to \infty$. Write $\mathcal{P}_k = \{x_j^k\}_{j=0}^{m_k}$. For each k define functions l_k, u_k by setting

$$l_k(x) = \inf_{t\in[x_{i-1}^k,x_i^k]} f(t),$$
$$u_k(x) = \sup_{t\in[x_{i-1}^k,x_i^k]} f(t),$$

whenever $x \in [x_{i-1}^k, x_i^k)$, $1 \leq i \leq m_k$. Then for each k the functions l_k, u_k are Lebesgue integrable and

$$\int_{[a,b]} l_k\,d\lambda = L(\mathcal{P}_k) \leq U(\mathcal{P}_k) = \int_{[a,b]} u_k\,d\lambda.$$

The sequence $\{l_k\}$ is monotonically increasing and bounded. Thus

$$l(x) := \lim_{k\to\infty} l_k(x)$$

exists for each $x \in [a,b]$, and l is a Lebesgue measurable function. Similarly, the function

$$u := \lim_{k\to\infty} u_k$$

is Lebesgue measurable and $l \le f \le u$ on $[a, b]$. Since f is Riemann integrable, it follows from Lebesgue's dominated convergence theorem (Theorem 6.17) and (6.2) that

$$\int_{[a,b]} (u - l)\, d\lambda = \lim_{k\to\infty} \int_{[a,b]} (u_k - l_k)\, d\lambda = \lim_{k\to\infty} (U(\mathcal{P}_k) - L(\mathcal{P}_k)) = 0. \tag{6.3}$$

Thus $l = f = u$ λ-a.e. on $[a, b]$ (see Exercise 4, Section 6.1), and invoking Theorem 6.17 once more, we have

$$\int_{[a,b]} f\, d\lambda = \lim_{k\to\infty} \int_{[a,b]} u_k\, d\lambda = \lim_{k\to\infty} U(\mathcal{P}_k) = (R)\int_a^b f(x)\, dx. \qquad \square$$

6.19. THEOREM. *A bounded function $f : [a, b] \to \mathbb{R}$ is Riemann integrable if and only if f is continuous λ-a.e. on $[a, b]$.*

PROOF. Suppose f is Riemann integrable and let $\{\mathcal{P}_k\}$ be a sequence of partitions of $[a, b]$ such that $\mathcal{P}_k \subset \mathcal{P}_{k+1}$ and $\lim_{k\to\infty} \|\mathcal{P}_k\| = 0$. Set $N = \cup_{k=1}^{\infty} \mathcal{P}_k$. Let l_k, u_k be as in the proof of Theorem 6.18. If $x \in [a, b]-N$, $l(x) = u(x)$ and $\varepsilon > 0$, then there is an integer k such that

$$u_k(x) - l_k(x) < \varepsilon.$$

Let $\mathcal{P}_k = \{x_j^k\}_{j=0}^{m_k}$. Then $x \in [x_{j-1}^k, x_j^k)$ for some $j \in \{1, 2, \ldots, m_k\}$. For every $y \in (x_{j-1}^k, x_j^k)$,

$$|f(y) - f(x)| \le u_k(x) - l_k(x) < \varepsilon.$$

Thus f is continuous at x. Since $l(x) = u(x)$ for λ-a.e. $x \in [a, b]$ and $\lambda(N) = 0$, we see that f is continuous at λ-a.e. point of $[a, b]$.

Now suppose f is bounded, $N \subset [a, b]$ with $\lambda(N) = 0$, and f is continuous at each point of $[a, b] - N$. Let $\{\mathcal{P}_k\}$ be a sequence of partitions of $[a, b]$ such that $\lim_{k\to\infty} \|\mathcal{P}_k\| = 0$. For each k define the Lebesgue integrable functions l_k and u_k as in the proof of Theorem 6.18. Then

$$L(\mathcal{P}_k) = \int_{[a,b]} l_k\, d\lambda \le \int_{[a,b]} u_k\, d\lambda = U(\mathcal{P}_k).$$

If $x \in [a, b] - N$ and $\varepsilon > 0$, then there is a $\delta > 0$ such that

$$|f(x) - f(y)| < \varepsilon/2$$

whenever $|y - x| < \delta$. There is a k_0 such that $\|\mathcal{P}_k\| < \frac{\delta}{2}$ whenever $k > k_0$. Thus

$$u_k(x) - l_k(x) \le 2 \sup\{|f(x) - f(y)| : |y - x| < \delta\} < \varepsilon$$

whenever $k > k_0$. Thus

$$\lim_{k\to\infty} (u_k(x) - l_k(x)) = 0$$

for each $x \in [a, b] - N$. By the dominated convergence theorem, Theorem 6.17, it follows that

$$\lim_{k\to\infty} (U(\mathcal{P}_k) - L(\mathcal{P}_k)) = \lim_{k\to\infty} \int_{[a,b]} (u_k - l_k)\, d\lambda = 0,$$

thus showing that f is Riemann integrable. □

Exercises for Section 6.3

1. Give an example of a nondecreasing sequence of functions mapping $[0,1]$ into $[0,1]$ such that each term in the sequence is Riemann integrable and such that the limit of the resulting sequence of Riemann integrals exists, but the limit of the sequence of functions is not Riemann integrable.
2. From here to Exercise 6 we outline a development of the **Riemann–Stieltjes integral** that is similar to that of the Riemann integral. Let f and g be two real-valued functions defined on a finite interval $[a,b]$. Given a partition $\mathcal{P}=\{x_i\}_{i=0}^m$ of $[a,b]$, for each $i\in\{1,2,\dots,m\}$ let x_i^* be an arbitrary point of the interval $[x_{i-1},x_i]$. We say that the Riemann–Stieltjes integral of f with respect to g exists if
$$\lim_{\|\mathcal{P}\|\to 0}\sum_{i=1}^m f(x_i^*)(g(x_i)-g(x_{i-1}))$$
exists, in which case the value is denoted by
$$\int_a^b f(x)\,dg(x).$$
Prove that if f is continuous and g is continuously differentiable on $[a,b]$, then
$$\int_a^b f\,dg=\int_a^b fg'\,dx.$$
3. Suppose f is a bounded function on $[a,b]$ and g is nondecreasing. Set
$$U_{RS}(\mathcal{P})=\sum_{i=1}^m\left[\sup_{x\in[x_{i-1},x_i]}f(x)\right](g(x_i)-g(x_{i-1})),$$
$$L_{RS}(\mathcal{P})=\sum_{i=1}^m\left[\inf_{x\in[x_{i-1},x_i]}f(x)\right](g(x_i)-g(x_{i-1})).$$
Prove that if $\mathcal{P}'$ is a refinement of $\mathcal{P}$, then $L_{RS}(\mathcal{P}')\geq L_{RS}(\mathcal{P})$ and $U_{RS}(\mathcal{P}')\leq U_{RS}(\mathcal{P})$. Also, if $\mathcal{P}_1$ and $\mathcal{P}_2$ are two partitions, then $L_{RS}(\mathcal{P}_1)\leq U_{RS}(\mathcal{P}_2)$.
4. If f is continuous and g nondecreasing, prove that
$$\int_a^b f\,dg$$
exists. Thus establish the same conclusion if g is assumed to be of bounded variation.
5. Prove the following integration by parts formula. If $\int_a^b f\,dg$ exists, then so does $\int_a^b g\,df$ and
$$f(b)g(b)-f(a)g(a)=\int_a^b f\,dg+\int_a^b g\,df.$$

6. Using the proof of Theorem 6.18 as a guide, show that the Riemann–Stieltjes and Lebesgue–Stieltjes integrals are in agreement. That is, if f is bounded, g is nondecreasing and right-continuous, and the Riemann–Stieltjes integral of f with respect to g exists, then

$$\int_a^b f\,dg = \int_{[a,b]} f\,d\lambda_g,$$

where λ_g is the Lebesgue–Stieltjes measure induced by g as in Section 4.6.

6.4. Improper Integrals

In this section we study the relation between Lebesgue integrals and improper integrals.

Let $a \in \mathbb{R}$ and let $f : [a, \infty) \to \mathbb{R}$ be a function that is Riemann integrable on each subinterval of $[a, \infty)$. The improper integral of f is defined as

$$(I) \int_a^\infty f(x)dx := \lim_{b\to\infty} (R) \int_a^b f(x)dx. \tag{6.4}$$

If the limit in (6.4) is finite, we say that the improper integral of f exist. We have the following result.

6.20. THEOREM. *Let $f : [a, \infty) \to \mathbb{R}$ be a nonnegative function that is Riemann integrable on each subinterval of $[a, \infty]$. Then*

$$\int_{[a,\infty)} f d\lambda = \lim_{b\to\infty} (R) \int_a^b f dx. \tag{6.5}$$

Thus, f is Lebesgue integrable on $[a, \infty)$ if and only if the improper integral $(I) \int_a^\infty f(x)dx$ exists. Moreover, in this case, $\int_{[a,\infty)} f(x)d\lambda = (I)$ $\int_a^\infty f(x)dx$.

PROOF. Let b_n, $n = 1, 2, 3, \ldots$ be any sequence with $b_n \to \infty$, $b_n > a$. We define $f_n = f\chi_{[a,b_n]}$. Then the monotone convergence theorem yields

$$\int_{[a,\infty)} f d\lambda = \lim_{n\to\infty} \int_{[a,\infty)} f_n d\lambda.$$

Thus

$$\int_{[a,\infty)} f d\lambda = \lim_{n\to\infty} \int_{[a,b_n]} f d\lambda.$$

Since f is Riemann integrable on each interval $[a, b_n]$, Theorem 6.18 yields $\int_{[a,b_n]} f d\lambda = (R) \int_a^{b_n} f dx$, and we conclude that

$$\int_{[a,\infty)} f d\lambda = \lim_{n\to\infty} (R) \int_a^{b_n} f dx. \tag{6.6}$$

The second part follows by noticing that the terms in (6.6) are both finite or ∞ at the same time. □

If $f : [a, \infty) \to \mathbb{R}$ takes also negative values, then we have the following result:

6.21. THEOREM. *Let* $f : [a, \infty) \to \mathbb{R}$ *be Riemann integrable on every subinterval of* $[a, \infty)$. *Then* f *is Lebesgue integrable if and only if the improper integral* $(I) \int_a^\infty |f(x)| dx$ *exists. Moreover, in this case,*

$$\int_{[a,\infty)} f d\lambda = (I) \int_a^\infty f(x) dx. \tag{6.7}$$

PROOF. Let $f = f^+ - f^-$. Assume that f is Lebesgue integrable on $[a, \infty)$. Thus f^+, f^- are both Lebesgue integrable. From Theorem 6.20 it follows that the improper integrals $(I) \int_a^\infty f^+(x) dx$ and $(I) \int_a^\infty f^-(x) dx$ exist. Moreover, for $b_n \to \infty$, $b_n > a$, we have

$$\int_{[a,\infty)} f^+ d\lambda = (I) \int_a^\infty f^+ dx = \lim_{n\to\infty} (R) \int_a^{b_n} f^+ dx \tag{6.8}$$

and

$$\int_{[a,\infty)} f^- d\lambda = (I) \int_a^\infty f^- dx = \lim_{n\to\infty} (R) \int_a^{b_n} f^- dx. \tag{6.9}$$

Note that

$$(R) \int_a^{b_n} f dx = (R) \int_a^{b_n} f^+ dx - (R) \int_a^{b_n} f^- dx.$$

Hence (6.8) and (6.9) imply that

$$\lim_{n\to\infty} (R) \int_a^{b_n} f dx = \lim_{n\to\infty} (R) \int_a^{b_n} f^+ dx - \lim_{n\to\infty} (R) \int_a^{b_n} f^- dx < \infty,$$

which means that the improper integral $(I) \int_a^\infty f dx$ exists. Moreover, (6.8) and (6.9) yield

$$(I) \int_a^\infty f dx = \int_{[a,\infty)} f^+ d\lambda - \int_{[a,\infty)} f^- d\lambda = \int_{[a,\infty)} f d\lambda,$$

which is (6.7). Analogously, using again (6.8) and (6.9), we have

$$\lim_{n\to\infty} (R) \int_a^{b_n} |f| dx = \lim_{n\to\infty} (R) \int_a^{b_n} f^+ dx + \lim_{n\to\infty} (R) \int_a^{b_n} f^- dx < \infty, \tag{6.10}$$

and hence the improper integral $(I) \int_a^\infty |f| dx$ exists with value

$$(I) \int_a^\infty |f| dx = \int_{[a,\infty)} f^+ d\lambda + \int_{[a,\infty)} f^- d\lambda = \int_{[a,\infty)} |f| d\lambda.$$

Conversely, if $(I) \int_a^\infty |f(x)| \, dx < \infty$, then (6.10) holds, and hence by (6.8) and (6.9) we have $\int_{[a,\infty)} f^+ d\lambda < \infty$ and $\int_{[a,\infty)} f^- d\lambda < \infty$, which yield $\int_{[a,\infty)} f d\lambda < \infty$, and hence f is Lebesgue integrable. □

Exercises for Section 6.4

1. Show that the improper integrals
$$(I)\int_0^\infty \cos(x^2)dx \quad \text{and} \quad (I)\int_0^\infty \sin(x^2)dx$$
both exist. Also, show that $\cos(x^2)$ and $\sin(x^2)$ are not Lebesgue integrable over $[0,\infty)$.
2. Let $f:[0,1]\to[0,\infty)$ be Riemann integrable on every closed subinterval of $(0,1]$. Show that f is Lebesgue integrable over $[0,1]$ if and only if $\lim_{\epsilon\to 0}(R)\int_\epsilon^1 f(x)dx$ exists in $\mathbb{R}$. Also, show that if this is the case, then we have
$$\int_{[0,1]} f d\lambda = \lim_{\epsilon\to 0}(R)\int_\epsilon^1 f(x)dx.$$
3. The gamma function for $t>0$ is defined by an integral as follows:
$$\Gamma(t) = (I)\int_0^\infty x^{t-1}e^{-x}dx.$$
Show that the integral $\int_0^\infty x^{t-1}e^{-x}dx = \lim_{\epsilon\to 0+, r\to\infty}(R)\int_\epsilon^r x^{t-1}e^{-x}dx$ exists as an improper integral (and hence as a Lebesgue integral).

6.5. L^p Spaces

The L^p spaces appear in many applications of analysis. They are also the prototypical examples of infinite-dimensional Banach spaces which will be studied in Chapter 8. It will be seen that there is a significant difference in these spaces when $p=1$ and $p>1$.

6.22. DEFINITION. For $1\le p\le\infty$ and $E\in\mathcal{M}$, let $L^p(E,\mathcal{M},\mu)$ denote the class of all measurable functions f on E such that $\|f\|_{p,E;\mu}<\infty$, where
$$\|f\|_{p,E;\mu} := \begin{cases} \left(\int_E |f|^p\, d\mu\right)^{1/p} & \text{if } 1\le p<\infty, \\ \inf\{M : |f|\le M\ \mu\text{-a.e. on } E\} & \text{if } p=\infty. \end{cases}$$

The quantity $\|f\|_{p,E;\mu}$ will be called the L^p **norm** of f on E and, for convenience, written $\|f\|_p$ when $E=X$ and the measure is clear from the context. The fact that it is a norm will be proved later in this section. We note immediately the following:

(i) $\|f\|_p\ge 0$ for every measurable f.
(ii) $\|f\|_p=0$ if and only if $f=0$ μ-a.e.
(iii) $\|cf\|_p=|c|\ \|f\|_p$ for all $c\in\mathbb{R}$.

For convenience, we will write $L^p(X)$ for the class $L^p(X,\mathcal{M},\mu)$. If X is a topological space, we let $L^p_{\text{loc}}(X)$ denote the class of functions f such that $f\in L^p(K)$ for each compact set $K\subset X$.

The next lemma shows that the classes $L^p(X)$ are **vector spaces** or, as is more commonly said in this context, **linear spaces**.

6.23. THEOREM. *Suppose* $1 \leq p \leq \infty$.
(i) *If* $f, g \in L^p(X)$, *then* $f + g \in L^p(X)$.
(ii) *If* $f \in L^p(X)$ *and* $c \in \mathbb{R}$, *then* $cf \in L^p(X)$.

PROOF. Assertion (ii) follows from property (iii) of the L^p norm noted above. If p is finite, assertion (i) follows from the inequality

$$|a + b|^p \leq 2^{p-1}(|a|^p + |b|^p), \tag{6.11}$$

which holds for all $a, b \in \mathbb{R}$, $1 \leq p < \infty$. For $p \geq 1$, inequality (6.11) follows from the fact that $t \mapsto t^p$ is a convex function on $(0, \infty)$ (note that $t \mapsto |t|$ is also convex according to the definition given in Exercise 6.8), and therefore

$$\left(\frac{a+b}{2}\right)^p \leq \frac{1}{2}(a^p + b^p).$$

If $p = \infty$, assertion (i) follows from the triangle inequality $|a + b| \leq |a| + |b|$, since if $|f(x)| \leq M$ μ-a.e. and $|g(x)| \leq N$ μ-a.e., then $|f(x) + g(x)| \leq M + N$ μ-a.e.. □

To deduce further properties of the L^p norms we will use the following arithmetic inequality.

6.24. LEMMA. *For* $a, b \geq 0$, $1 < p < \infty$, *and* p' *determined by the equation*

$$\frac{1}{p} + \frac{1}{p'} = 1,$$

we have

$$ab \leq \frac{a^p}{p} + \frac{b^{p'}}{p'}.$$

Equality holds if and only if $a^p = b^{p'}$.

PROOF. Recall that $\ln(x)$ is an increasing, strictly concave function on $(0, \infty)$, i.e.,

$$\ln(\lambda x + (1 - \lambda)y) > \lambda \ln(x) + (1 - \lambda)\ln(y)$$

for $x, y \in (0, \infty)$, $x \neq y$, and $\lambda \in [0, 1]$.

Set $x = a^p$, $y = b^{p'}$, and $\lambda = \frac{1}{p}$ (thus $(1 - \lambda) = \frac{1}{p'}$) to obtain

$$\ln(\frac{1}{p}a^p + \frac{1}{p'}b^{p'}) > \frac{1}{p}\ln(a^p) + \frac{1}{p'}\ln(b^{p'}) = \ln(ab).$$

Clearly equality holds in this inequality if and only if $a^p = b^{p'}$. □

For $p \in [1, \infty]$ the number p' defined by $\frac{1}{p} + \frac{1}{p'} = 1$ is called the **Lebesgue conjugate** of p. We adopt the convention that $p' = \infty$ when $p = 1$ and $p' = 1$ when $p = \infty$.

6.25. THEOREM (Hölder's inequality). *If $1 \le p \le \infty$ and f, g are measurable functions, then*

$$\int_X |fg| \, d\mu = \int_X |f| \, |g| \, d\mu \le \|f\|_p \|g\|_{p'}.$$

Equality holds, for $1 < p < \infty$, if and only if

$$\|f\|_p^p \, |g|^{p'} = \|g\|_{p'}^{p'} \, |f|^p \ \mu\text{-a.e.}$$

(Recall the convention that $0 \cdot \infty = \infty \cdot 0 = 0$.)

PROOF. If $p = 1$, then

$$\int_X |f| \, |g| \, d\mu \le \|g\|_\infty \int_X |f| \, d\mu = \|g\|_\infty \|f\|_1,$$

and an analogous inequality holds if $p = \infty$.

If $1 < p < \infty$, the assertion is clear unless $0 < \|f\|_p, \|g\|_{p'} < \infty$. In this case, set

$$\tilde{f} = \frac{f}{\|f\|_p} \quad \text{and} \quad \tilde{g} = \frac{g}{\|g\|_{p'}},$$

so that $\|\tilde{f}\|_p = 1$ and $\|\tilde{g}\|_{p'} = 1$, and apply Lemma 6.24 to obtain

$$\frac{1}{\|f\|_p \|g\|_{p'}} \int_X |f| \, |g| \, d\mu = \int_X \left|\tilde{f}\right| \, |\tilde{g}| \, d\mu \le \frac{1}{p} \left\|\tilde{f}\right\|_p^p + \frac{1}{p'} \|\tilde{g}\|_{p'}^{p'}. \qquad \square$$

The statement concerning equality follows immediately from the preceding lemma when $1 < p < \infty$.

6.26. THEOREM. *Suppose $(X, \mathcal{M}, \mu)$ is a σ-finite measure space. If f is measurable, $1 \le p \le \infty$, and $\frac{1}{p} + \frac{1}{p'} = 1$, then*

$$\|f\|_p = \sup \left\{ \int_X fg \, d\mu : \|g\|_{p'} \le 1 \right\}. \tag{6.12}$$

PROOF. Suppose f is measurable. If $g \in L^{p'}(X)$ with $\|g\|_{p'} \le 1$, then by Hölder's inequality,

$$\int_X fg \, d\mu \le \|f\|_p \|g\|_{p'} \le \|f\|_p.$$

Thus

$$\sup \left\{ \int_X fg \, d\mu : \|g\|_{p'} \le 1 \right\} \le \|f\|_p,$$

and it remains to prove the opposite inequality.

If $p = 1$, set $g = \operatorname{sign}(f)$; then $\|g\|_\infty \le 1$ and

$$\int_X fg \, d\mu = \int_X |f| \, d\mu = \|f\|_1.$$

Now consider the case $1 < p < \infty$. If $\|f\|_p = 0$, then $f = 0$ a.e., and the desired inequality is clear. If $0 < \|f\|_p < \infty$, set

$$g = \frac{|f|^{p/p'}\operatorname{sign}(f)}{\|f\|_p^{p/p'}}.$$

Then $\|g\|_{p'} = 1$ and

$$\int fg\,d\mu = \frac{1}{\|f\|_p^{p/p'}}\int |f|^{p/p'+1}\,d\mu = \frac{\|f\|_p^p}{\|f\|_p^{p/p'}} = \|f\|_p.$$

If $\|f\|_p = \infty$, let $\{X_k\}_{k=1}^\infty$ be an increasing sequence of measurable sets such that $\mu(X_k) < \infty$ for each k and $X = \cup_{k=1}^\infty X_k$. For each k set

$$h_k(x) = \chi_{X_k}\min(|f(x)|, k)$$

for $x \in X$. Then $h_k \in L^p(X)$, $h_k \leq h_{k+1}$, and $\lim_{k\to\infty} h_k = |f|$. By the monotone convergence theorem, $\lim_{k\to\infty}\|h_k\|_p = \infty$. Since we may assume without loss of generality that $\|h_k\|_p > 0$ for each k, there exist, by the result just proved, $g_k \in L^{p'}(X)$ such that $\|g_k\|_{p'} = 1$ and

$$\int h_k g_k\,d\mu = \|h_k\|_p.$$

Since $h_k \geq 0$, we have $g_k \geq 0$ and hence

$$\int f(\operatorname{sign}(f)g_k)\,d\mu = \int |f|\,g_k\,d\mu \geq \int h_k g_k d\mu = \|h_k\|_p \to \infty$$

as $k \to \infty$. Thus

$$\sup\{\int fg\,d\mu : \|g\|_{p'} \leq 1\} = \infty = \|f\|_p.$$

Finally, for the case $p = \infty$, suppose $M := \sup\{\int fg d\mu : \|g\|_1 \leq 1\} < \|f\|_\infty$. Thus there exists $\varepsilon > 0$ such that $0 < M + \varepsilon < \|f\|_\infty$. Then the set $E_\varepsilon := \{x : |f(x)| \geq M + \varepsilon\}$ has positive measure, since otherwise, we would have $\|f\|_\infty \leq M + \varepsilon$. Since μ is σ-finite, there is a measurable set E such that $0 < \mu(E_\varepsilon \cap E) < \infty$. Set

$$g_\varepsilon = \frac{1}{\mu(E_\varepsilon \cap E)}\chi_{E_\varepsilon\cap E}\operatorname{sign}(f).$$

Then $\|g_\varepsilon\|_1 = 1$ and

$$\int fg_\varepsilon\,d\mu = \frac{1}{\mu(E_\varepsilon \cap E)}\int_{E_\varepsilon\cap E}|f|\,d\mu \geq M + \varepsilon.$$

Thus, $M + \varepsilon \leq \sup\{fgd\mu : \|g\|_1 \leq 1\} \leq \|f\|_\infty$, which contradicts that $\sup\{fgd\mu : \|g\|_1 \leq 1\} = M$. We conclude that

$$\sup\{\int fg\,d\mu : \|g\|_1 \leq 1\} = \|f\|_\infty. \qquad \square$$

6.27. THEOREM (Minkowski's inequality). *Suppose* $1 \leq p \leq \infty$ *and* $f,g \in L^p(X)$. *Then*

$$\|f+g\|_p \leq \|f\|_p + \|g\|_p.$$

PROOF. The assertion is clear if $p = 1$ or $p = \infty$, so suppose $1 < p < \infty$. Then, applying first the triangle inequality and then Hölder's inequality, we obtain

$$\begin{aligned}
\|f+g\|_p^p &= \int |f+g|^p \, d\mu = \int |f+g|^{p-1} |f+g| \\
&\leq \int |f+g|^{p-1} |f| \, d\mu + \int |f+g|^{p-1} |g| \, d\mu \\
&\leq \left(\int (|f+g|^{p-1})^{p'}\right)^{1/p'} \left(\int |f|^p \, d\mu\right)^{1/p} \\
&\quad + \left(\int (|f+g|^{p-1})^{p'}\right)^{1/p'} \left(\int |g|^p \, d\mu\right)^{1/p} \\
&\leq \left(\int |f+g|^p\right)^{(p-1)/p} \left(\int |f|^p \, d\mu\right)^{1/p} \\
&\quad + \left(\int |f+g|^p\right)^{(p-1)/p} \left(\int |g|^p \, d\mu\right)^{1/p} \\
&= \|f+g\|_p^{p-1} \|f\|_p + \|f+g\|_p^{p-1} \|g\|_p \\
&= \|f+g\|_p^{p-1} (\|f\|_p + \|g\|_p).
\end{aligned}$$

The assertion is clear if $\|f+g\|_p = 0$. Otherwise, we divide by $\|f+g\|_p^{p-1}$ to obtain

$$\|f+g\|_p \leq \|f\|_p + \|g\|_p. \qquad \square$$

As a consequence of Theorem 6.27 and the remarks following Definition 6.22 we can say that for $1 \leq p \leq \infty$ the spaces $L^p(X)$ are, in the terminology of Chapter 8, **normed linear spaces**, provided we agree to identify functions that are equal μ-a.e. The norm $\|\cdot\|_p$ induces a metric ρ on $L^p(X)$ if we define

$$\rho(f,g) := \|f-g\|_p$$

for $f, g \in L^p(X)$ and agree to interpret the statement "$f = g$" as $f = g$ μ-a.e.

6.28. DEFINITIONS. A sequence $\{f_k\}_{k=1}^{\infty}$ is a **Cauchy sequence** in $L^p(X)$ if given $\varepsilon > 0$, there is a positive integer N such that

$$\|f_k - f_m\|_p < \varepsilon$$

whenever $k, m > N$. The sequence $\{f_k\}_{k=1}^{\infty}$ **converges in** $L^p(X)$ to $f \in L^p(X)$ if

$$\lim_{k\to\infty} \|f_k - f\|_p = 0.$$

6.29. THEOREM. *If $1 \le p \le \infty$, then $L^p(X)$ is a complete metric space under the metric ρ, i.e., if $\{f_k\}_{k=1}^{\infty}$ is a Cauchy sequence in $L^p(X)$, then there is an $f \in L^p(X)$ such that*

$$\lim_{k\to\infty} \|f_k - f\|_p = 0.$$

PROOF. Suppose $\{f_k\}_{k=1}^{\infty}$ is a Cauchy sequence in $L^p(X)$. There is an integer N such that $\|f_k - f_m\|_p < 1$ whenever $k, m \ge N$. By Minkowski's inequality,

$$\|f_k\|_p \le \|f_N\|_p + \|f_k - f_N\|_p \le \|f_N\|_p + 1$$

whenever $k \ge N$. Thus the sequence $\{\|f_k\|_p\}_{k=1}^{\infty}$ is bounded.

Consider the case $1 \le p < \infty$. For $\varepsilon > 0$, let $A_{k,m} := \{x : |f_k(x) - f_m(x)| \ge \varepsilon\}$. Then,

$$\int_{A_{k,m}} |f_k - f_m|^p \, d\mu \ge \varepsilon^p \mu(A_{k,m});$$

that is,

$$\varepsilon^p \mu(\{x : |f_k(x) - f_m(x)| \ge \varepsilon\}) \le \|f_k - f_m\|_p^p.$$

Thus $\{f_k\}$ is fundamental in measure, and consequently by Exercise 6, Section 5.2, and Theorem 5.26, there exists a subsequence $\{f_{k_j}\}_{j=1}^{\infty}$ that converges μ-a.e. to a measurable function f. By Fatou's lemma,

$$\|f\|_p^p = \int |f|^p \, d\mu \le \liminf_{j\to\infty} \int |f_{k_j}|^p \, d\mu < \infty.$$

Thus $f \in L^p(X)$.

Let $\varepsilon > 0$ and let M be such that $\|f_k - f_m\|_p < \varepsilon$ whenever $k, m > M$. Using Fatou's lemma again we see that

$$\|f_k - f\|_p^p = \int |f_k - f|^p \, d\mu \le \liminf_{j\to\infty} \int |f_k - f_{k_j}|^p \, d\mu < \varepsilon^p$$

whenever $k > M$. Thus f_k converges to f in $L^p(X)$.

The case $p = \infty$ is left as Exercise 2, Section 6.5. □

As a consequence of Theorem 6.29 we see for $1 \le p \le \infty$ that $L^p(X)$ is a **Banach space**; i.e., a normed linear space that is complete with respect to the metric induced by the norm.

Here we include a useful result relating norm convergence in L^p and pointwise convergence.

6.30. THEOREM (Vitali's convergence theorem). *Suppose $\{f_k\}, f \in L^p(X)$, $1 \le p < \infty$. Then $\|f_k - f\|_p \to 0$ if the following three conditions hold:*

(i) *$f_k \to f$ μ-a.e.*

(ii) *For each $\varepsilon > 0$, there exists a measurable set E such that $\mu(E) < \infty$ and*

$$\int_{\widetilde{E}} |f_k|^p \, d\mu < \varepsilon, \quad \text{for all} \quad k \in \mathbb{N}.$$

(iii) *For each $\varepsilon > 0$, there exists $\delta > 0$ such that $\mu(E) < \delta$ implies*

$$\int_E |f_k|^p \, d\mu < \varepsilon \quad \textit{for all} \quad k \in \mathbb{N}.$$

Conversely, if $\|f_k - f\|_p \to 0$, then (ii) and (iii) hold. Furthermore, (i) holds for a subsequence.

PROOF. Assume that the three conditions hold. Choose $\varepsilon > 0$ and let $\delta > 0$ be the corresponding number given by (iii). Condition (ii) provides a measurable set E with $\mu(E) < \infty$ such that

$$\int_{\tilde{E}} |f_k|^p \, d\mu < \varepsilon$$

for all positive integers k. Since $\mu(E) < \infty$, we can apply Egorov's theorem to obtain a measurable set $B \subset E$ with $\mu(E - B) < \delta$ such that f_k converges uniformly to f on B. Now write

$$\int_X |f_k - f|^p \, d\mu = \int_B |f_k - f|^p \, d\mu + \int_{E-B} |f_k - f|^p \, d\mu + \int_{\tilde{E}} |f_k - f|^p \, d\mu.$$

The first integral on the right can be made arbitrarily small for large k, because of the uniform convergence of f_k to f on B. The second and third integrals will be estimated with the help of the inequality

$$|f_k - f|^p \leq 2^{p-1}(|f_k|^p + |f|^p);$$

see (6.11). From (iii) we have $\int_{E-B} |f_k|^p < \varepsilon$ for all $k \in \mathbb{N}$, and then Fatou's lemma shows that $\int_{E-B} |f|^p < \varepsilon$ as well. The third integral can be handled in a similar way using (ii). Thus, it follows that $\|f_k - f\|_p \to 0$.

Now suppose $\|f_k - f\|_p \to 0$. Then for each $\varepsilon > 0$ there exists a positive integer k_0 such that $\|f_k - f\|_p < \varepsilon/2$ for $k > k_0$. With the help of Exercise 3, Section 6.5, there exist measurable sets A and B of finite measure such that

$$\int_{\tilde{A}} |f|^p \, d\mu < (\varepsilon/2)^p \quad \text{and} \quad \int_{\tilde{B}} |f_k|^p \, d\mu < (\varepsilon)^p \quad \text{for} \quad k = 1, 2, \ldots, k_0.$$

Minkowski's inequality implies that

$$\|f_k\|_{p,\tilde{A}} \leq \|f_k - f\|_{p,\tilde{A}} + \|f\|_{p,\tilde{A}} < \varepsilon \quad \text{for} \quad k > k_0.$$

Then set $E = A \cup B$ to obtain the necessity of (ii).

Similar reasoning establishes the necessity of (iii).

According to Exercise 4, Section 6.5, convergence in L^p implies convergence in measure. Hence, (i) holds for a subsequence. □

Finally, we conclude this section by considering how $L^p(X)$ compares with $L^q(X)$ for $1 \leq p < q \leq \infty$. For example, let $X = [0, 1]$ and let $\mu := \lambda$.

In this case it is easy to see that $L^q \subset L^p$, for if $f \in L^q$, then $|f(x)|^q \geq |f(x)|^p$ if $x \in A := \{x : |f(x)| \geq 1\}$. Therefore,

$$\int_A |f|^p \, d\lambda \leq \int_A |f|^q \, d\lambda < \infty,$$

while

$$\int_{[0,1]\setminus A} |f|^p \, d\lambda \leq 1 \cdot \lambda([0,1]) < \infty.$$

This observation extends to a more general situation via Hölder's inequality.

6.31. THEOREM. *If $\mu(X) < \infty$ and $1 \leq p \leq q \leq \infty$, then $L^q(X) \subset L^p(X)$ and*

$$\|f\|_{q;\mu} \leq \mu(X)^{\frac{1}{p}-\frac{1}{q}} \|f\|_{p;\mu} \, .$$

PROOF. If $q = \infty$, then the result is immediate:

$$\|f\|_p^p = \int_X |f|^p \, d\mu \leq \|f\|_\infty^p \int_X 1 \, d\mu = \|f\|_\infty^p \, \mu(X) < \infty.$$

If $q < \infty$, then Hölder's inequality with conjugate exponents q/p and $q/(q-p)$ implies that

$$\|f\|_p^p = \int_X |f|^p \cdot 1 \, d\mu \leq \||f|^p\|_{q/p} \|1\|_{q/(q-p)} = \|f\|_q^p \, \mu(X)^{(q-p)/q} < \infty. \quad \square$$

6.32. THEOREM. *If $0 < p < q < r \leq \infty$, then $L^p(X) \cap L^r(X) \subset L^q(X)$ and*

$$\|f\|_{q;\mu} \leq \|f\|_{p;\mu}^{\lambda} \|f\|_{r;\mu}^{1-\lambda},$$

where $0 < \lambda < 1$ is defined by the equation

$$\frac{1}{q} = \frac{\lambda}{p} + \frac{1-\lambda}{r}.$$

PROOF. If $r < \infty$, use Hölder's inequality with conjugate indices $p/\lambda q$ and $r/(1-\lambda)q$ to obtain

$$\begin{aligned} \int_X |f|^q = \int_X |f|^{\lambda q} |f|^{(1-\lambda)q} &\leq \left\| |f|^{\lambda q} \right\|_{p/\lambda q} \left\| |f|^{(1-\lambda)q} \right\|_{r/(1-\lambda)q} \\ &= \left(\int_X |f|^p \right)^{\lambda q/p} \left(\int_X |f|^r \right)^{(1-\lambda)q/r} \\ &= \|f\|_p^{\lambda q} \|f\|_r^{(1-\lambda)q}. \end{aligned}$$

We obtain the desired result by taking qth roots of both sides.

When $r = \infty$, we have

$$\int_X |f|^q \leq \|f\|_\infty^{q-p} \int_X |f|^p,$$

and so

$$\|f\|_q \leq \|f\|_p^{p/q} \|f\|_\infty^{1-(p/q)} = \|f\|_p^{\lambda} \|f\|_\infty^{1-\lambda}. \quad \square$$

Exercises for Section 6.5

1. Use Theorem 5.27 to show that if $f \in L^p(X)$ $(1 \le p < \infty)$, then there is a sequence $\{f_k\}$ of measurable simple functions such that $|f_k| \le |f|$ for each k and
$$\lim_{k\to\infty} \|f - f_k\|_{L^p(X)} = 0.$$
2. Prove Theorem 6.29 for the case $p = \infty$.
3. Suppose $(X, \mathcal{M}, \mu)$ is an arbitrary measure space, $\|f\|_p < \infty, 1 \le p < \infty$, and $\varepsilon > 0$. Prove that there is a measurable set E with $\mu(E) < \infty$ such that
$$\int_{\tilde{E}} |f|^p \, d\mu < \varepsilon.$$
4. Prove that convergence in L^p, $1 \le p < \infty$, implies convergence in measure.
5. Let $(X, \mathcal{M}, \mu)$ be a σ-finite measure space. Prove that there is a function $f \in L^1(\mu)$ such that $0 < f < 1$ everywhere on X.
6. Suppose μ and ν are measures on $(X, \mathcal{M})$ with the property that $\mu(E) \le \nu(E)$ for each $E \in \mathcal{M}$. For $p \ge 1$ and $f \in L^p(X, \nu)$, show that $f \in L^p(X, \mu)$ and that
$$\int_X |f|^p \, d\mu \le \int_X |f|^p \, d\nu.$$
7. Suppose $f \in L^p(X, \mathcal{M}, \mu)$, $1 \le p < \infty$. Then for all $t > 0$,
$$\mu(\{|f| > t\}) \le t^{-p} \|f\|_{p;\mu}^p .$$
This is known as **Chebyshev's inequality**.
8. Prove that a differentiable function f on (a, b) is convex if and only if f' is monotonically increasing.
9. Prove that a convex function is continuous.
10. (a) Prove **Jensen's inequality**: Let $f \in L^1(X, \mathcal{M}, \mu)$, where $\mu(X) < \infty$ and suppose $f(X) \subset [a, b]$. If φ is a convex function on $[a, b]$, then
$$\varphi\left(\frac{1}{\mu(X)} \int_X f \, d\mu\right) \le \frac{1}{\mu(X)} \int_X (\varphi \circ f) \, d\mu.$$
Thus, $\varphi(\text{average}(f)) \le \text{average}(\varphi \circ f)$. Hint: Let
$$t_0 = [\mu(X)^{-1}] \int_X f \, d\mu. \quad \text{Then} \quad t_0 \in (a, b).$$
Furthermore, with
$$\alpha := \sup_{t\in(a,t_0)} \frac{\varphi(t_0) - \varphi(t)}{t_0 - t},$$
we have $\varphi(t) - \varphi(t_0) \ge \alpha(t - t_0)$ for all $t \in (a, b)$. In particular, $\varphi(f(x)) - \varphi(t_0) \ge \alpha(f(x) - t_0)$ for all $x \in X$. Now integrate.

(b) Observe that if $\varphi(t) = t^p$, $1 \le p < \infty$, then Jensen's inequality follows from Hölder's inequality:

$$[\mu(X)]^{-1} \int_X f \cdot 1 \, d\mu \le \|f\|_p [\mu(X)]^{\frac{1}{p'}-1} = \|f\|_p [\mu(X)]^{-1/p}$$

$$\implies$$

$$\left(\frac{1}{\mu(X)} \int_X f \, d\mu \right)^p \le \frac{1}{\mu(X)} \int_X (|f|^p) \, d\mu.$$

(c) However, Jensen's inequality is stronger than Hölder's inequality in the following sense: If f is defined on $[0, 1]$, then

$$e^{\int_X f \, d\lambda} \le \int_X e^{f(x)} \, d\lambda.$$

(d) Suppose $\varphi \colon \mathbb{R} \to \mathbb{R}$ is such that

$$\varphi\left(\int_0^1 f \, d\lambda \right) \le \int_0^1 \varphi(f) \, d\lambda$$

for every real bounded measurable function f. Prove that φ is convex.

(e) Thus, we have

$$\varphi\left(\int_0^1 f \, d\lambda \right) \le \int_0^1 \varphi(f) \, d\lambda$$

for each bounded measurable f if and only if φ is convex.

11. In the context of a measure space $(X, \mathcal{M}, \mu)$, suppose f is a bounded measurable function with $a \le f(x) \le b$ for μ-a.e. $x \in X$. Prove that for each integrable function g, there exists a number $c \in [a, b]$ such that

$$\int_X f \, |g| \, d\mu = c \int_X |g| \, d\mu.$$

12. (a) Suppose f is a Lebesgue integrable function on $\mathbb{R}^n$. Prove that for each $\varepsilon > 0$ there is a continuous function g with compact support on $\mathbb{R}^n$ such that

$$\int_{\mathbb{R}^n} |f(y) - g(y)| \, d\lambda(y) < \varepsilon.$$

(b) Show that the above result is true for $f \in L^p(\mathbb{R}^n)$, $1 \le p < \infty$. That is, show that the continuous functions with compact support are dense in $L^p(\mathbb{R}^n)$. Hint: Use Corollary 5.28 to show that step functions are dense in $L^p(\mathbb{R}^n)$.

13. If $f \in L^p(\mathbb{R}^n)$, $1 \le p < \infty$, then prove

$$\lim_{|h| \to 0} \|f(x+h) - f(x)\|_p = 0.$$

Also, show that this result fails when $p = \infty$.

14. Let $p_1, p_2, \ldots, p_m$ be positive real numbers such that

$$\sum_{i=1}^{m} p_i = 1.$$

For $f_1, f_2, \ldots, f_m \in L^1(X, \mu)$, prove that

$$f_1^{p_1} f_2^{p_2} \cdots f_m^{p_m} \in L^1(X, \mu)$$

and

$$\int_X (f_1^{p_1} f_2^{p_2} \cdots f_m^{p_m})\, d\mu \leq \|f_1\|_1^{p_1} \|f_2\|_1^{p_2} \cdots \|f_m\|_1^{p_m}.$$

6.6. Signed Measures

We develop the basic properties of countably additive set functions of arbitrary sign, or signed measures. In particular, we establish the decomposition theorems of Hahn and Jordan, which show that signed measures and (positive) measures are closely related.

Let $(X, \mathcal{M}, \mu)$ be a measure space. Suppose f is measurable, at least one of f^+, f^- is integrable, and set

$$\nu(E) = \int_E f \, d\mu \tag{6.13}$$

for $E \in \mathcal{M}$. (Recall from Corollary 6.10, that the integral in (6.13) exists.) Then ν is an extended real-valued function on $\mathcal{M}$ with the following properties:

(i) ν assumes at most one of the values $+\infty$, $-\infty$.
(ii) $\nu(\emptyset) = 0$.
(iii) If $\{E_k\}_{k=1}^{\infty}$ is a disjoint sequence of measurable sets, then

$$\nu\Big(\bigcup_{k=1}^{\infty} E_k\Big) = \sum_{k=1}^{\infty} \nu(E_k),$$

where the series on the right either converges absolutely or diverges to $\pm\infty$ (see Exercise 1, Section 6.6).
(iv) $\nu(E) = 0$ whenever $\mu(E) = 0$.

In Section 6.6 we will show that the properties (i)–(iv) characterize set functions of the type (6.13).

6.33. Remark. An extended real-valued function ν defined on $\mathcal{M}$ is a **signed measure** if it satisfies properties (i)–(iii) above. If in addition it satisfies (iv), the signed measure ν is said to be **absolutely continuous** with respect to μ, written $\nu \ll \mu$. In some contexts, we will underscore that a measure μ is not a signed measure by saying that it is a **positive measure**. In other words, a positive measure is merely a measure in the sense defined in Definition 4.47.

6.34. DEFINITION. Let ν be a signed measure on $\mathcal{M}$. A set $A \in \mathcal{M}$ is a **positive set** for ν if $\nu(E) \geq 0$ for each measurable subset E of A. A set $B \in \mathcal{M}$ is a **negative set** for ν if $\nu(E) \leq 0$ for each measurable subset E of B. A set $C \in \mathcal{M}$ is a **null set** for ν if $\nu(E) = 0$ for each measurable subset E of C.

Note that every measurable subset of a positive set for ν is also a positive set and that analogous statements hold for negative sets and null sets. It follows that every countable union of positive sets is a positive set. To see this, suppose $\{P_k\}_{k=1}^{\infty}$ is a sequence of positive sets. Then there exist disjoint measurable sets $P_k^* \subset P_k$ such that $P := \bigcup_{k=1}^{\infty} P_k = \bigcup_{k=1}^{\infty} P_k^*$ (Lemma 4.7). If E is a measurable subset of P, then

$$\nu(E) = \sum_{k=1}^{\infty} \nu(E \cap P_k^*) \geq 0,$$

since each P_k^* is positive for ν.

It is important to observe the distinction between measurable sets E such that $\nu(E) = 0$ and null sets for ν. If E is a null set for ν, then $\nu(E) = 0$, but the converse is not generally true.

6.35. THEOREM. *If ν is a signed measure on $\mathcal{M}$, $E \in \mathcal{M}$, and $0 < \nu(E) < \infty$, then E contains a positive set A with $\nu(A) > 0$.*

PROOF. If E is positive, then the conclusion holds for $A = E$. Assume that E is not positive, and inductively construct a sequence of sets E_k as follows. Set

$$c_1 := \inf\{\nu(B) : B \in \mathcal{M}, B \subset E\} < 0.$$

There exists a measurable set $E_1 \subset E$ such that

$$\nu(E_1) < \frac{1}{2}\max(c_1, -1) < 0.$$

For $k \geq 1$, if $E \setminus \bigcup_{j=1}^{k} E_j$ is not positive, then

$$c_{k+1} := \inf\{\nu(B) : B \in \mathcal{M}, B \subset E \setminus \bigcup_{j=1}^{k} E_j\} < 0,$$

and there is a measurable set $E_{k+1} \subset E \setminus \bigcup_{j=1}^{k} E_j$ such that

$$\nu(E_{k+1}) < \frac{1}{2}\max(c_{k+1}, -1) < 0.$$

Note that if $c_k = -\infty$, then $\nu(E_k) < -1/2$.

If at any stage $E \setminus \bigcup_{j=1}^{k} E_j$ is a positive set, let $A = E \setminus \bigcup_{j=1}^{k} E_j$ and observe that

$$\nu(A) = \nu(E) - \sum_{j=1}^{k} \nu(E_j) > \nu(E) > 0.$$

Otherwise, set $A = E \setminus \bigcup_{k=1}^{\infty} E_k$ and observe that

$$\nu(E) = \nu(A) + \sum_{k=1}^{\infty} \nu(E_k).$$

Since $\nu(E) > 0$, we have $\nu(A) > 0$. Since $\nu(E)$ is finite, the series converges absolutely, $\nu(E_k) \to 0$, and therefore $c_k \to 0$ as $k \to \infty$. If B is a measurable subset of A, then $B \bigcap E_k = \emptyset$ for $k = 1, 2, \ldots$, and hence

$$\nu(B) \geq c_k$$

for $k = 1, 2, \ldots$. Thus $\nu(B) \geq 0$. This shows that A is a positive set and the lemma is proved. □

6.36. THEOREM (Hahn decomposition). *If ν is a signed measure on $\mathcal{M}$, then there exist disjoint sets P and N such that P is a positive set, N is a negative set, and $X = P \cup N$.*

PROOF. By considering $-\nu$ in place of ν if necessary, we may assume $\nu(E) < \infty$ for each $E \in \mathcal{M}$. Set

$$\lambda := \sup\{\nu(A) : A \text{ is a positive set for } \nu\}.$$

Since $\emptyset$ is a positive set, $\lambda \geq 0$. Let $\{A_k\}_{k=1}^{\infty}$ be a sequence of positive sets for which

$$\lim_{k\to\infty} \nu(A_k) = \lambda.$$

Set $P = \bigcup_{k=1}^{\infty} A_k$. Then P is positive and hence $\nu(P) \leq \lambda$. On the other hand, each $P \setminus A_k$ is positive, and hence

$$\nu(P) = \nu(A_k) + \nu(P \setminus A_k) \geq \nu(A_k).$$

Thus $\nu(P) = \lambda < \infty$.

Set $N = X \setminus P$. We have only to show that N is negative. Suppose B is a measurable subset of N. If $\nu(B) > 0$, then by Theorem 6.35, B must contain a positive set B^* such that $\nu(B^*) > 0$. But then $B^* \bigcup P$ is positive and

$$\nu(B^* \cup P) = \nu(B^*) + \nu(P) > \lambda,$$

contradicting the choice of λ. □

Note that the Hahn decomposition above is not unique if ν has a nonempty null set.

The following definition describes a relation between measures that is the antithesis of absolute continuity.

6.37. DEFINITION. Two measures μ_1 and μ_2 defined on a measure space $(X, \mathcal{M})$ are said to be **mutually singular** (written $\mu_1 \perp \mu_2$) if there exists a measurable set E such that

$$\mu_1(E) = 0 = \mu_2(X - E).$$

6.38. THEOREM (Jordan decomposition). *If ν is a signed measure on $\mathcal{M}$, then there exists a unique pair of mutually singular measures ν^+ and ν^-, at least one of which is finite, such that*

$$\nu(E) = \nu^+(E) - \nu^-(E)$$

for each $E \in \mathcal{M}$.

PROOF. Let $P \cup N$ be a Hahn decomposition of X with $P \cap N = \emptyset$, P positive, and N negative for ν. Set

$$\nu^+(E) = \nu(E \cap P),$$
$$\nu^-(E) = -\nu(E \cap N),$$

for $E \in \mathcal{M}$. Clearly ν^+ and ν^- are measures on $\mathcal{M}$ and $\nu = \nu^+ - \nu^-$. The measures ν^+ and ν^- are mutually singular, since $\nu^+(N) = 0 = \nu^-(X - N)$. That at least one of the measures ν^+, ν^- is finite follows immediately from the fact that $\nu^+(X) = \nu(P)$ and $\nu^-(X) = -\nu(N)$, at least one of which is finite.

If ν_1 and ν_2 are positive measures such that $\nu = \nu_1 - \nu_2$, and $A \in \mathcal{M}$ is such that $\nu_1(X - A) = 0 = \nu_2(A)$, then

$$\begin{aligned}\nu_1(X - P) &= \nu_1((X \cap A) \setminus P) \\ &= \nu((X \cap A) \setminus P) + \nu_2((X \cap A) \setminus P) \\ &= -\nu^-(X \cap A) \leq 0.\end{aligned}$$

Thus $\nu_1(X \setminus P) = 0$. Similarly $\nu_2(P) = 0$. For every $E \in \mathcal{M}$ we have

$$\nu^+(E) = \nu(E \cap P) = \nu_1(E \cap P) - \nu_2(E \cap P) = \nu_1(E).$$

Analogously $\nu^- = \nu_2$. □

Note that if ν is the signed measure defined by (6.13), then the sets $P = \{x : f(x) > 0\}$ and $N = \{x : f(x) \leq 0\}$ form a Hahn decomposition of X for ν, and

$$\nu^+(E) = \int_{E \cap P} f \, d\mu = \int_E f^+ \, d\mu$$

and

$$\nu^-(E) = -\int_{E \cap N} f \, d\mu = \int_E f^- \, d\mu$$

for each $E \in \mathcal{M}$.

6.39. DEFINITION. The **total variation** of a signed measure ν is denoted by $\|\nu\|$ and is defined as

$$\|\nu\| = \nu^+ + \nu^-.$$

We conclude this section by examining alternative characterizations of absolutely continuous measures. We leave it as an exercise to prove that the following three conditions are equivalent:

$$\begin{aligned} &\text{(i) } \nu \ll \mu, \\ (6.14) \qquad &\text{(ii) } \|\nu\| \ll \mu, \\ &\text{(iii) } \nu^+ \ll \mu \quad \text{and} \quad \nu^- \ll \mu. \end{aligned}$$

6.40. THEOREM. *Let ν be a finite signed measure and μ a positive measure on $(X, \mathcal{M})$. Then $\nu \ll \mu$ if and only if for every $\varepsilon > 0$ there exists $\delta > 0$ such that $|\nu(E)| < \varepsilon$ whenever $\mu(E) < \delta$.*

PROOF. Because of condition (ii) in (6.14) and the fact that $|\nu(E)| \leq \|\nu\|(E)$, we may assume that ν is a finite positive measure. Since the ε, δ condition is easily seen to imply that $\nu \ll \mu$, we will prove only the converse. Proceeding by contradiction, suppose then there exist $\varepsilon > 0$ and a sequence of measurable sets $\{E_k\}$ such that $\mu(E_k) < 2^{-k}$ and $\nu(E_k) > \varepsilon$ for all k. Set

$$F_m = \bigcup_{k=m}^{\infty} E_k$$

and

$$F = \bigcap_{m=1}^{\infty} F_m.$$

Then $\mu(F_m) < 2^{1-m}$, so $\mu(F) = 0$. But $\nu(F_m) \geq \varepsilon$ for each m, and since ν is finite, we have

$$\nu(F) = \lim_{m\to\infty} \nu(F_m) \geq \varepsilon,$$

thus reaching a contradiction. □

Exercises for Section 6.6

1. Prove property (iii) that follows (6.13).
2. Prove that the three conditions in (6.14) are equivalent.
3. Let $(X, \mathcal{M}, \mu)$ be a finite measure space, and let $f \in L^1(X, \mu)$. In particular, f is $\mathcal{M}$-measurable. Suppose $\mathcal{M}_0 \subset \mathcal{M}$ us a σ-algebra. Of course, f may not be $\mathcal{M}_0$-measurable. However, prove that there is a unique $\mathcal{M}_0$-measurable function f_0 such that
$$\int_X fg\, d\mu = \int_X f_0 g\, d\tilde{\mu}$$
for each $\mathcal{M}_0$-measurable g for which the integrals are finite and $\tilde{\mu} = \mu \llcorner \mathcal{M}_0$. Hint: Use the Radon–Nikodym theorem.

4. Show that the total variation of the measure ν satisfies
$$\|\nu\|(A) = \sup \left| \int_A f\, d\nu : f \in C_c(A), |f| \leq 1 \right|$$
for each open set A.
5. Suppose that μ and ν are σ-finite measures on $(X, \mathcal{M})$ such that $\mu \ll \nu$ and $\nu \ll \mu$. Prove that
$$\frac{d\nu}{d\mu} \neq 0$$
almost everywhere and
$$\frac{d\mu}{d\nu} = 1/\frac{d\nu}{d\mu}.$$
6. Let $f: \mathbb{R} \to \mathbb{R}$ be a nondecreasing continuously differentiable function and let λ_f be the corresponding Lebesgue–Stieltjes measure; see Definition 4.29. Prove:
   (a) $\lambda_f << \lambda$.
   (b) $\frac{d\lambda_f}{d\lambda} = f'$.
7. Let $(X, \mathcal{M}, \mu)$ be a finite measure space with $\mu(X) < \infty$. Let ν_k be a sequence of finite measures on $\mathcal{M}$ (that is, $\nu_k(X) < \infty$ for all k) with the property that they are **uniformly absolutely continuous** with respect to μ; that is, for each $\varepsilon > 0$, there exist $\delta > 0$ and a positive integer K such that $\nu_k(E) < \varepsilon$ for all $k \geq K$ and all $E \in \mathcal{M}$ for which $\mu(E) < \delta$. Assume that the limit
$$\nu(E) := \lim_{k\to\infty} \nu_k(E), \; E \in \mathcal{M}$$
exists. Prove that ν is a σ-finite measure on $\mathcal{M}$.
8. Let $(X, \mathcal{M}, \mu)$ be a finite measure space and define a metric space $\widetilde{\mathcal{M}}$ as follows: for $A, B \in \mathcal{M}$, define
$$\mathrm{d}(A, B) := \mu(A \Delta B), \quad \text{where } A\Delta B \text{ denotes symmetric difference.}$$
The space $\widetilde{\mathcal{M}}$ is defined to comprise all sets in $\mathcal{M}$ with sets A and B are identified if $\mu(A \Delta B) = 0$.
   (a) Prove that $(\widetilde{\mathcal{M}}, \mathrm{d})$ is a complete metric space.
   (b) Prove that $(\widetilde{\mathcal{M}}, \mathrm{d})$ is separable if and only if $L^p(X, \widetilde{\mathcal{M}}, \mu)$ is, $1 \leq p < \infty$.
9. Show that the space above is not compact when $X = [0, 1]$, $\mathcal{M}$ is the family of Borel sets on $[0, 1]$, and μ is Lebesgue measure.
10. Let $\{\nu_k\}$ be a sequence of measures on the finite measure space $(X, \mathcal{M}, \mu)$ such that
   - $\nu_k(X) < \infty$ for each k,
   - the limit exists and is finite for each $E \in \mathcal{M}$,
$$\nu(E) := \lim_{k\to\infty} \nu_k(E),$$
   - $\nu_k << \mu$ for each k.

(a) Prove that each ν_k is well defined and continuous on the space $(\widetilde{\mathcal{M}}, \mathrm{d})$.

(b) For $\varepsilon > 0$, let

$$\mathcal{M}_{i,j} := \{E \in \mathcal{M} : |\nu_i(E) - \nu_j(E)| \le \frac{\varepsilon}{3}\},\ i, j = 1, 2, \ldots,$$

and

$$\mathcal{M}_p := \bigcap_{i,j \ge p} \mathcal{M}_{i,j},\ p = 1, 2, \ldots.$$

Prove that $\mathcal{M}_p$ is a closed set in $(\widetilde{\mathcal{M}}, \mathrm{d})$.

(c) Prove that there is some q such that $\mathcal{M}_q$ contains an open set; call it U.

(d) Prove that the $\{\nu_k\}$ are uniformly absolutely continuous with respect to μ, as in the previous Problem 7.
Hint: Let A be an interior point in U and for $B \in \mathcal{M}$, write

$$\nu_k(B) = \nu_q(B) + [\nu_k(B) - \nu_q(B)]$$

and use the identity $\nu_i(B) = \nu_i(B \cap A) + \nu_i(B \setminus A)$, $i = 1, 2 \ldots$, to estimate $\nu_i(B)$.

(e) Prove that ν is a finite measure.

6.7. The Radon–Nikodym Theorem

If f is an integrable function on the measure space $(X, \mathcal{M}, \mu)$, then the signed measure

$$\nu(E) = \int_E f\, d\mu$$

defined for all $E \in \mathcal{M}$ is absolutely continuous with respect to μ. The Radon–Nikodym theorem states that essentially every signed measure ν, absolutely continuous with respect to μ, is of this form. The proof of Theorem 6.41 below is due to A. Schep [46].

6.41. THEOREM. *Suppose $(X, \mathcal{M}, \mu)$ is a finite measure space and ν is a measure on $(X, \mathcal{M})$ with the property*

$$\nu(E) \le \mu(E)$$

for each $E \in \mathcal{M}$. Then there is a measurable function $f\colon X \to [0, 1]$ such that

$$\nu(E) = \int_E f\; d\mu, \quad \textit{for each} \quad E \in \mathcal{M}. \tag{6.15}$$

More generally, if g is a nonnegative measurable function on X, then

$$\int_X g\; d\nu = \int_X gf\; d\mu.$$

PROOF. Let

$$H := \left\{ f : \; f \text{ measurable}, 0 \le f \le 1, \; \int_E f \, d\mu \le \nu(E) \text{ for all } E \in \mathcal{M} \right\},$$

and let

$$M := \sup \left\{ \int_X f \, d\mu : f \in H \right\}.$$

Then there exist functions $f_k \in H$ such that

$$\int_X f_k \, d\mu > M - k^{-1}.$$

Observe that we may assume $0 \le f_1 \le f_2 \le \ldots$, because if $f, g \in H$, then so is $\max\{f, g\}$ in view of the following:

$$\begin{aligned}\int_E \max\{f, g\} \, d\mu &= \int_{E \cap A} \max\{f, g\} \, d\mu + \int_{E \cap (X \setminus A)} \max\{f, g\} \, d\mu \\ &= \int_{E \cap A} f \, d\mu + \int_{X \setminus A} g \, d\mu \\ &\le \nu(E \cap A) + \nu(E \cap (X \setminus A)) = \nu(E),\end{aligned}$$

where $A := \{x : f(x) \ge g(x)\}$. Therefore, since $\{f_k\}$ is an increasing sequence, the limit below exists:

$$f_\infty(x) := \lim_{k \to \infty} f_k(x).$$

Note that f_∞ is a measurable function. Clearly $0 \le f_\infty \le 1$, and the monotone convergence theorem implies

$$\int_X f_\infty \, d\mu = \lim_{k \to \infty} \int_X f_k = M$$

for each $E \in \mathcal{M}$ and

$$\int_E f_k \, d\mu \le \nu(E)$$

for each k and $E \in \mathcal{M}$. So $f_\infty \in H$.

The proof of the theorem will be concluded by showing that (6.15) is satisfied by taking f as f_∞. For this purpose, assume for the sake of obtaining a contradiction that

$$\int_E f_\infty \, d\mu < \nu(E) \tag{6.16}$$

for some $E \in \mathcal{M}$. Let

$$\begin{aligned}E_0 &= \{x \in E : f_\infty(x) < 1\}, \\ E_1 &= \{x \in E : f_\infty(x) = 1\}.\end{aligned}$$

Then

$$\begin{aligned}\nu(E) &= \nu(E_0) + \nu(E_1) \\ &> \int_E f_\infty \, d\mu \\ &= \int_{E_0} f_\infty \, d\mu + \mu(E_1) \\ &\geq \int_{E_0} f_\infty \, d\mu + \nu(E_1),\end{aligned}$$

which implies

$$\nu(E_0) > \int_{E_0} f_\infty \, d\mu.$$

Let $\varepsilon^* > 0$ be such that

$$\int_{E_0} f_\infty + \varepsilon^* \chi_{E_0} \, d\mu < \nu(E_0) \tag{6.17}$$

and let $F_k := \{x \in E_0 : f_\infty(x) < 1 - 1/k\}$. Observe that $F_1 \subset F_2 \subset \ldots$, and since $f_\infty < 1$ on E_0, we have $\cup_{k=1}^\infty F_k = E_0$ and therefore that $\nu(F_k) \uparrow \nu(E_0)$. Furthermore, it follows from (6.17) that

$$\int_{E_0} f_\infty + \varepsilon^* \chi_{E_0} \, d\mu < \nu(E_0) - \eta$$

for some $\eta > 0$. Therefore, since $[f_\infty + \varepsilon^* \chi_{F_k}]\chi_{F_k} \uparrow [f_\infty + \varepsilon^* \chi_{E_0}]\chi_{E_0}$, there exists k^* such that

$$\begin{aligned}&\int_{F_k} f_\infty + \varepsilon^* \chi_{F_k} \, d\mu \\ &= \int_X [f_\infty + \varepsilon^* \chi_{F_k}]\chi_{F_k} \, d\mu \\ &\to \int_{E_0} f_\infty + \varepsilon^* \chi_{E_0} \, d\mu \qquad \text{by the monotone convergence theorem} \\ &< \nu(E_0) - \eta \\ &< \nu(F_k) \qquad\qquad\qquad\qquad \text{for all } k \geq k^*.\end{aligned}$$

For all such $k \geq k^*$ and $\varepsilon := \min(\varepsilon^*, 1/k)$, we claim that

$$f_\infty + \varepsilon \chi_{F_k} \in H. \tag{6.18}$$

The validity of this claim would imply that $\int f_\infty + \varepsilon \chi_{F_k} \, d\mu = M + \varepsilon\mu(F_k) > M$, contradicting the definition of M, which would mean that our contradiction hypothesis, (6.16), is false, thus establishing our theorem.

So, to finish the proof, it suffices to prove (6.18) for some $k \geq k^*$, which will remain fixed throughout the remainder of the proof. For this, first note that $0 \leq f_\infty + \varepsilon \chi_{F_k} \leq 1$. To show that

$$\int_E f_\infty + \varepsilon \chi_{F_k} \, d\mu \leq \nu(E) \quad \text{for all } E \in \mathcal{M},$$

we proceed by contradiction; if not, there would exist a measurable set $G \subset X$ such that

$$\begin{aligned}\nu(G \setminus F_k) &+ \int_{G\cap F_k} f_\infty + \varepsilon\chi_{F_k}\, d\mu \\ &\geq \int_{G\setminus F_k} f_\infty\, d\mu + \int_{G\cap F_k} f_\infty + \varepsilon\chi_{F_k}\, d\mu \\ &= \int_{G\setminus F_k} f_\infty + \varepsilon\chi_{F_k}\, d\mu + \int_{G\cap F_k} f_\infty + \varepsilon\chi_{F_k}\, d\mu \\ &= \int_G f_\infty + \varepsilon\chi_{F_k}\, d\mu > \nu(G).\end{aligned}$$

This implies

$$\int_{G\cap F_k} (f_\infty + \varepsilon\chi_{F_k})\, d\mu > \nu(G) - \nu(G \setminus F_k) = \nu(G \cap F_k). \tag{6.19}$$

Hence, we may assume $G \subset F_k$.

Let $\mathcal{F}_1$ be the collection of all measurable sets $G \subset F_k$ such that (6.19) holds. Define $\alpha_1 := \sup\{\mu(G) : G \in \mathcal{F}_1\}$ and let $G_1 \in \mathcal{F}_1$ be such that $\mu(G_1) > \alpha_1 - 1$. Similarly, let $\mathcal{F}_2$ be the collection of all measurable sets $G \subset F_k \setminus G_1$ such that (6.19) holds for G. Define $\alpha_2 := \sup\{\mu(G) : G \in \mathcal{F}_2\}$ and let $G_2 \in \mathcal{F}_2$ be such that $\mu(G_2) > \alpha_2 - \frac{1}{2^2}$. Proceeding inductively, we obtain a decreasing sequence α_k and disjoint measurable sets G_j, where $\mu(G_j) > \alpha_j - \frac{1}{j^2}$. Observe that $\alpha_j \to 0$, for if $\alpha_j \to a > 0$, then $\mu(G_j) \downarrow a$. Since $\sum_{j=1}^\infty \frac{1}{j^2} < \infty$, this would imply

$$\mu(\bigcup G_j) = \sum_{j=1}^\infty \mu(G_j) > \sum_{j=1}^\infty \left(\alpha_j - \frac{1}{j^2}\right) = \infty,$$

contrary to the finiteness of μ. This implies

$$\mu(F_k \setminus \bigcup_{j=1}^\infty G_j) = 0,$$

for if not, there would be two possibilities:

(i) There would exist a set $T \subset F_k \setminus \bigcup_{j=1}^\infty G_j$ of positive μ measure for which (6.19) would hold with T replacing G. Since $\mu(T) > 0$, there would exist α_k such that

$$\alpha_k < \mu(T),$$

and since

$$T \subset F_k \setminus \bigcup_{j=1}^\infty G_j \subset F_k \setminus \bigcup_{i=1}^{j-1} G_j,$$

this would contradict the definition of α_j. Hence,

$$\begin{aligned}\nu(F_k) &> \int_{F_k} f_\infty + \varepsilon\chi_{F_k} d\,\mu \\ &= \sum_j \int_{G_j} f_\infty + \varepsilon\chi_{F_k} d\,\mu \\ &> \sum_j \nu(G_j) = \nu(F_k),\end{aligned}$$

which is impossible, and therefore (i) cannot occur.

(ii) If there were no set T as in (i), then $F_k \setminus \bigcup_j^\infty G_j$ could not satisfy (6.18) and thus

$$\int_{F_k \setminus \cup_j G_{j=1}^\infty} f_\infty + \varepsilon\chi_{F_k}\, d\mu \le \nu(F_k \setminus \cup_{j=1}^\infty G_j). \tag{6.20}$$

With $S := F_k \setminus \bigcup_{j=1}^\infty G_j$ we have

$$\int_S f_\infty + \varepsilon\chi_{F_k}\, d\mu \le \nu(S).$$

Since

$$\int_{G_j} f_\infty + \varepsilon\chi_{F_k}\, d\mu > \nu(G_j)$$

for each $j \in \mathbb{N}$, it follows that $F_k \setminus S$ must also satisfy (6.18), which contradicts (6.20). Hence, both (i) and (ii) do not occur, and thus we conclude that

$$\mu(F_k \setminus \bigcup_{j=1}^\infty G_j) = 0,$$

as desired. □

6.42. Notation. The function f in (6.15) (and also in (6.22) below) is called the **Radon–Nikodym derivative** of ν with respect to μ and is denoted by

$$f := \frac{d\nu}{d\mu}.$$

The previous theorem yields the notationally convenient result

$$\int_X g\, d\nu = \int_X g \frac{d\nu}{d\mu}\, d\mu. \tag{6.21}$$

6.43. Theorem (Radon–Nikodym). *If $(X, \mathcal{M}, \mu)$ is a σ-finite measure space and ν is a σ-finite signed measure on $\mathcal{M}$ that is absolutely continuous with respect to μ, then there exists a measurable function f such that either f^+ or f^- is integrable and*

$$\nu(E) = \int_E f\, d\mu \tag{6.22}$$

for each $E \in \mathcal{M}$.

PROOF. We first assume, temporarily, that μ and ν are finite measures. Referring to Theorem 6.41, there exist Radon–Nikodym derivatives

$$f_\nu := \frac{d\nu}{d(\nu+\mu)} \quad \text{and} \quad f_\mu := \frac{d\mu}{d(\nu+\mu)}.$$

Define $A = X \cap \{f_\mu(x) > 0\}$ and $B = X \cap \{f_\mu(x) = 0\}$. Then

$$\mu(B) = \int_B f_\mu \, d(\nu+\mu) = 0,$$

and therefore $\nu(B) = 0$, since $\nu \ll \mu$. Now define

$$f(x) = \begin{cases} \frac{f_\nu(x)}{f_\mu(x)} & \text{if } x \in A, \\ 0 & \text{if } x \in B. \end{cases}$$

If E is a measurable subset of A, then

$$\nu(E) = \int_E f_\nu \, d(\nu+\mu) = \int_E f \cdot f_\mu \, d(\nu+\mu) = \int_E f \; d\mu$$

by (6.21). Since both ν and μ are 0 on B, we have

$$\nu(E) = \int_E f \; d\mu$$

for all measurable E.

Next, consider the case that μ and ν are σ-finite measures. There is a sequence of disjoint measurable sets $\{X_k\}_{k=1}^\infty$ such that $X = \cup_{k=1}^\infty X_k$ and both $\mu(X_k)$ and $\nu(X_k)$ are finite for each k. Set $\mu_k := \mu \llcorner X_k, \nu_k := \nu \llcorner X_k$ for $k = 1, 2, \ldots$. Clearly μ_k and ν_k are finite measures on $\mathcal{M}$ and $\nu_k \ll \mu_k$. Thus there exist nonnegative measurable functions f_k such that for all $E \in \mathcal{M}$,

$$\nu(E \cap X_k) = \nu_k(E) = \int_E f_k \; d\mu_k = \int_{E \cap X_k} f_k \; d\mu.$$

It is clear that we may assume $f_k = 0$ on $X - X_k$. Set $f := \sum_{k=1}^\infty f_k$. Then for all $E \in \mathcal{M}$,

$$\nu(E) = \sum_{k=1}^\infty \nu(E \cap X_k) = \sum_{k=1}^\infty \int_{E \cap X_k} f \; d\mu = \int_E f \; d\mu.$$

Finally, suppose that ν is a signed measure and let $\nu = \nu^+ - \nu^-$ be the Jordan decomposition of ν. Since the measures are mutually singular, there is a measurable set P such that $\nu^+(X - P) = 0 = \nu^-(P)$. For all $E \in \mathcal{M}$ such that $\mu(E) = 0$,

$$\nu^+(E) = \nu(E \cap P) = 0,$$
$$\nu^-(E) = -\nu(E - P) = 0,$$

since $\mu(E\cap P)+\mu(E-P) = \mu(E) = 0$. Thus ν^+ and ν^- are absolutely continuous with respect to μ, and consequently, there exist nonnegative measurable functions f^+ and f^- such that

$$\nu^{\pm}(E) = \int_E f^{\pm}\, d\mu$$

for each $E \in \mathcal{M}$. Since at least one of the measures $\nu^{\pm}$ is finite, it follows that at least one of the functions $f^{\pm}$ is μ-integrable. Set $f = f^+ - f^-$. In view of Theorem 6.9, p. 145,

$$\nu(E) = \int_E f^+\, d\mu - \int_E f^-\, d\mu = \int_E f\, d\mu$$

for each $E \in \mathcal{M}$. □

An immediate consequence of this result is the following.

6.44. THEOREM (Lebesgue decomposition). *Let μ and ν be σ-finite measures defined on the measure space $(X, \mathcal{M})$. Then there is a decomposition of ν such that $\nu = \nu_0 + \nu_1$, where $\nu_0 \perp \mu$ and $\nu_1 \ll \mu$. The measures ν_0 and ν_1 are unique.*

PROOF. We employ the same device as in the proof of the preceding theorem by considering the Radon–Nikodym derivatives

$$f_\nu := \frac{d\nu}{d(\nu+\mu)} \quad \text{and} \quad f_\mu := \frac{d\mu}{d(\nu+\mu)}.$$

Define $A = X \cap \{f_\mu(x) > 0\}$ and $B = X \cap \{f_\mu(x) = 0\}$. Then X is the disjoint union of A and B. With $\gamma := \mu + \nu$ we will show that the measures

$$\nu_0(E) := \nu(E \cap B) \qquad \text{and} \qquad \nu_1(E) := \nu(E \cap A) = \int_{E\cap A} f_\nu\, d\gamma$$

provide our desired decomposition. First, note that $\nu = \nu_0 + \nu_1$. Next, we have $\nu_0(A) = 0$, and so $\nu_0 \perp \mu$. Finally, to show that $\nu_1 \ll \mu$, consider E with $\mu(E) = 0$. Then

$$0 = \mu(E) = \int_E f_\mu\, d\gamma = \int_{A\cap E} f_\mu\, d\lambda.$$

Thus, $f_\mu = 0$ γ-a.e. on E. Then, since $f_\mu > 0$ on A, we must have $\gamma(A\cap E) = 0$. This implies $\nu(A \cap E) = 0$ and therefore $\nu_1(E) = 0$, which establishes $\nu_1 \ll \mu$. The proof of uniqueness is left as an exercise. □

Exercises for Section 6.7

1. Prove the uniqueness assertion in Theorem 6.44.
2. Use the following example to show that the hypothesis in the Radon–Nikodym theorem that μ is σ-finite cannot be omitted. Let $X = [0, 1]$, let $\mathcal{M}$ denote the class of Lebesgue measurable subsets of $[0, 1]$, and take ν to be the Lebesgue measure and μ to be the counting measure on $\mathcal{M}$. Then ν is finite and absolutely continuous with respect to μ, but there is

no function f such that $\nu(E) = \int_E f d\mu$ for all $E \in \mathcal{M}$. At what point does the proof of Theorem 6.43 break down for this example?

6.8. The Dual of L^p

Using the Radon–Nikodym theorem, we completely characterize the continuous linear mappings of $L^p(X)$ into $\mathbb{R}$.

6.45. Definitions. Let $(X, \mathcal{M}, \mu)$ be a measure space. A **linear functional** on $L^p(X) = L^p(X, \mathcal{M}, \mu)$ is a real-valued linear function on $L^p(X)$, i.e., a function $F : L^p(X) \to \mathbb{R}$ such that

$$F(af + bg) = aF(f) + bF(g)$$

whenever $f, g \in L^p(X)$ and $a, b \in \mathbb{R}$. Set

$$\|F\| \equiv \sup\{|F(f)| : f \in L^p(X), \|f\|_p \leq 1\}.$$

A linear functional F on $L^p(X)$ is said to be **bounded** if $\|F\| < \infty$.

6.46. Theorem. *A linear functional on $L^p(X)$ is bounded if and only if it is continuous with respect to (the metric induced by) the norm $\|\cdot\|_p$.*

Proof. Let F be a linear functional on $L^p(X)$.

If F is bounded, then $\|F\| < \infty$, and if $0 \neq f \in L^p(X)$, then

$$\left| F\left(\frac{f}{\|f\|_p}\right) \right| \leq \|F\|,$$

i.e.,

$$|F(f)| \leq \|F\| \; \|f\|_p$$

whenever $f \in L^p(X)$. In particular, for all $f, g \in L^p(X)$, we have

$$|F(f - g)| \leq \|F\| \; \|f - g\|_p,$$

and hence F is uniformly continuous on $L^p(X)$.

On the other hand, if F is continuous at 0, then there exists a $\delta > 0$ such that

$$|F(f)| \leq 1$$

whenever $\|f\|_p \leq \delta$. Thus if $f \in L^p(X)$ with $\|f\|_p > 0$, then

$$|F(f)| = \left| \frac{\|f\|_p}{\delta} F\left(\frac{\delta}{\|f\|_p} f\right) \right| \leq \frac{1}{\delta} \|f\|_p,$$

whence $\|F\| \leq \frac{1}{\delta}$. □

6.47. Theorem. *If $1 \leq p \leq \infty$, $\frac{1}{p} + \frac{1}{p'} = 1$, and $g \in L^{p'}(X)$, then*

$$F(f) = \int fg \, d\mu$$

defines a bounded linear functional on $L^p(X)$ with

$$\|F\| = \|g\|_{p'}.$$

PROOF. That F is a bounded linear functional on $L^p(X)$ follows immediately from Hölder's inequality and the elementary properties of the integral. The rest of the assertion follows from Theorem 6.26, since

$$\|F\| = \sup\left\{\int fg\,d\mu : \|f\|_p \le 1\right\} = \|g\|_{p'}.$$

Note that while the hypotheses of Theorem 6.26 include a σ-finiteness condition, that assumption is not needed to establish (6.12) if the function is integrable. That is the situation we have here, since it is assumed that $g \in L^{p'}$. □

The next theorem shows that all bounded linear functionals on $L^p(X)$ $(1 \le p < \infty)$ are of this form.

6.48. THEOREM. *If* $1 < p < \infty$ *and* F *is a bounded linear functional on* $L^p(X)$, *then there is a* $g \in L^{p'}(X)$, $(\frac{1}{p} + \frac{1}{p'} = 1)$ *such that*

$$F(f) = \int fg\,d\mu \tag{6.23}$$

for all $f \in L^p(X)$. *Moreover,* $\|g\|_{p'} = \|F\|$, *and the function* g *is unique in the sense that if* (6.23) *holds with* $\tilde{g} \in L^{p'}(X)$, *then* $\tilde{g} = g$ μ-*a.e. If* $p = 1$, *the same conclusion holds under the additional assumption that* μ *is* σ-*finite.*

PROOF. Assume first $\mu(X) < \infty$. Note that our assumption implies that $\chi_E \in L^p(X)$ whenever $E \in \mathcal{M}$. Set

$$\nu(E) = F(\chi_E)$$

for $E \in \mathcal{M}$. Suppose $\{E_k\}_{k=1}^{\infty}$ is a sequence of disjoint measurable sets and let $E := \bigcup_{k=1}^{\infty} E_k$. Then for every positive integer N,

$$\begin{aligned}
\left|\nu(E) - \sum_{k=1}^{N} \nu(E_k)\right| &= \left|F(\chi_E - \sum_{k=1}^{N} \chi_{E_k})\right| \\
&= \left|F(\sum_{k=N+1}^{\infty} \chi_{E_k})\right| \\
&\le \|F\|\,(\mu(\bigcup_{k=N+1}^{\infty} E_k))^{\frac{1}{p}}
\end{aligned}$$

and $\mu(\bigcup_{k=N+1}^{\infty} E_k) = \sum_{k=N+1}^{\infty} \mu(E_k) \to 0$ as $N \to \infty$, since

$$\mu(E) = \sum_{k=1}^{\infty} \mu(E_k) < \infty.$$

Thus,

$$\nu(E) = \sum_{k=1}^{\infty} \nu(E_k),$$

and since the same result holds for every rearrangement of the sequence $\{E_k\}_{k=1}^{\infty}$, the series converges absolutely. It follows that ν is a signed measure, and since $|\nu(E)| \leq \|F\| \, (\mu(E))^{\frac{1}{p}}$, we see that $\nu \ll \mu$. By the Radon–Nikodym theorem there is a $g \in L^1(X)$ such that

$$F(\chi_E) = \nu(E) = \int \chi_E g \, d\mu$$

for each $E \in \mathcal{M}$. From the linearity of both F and the integral, it is clear that

$$F(f) = \int f g \, d\mu \tag{6.24}$$

whenever f is a simple function.

Step 1: Assume $\mu(X) < \infty$ and $p = 1$.
We proceed to show that $g \in L^\infty(X)$. Assume that $\|g^+\|_\infty > \|F\|$. Let $M > 0$ be such that $\|g^+\|_\infty > M > \|F\|$ and set $E_M := \{x : g(x) > M\}$. We have $\mu(E_M) > 0$, since otherwise, we would have $\|g^+\|_\infty \leq M$. Then

$$M\mu(E_M) \leq \int \chi_{E_M} g \, d\mu = F(\chi_{E_M}) \leq \|F\| \mu(E_M).$$

Thus $\mu(E_M) > 0$ yields $M \leq \|F\|$, which is a contradiction. We conclude that $\|g^+\|_\infty \leq \|F\|$. Similarly, $\|g^-\|_\infty \leq \|F\|$, and hence $\|g\|_\infty \leq \|F\|$.

If f is an arbitrary function in L^1, then we know by Theorem 5.27 that there exist simple functions f_k with $|f_k| \leq |f|$ such that $\{f_k\} \to f$ pointwise and $\|f_k - f\|_1 \to 0$. Therefore, by Lebesgue's dominated convergence theorem,

$$|F(f_k) - F(f)| = |F(f_k - f)| \leq \|F\| \, \|f - f_k\|_1 \to 0,$$

and

$$\begin{aligned} \left| F(f) - \int f g \, d\mu \right| &\leq |F(f - f_k)| + \left| \int (f_k - f) g \, d\mu \right| \\ &\leq \|F\| \;\; \|f - f_k\|_1 + \|f - f_k\|_1 \|g\|_\infty \\ &\leq 2 \, \|F\| \;\; \|f - f_k\|_1, \end{aligned}$$

and thus we have our desired result when $p = 1$ and $\mu(X) < \infty$. By Theorem 6.26, we have $\|g\|_\infty = \|F\|$.

Step 2. Assume $\mu(X) < \infty$ and $1 < p < \infty$.

Let $\{h_k\}_{k=1}^{\infty}$ be an increasing sequence of nonnegative simple functions such that $\lim_{k\to\infty} h_k = |g|$. Set $g_k = h_k^{p'-1}\operatorname{sign}(g)$. Then

$$\begin{aligned} \|h_k\|_{p'}^{p'} &= \int |h_k|^{p'}\, d\mu \le \int g_k g\, d\mu = F(g_k) \\ &\le \|F\|\ \|g_k\|_p = \|F\| \left(\int h_k^{p'}\, d\mu\right)^{\frac{1}{p}} = \|F\|\ \|h_k\|_{p'}^{\frac{p'}{p}}. \end{aligned} \tag{6.25}$$

We wish to conclude that $g \in L^{p'}(X)$. For this we may assume that $\|g\|_{p'} > 0$ and hence that $\|h_k\|_{p'} > 0$ for large k. It then follows from (6.25) that $\|h_k\|_{p'} \le \|F\|$ for all k, and thus by Fatou's lemma we have

$$\|g\|_{p'} \le \liminf_{k\to\infty} \|h_k\|_{p'} \le \|F\|,$$

which shows that $g \in L^{p'}(X)$, $1 < p < \infty$.

Now let $f \in L^p(X)$ and let $\{f_k\}$ be a sequence of simple functions such that $\|f - f_k\|_p \to 0$ as $k \to \infty$ (Exercise 1, Section 6.5). Then

$$\begin{aligned} \left|F(f) - \int fg\, d\mu\right| &\le |F(f - f_k)| + \left|\int (f_k - f)g\, d\mu\right| \\ &\le \|F\|\ \|f - f_k\|_p + \|f - f_k\|_p \|g\|_{p'} \\ &\le 2\,\|F\|\ \|f - f_k\|_p \end{aligned}$$

for all k, whence

$$F(f) = \int fg\, d\mu$$

for all $f \in L^p(X)$. By Theorem 6.26, we have $\|g\|_{p'} = \|F\|$. Thus, using step 1 also, we conclude that the proof is complete under the assumptions that $\mu(X) < \infty$, $1 \le p < \infty$.

Step 3. Assume that μ is σ-finite and $1 \le p < \infty$.

Suppose $Y \in \mathcal{M}$ is σ-finite. Let $\{Y_k\}$ be an increasing sequence of measurable sets such that $\mu(Y_k) < \infty$ for each k and such that $Y = \bigcup_{k=1}^{\infty} Y_k$. Then, from Steps 1 and 2 above (see also Exercise 7, Section 6.1), for each k there is a measurable function g_k such that $\left\|g_k \chi_{Y_k}\right\|_{p'} \le \|F\|$ and

$$F(f\chi_{Y_k}) = \int f\chi_{Y_k} g_k\, d\mu$$

for each $f \in L^p(X)$. We may assume $g_k = 0$ on $Y - Y_k$. If $k < m$, then

$$F(f\chi_{Y_k}) = \int f g_m \chi_{Y_k}\, d\mu$$

for each $f \in L^p(X)$. Thus

$$\int f(g_k - g_m\chi_{Y_k})\, d\mu = 0$$

for each $f \in L^p(X)$. By Theorem 6.26, this implies that $g_k = g_m$ μ-a.e. on Y_k. Thus $\{g_k\}$ converges μ-a.e. to a measurable function g, and by Fatou's lemma,

$$\|g\|_{p'} \le \liminf_{k\to\infty} \|g_k\|_{p'} \le \|F\|, \tag{6.26}$$

which shows that $g \in L^{p'}$, $1 < p' < \infty$. For $p' = \infty$, we also have $\|g\|_\infty \le \|F\|$.

Fix $f \in L^p(X)$ and set $f_k := f\chi_{Y_k}$. Then f_k converges to $f\chi_Y$ and $|f - f_k| \le 2\,|f|$. By the dominated convergence theorem, $\|(f\chi_Y - f_k)\|_p \to 0$ as $k \to \infty$. Thus, since $g_k = g$ μ-a.e. on Y_k,

$$\begin{aligned}\left|F(f\chi_Y) - \int fg\,d\mu\right| &\le |F(f\chi_Y - f_k)| + \int |f - f_k|\,|g|\,d\mu \\ &\le 2\,\|F\|\,\,\|f\chi_Y - f_k\|_p\end{aligned}$$

and therefore

$$F(f\chi_Y) = \int fg\,d\mu \tag{6.27}$$

for each $f \in L^p(X)$.

If μ is σ-finite, we may set $Y = X$ and deduce that

$$F(f) = \int fg\,d\mu$$

whenever $f \in L^p(X)$. Thus, in view of Theorem 6.47,

$$\|F\| = \sup\{|F(f)| : \|f\|_p = 1\} = \|g\|_{p'}\,.$$

Step 4. Assume that μ is not σ-finite and $1 < p < \infty$.

When $1 < p < \infty$ and μ is not assumed to be σ-finite, we will conclude the proof by making a judicious choice for Y in (6.27) so that

$$F(f) = F(f\chi_Y) \quad \text{for each } f \in L^p(X) \text{ and } \|g\|_{p'} = \|F\|\,. \tag{6.28}$$

For each positive integer k there is $h_k \in L^p(X)$ such that $\|h_k\|_p \le 1$ and

$$\|F\| - \frac{1}{k} < |F(h_k)|\,.$$

Set

$$Y = \bigcup_{k=1}^{\infty} \{x : h_k(x) \ne 0\}.$$

Then Y is a measurable σ-finite subset of X, and thus by Step 3, there is a $g \in L^{p'}(X)$ such that $g = 0$ μ-a.e. on $X - Y$ and

$$F(f\chi_Y) = \int fg\,d\mu \tag{6.29}$$

for each $f \in L^p(X)$. Since for each k,

$$\|F\| - \frac{1}{k} < F(h_k) = \int h_k g\,d\mu \le \|g\|_{p'}\,,$$

we see that $\|g\|_{p'} \geq \|F\|$. On the other hand, appealing to Theorem 6.26 again,

$$\|F\| \geq \sup_{\|f\|_p \leq 1} F(f\chi_Y) = \sup_{\|f\|_p \leq 1} \int fg\,d\mu = \|g\|_{p'},$$

which establishes the second part of (6.28),

To establish the first part of (6.28), with the help of (6.29), it suffices to show that $F(f) = F(f\chi_Y)$ for each $f \in L^p(X)$. For the sake of obtaining a contradiction, suppose there is a function $f_0 \in L^p(X)$ such that $F(f_0) \neq F(f_0\chi_Y)$. Set $Y_0 = \{x : f_0(x) \neq 0\} - Y$. Then, since Y_0 is σ-finite, there is a $g_0 \in L^{p'}(X)$ such that $g_0 = 0$ μ-a.e. on $X - Y_0$ and

$$F(f\chi_{Y_0}) = \int fg_0\,d\mu$$

for each $f \in L^p(X)$. Since g and g_0 are nonzero on disjoint sets, note that

$$\|g + g_0\|_{p'}^{p'} = \|g\|_{p'}^{p'} + \|g_0\|_{p'}^{p'}$$

and that $\|g_0\|_{p'} > 0$, since

$$\int f_0 g_0\,d\mu = F(f_0[1 - \chi_Y]) = F(f_0) - F(f_0\chi_Y) \neq 0.$$

Moreover, since

$$F(f\chi_{Y\cup Y_0}) = \int f(g + g_0)\,d\mu$$

for each $f \in L^p(X)$, we see that $\|g + g_0\|_{p'} \leq \|F\|$. Thus

$$\begin{aligned}\|F\|^{p'} &= \|g\|_{p'}^{p'} \\ &< \|g\|_{p'}^{p'} + \|g_0\|_{p'}^{p'} \\ &= \|g + g_0\|_{p'}^{p'} \\ &\leq \|F\|^{p'}.\end{aligned}$$

This contradiction implies that

$$F(f) = \int fg\,d\mu$$

for each $f \in L^p(X)$, which establishes the first part of (6.28), as desired.

Step 5. Uniqueness of g.

If $\tilde{g} \in L^{p'}(X)$ is such that

$$\int f(g - \tilde{g})\,d\mu = 0$$

for all $f \in L^p(X)$, then by Theorem 6.26, $\|g - \tilde{g}\|_{p'} = 0$, and thus $g = \tilde{g}$ μ-a.e., thus establishing the uniqueness of g. □

Exercises for Section 6.8

1. Suppose f is a nonnegative measurable function. Set
$$E_t = \{x : f(x) > t\}$$
and
$$g(t) = -\mu(E_t)$$
for $t \in \mathbb{R}$. Show that
$$\int f\, d\mu = \int_0^\infty t\, d\lambda_g(t),$$
where λ_g is the Lebesgue–Stieltjes measure induced by g as in Section 4.6.
2. Let f and g be integrable functions on a measure space $(X, \mathcal{M}, \mu)$ with the property that
$$\mu[\{f > t\}\Delta\{g > t\}] = 0$$
for λ-a.e. t. Prove that $f = g$ μ-a.e.
3. Let f be a Lebesgue measurable function on $[0,1]$ and let $Q := [0,1]\times[0,1]$.
   (a) Show that $F(x,y) := f(x) - f(y)$ is measurable with respect to Lebesgue measure in $\mathbb{R}^2$.
   (b) If $F \in L^1(Q)$, show that $f \in L^1([0,1])$.

6.9. Product Measures and Fubini's Theorem

In this section we introduce product measures and prove Fubini's theorem, which generalizes the notion of iterated integration of Riemannian calculus.

Let $(X, \mathcal{M}_X, \mu)$ and $(Y, \mathcal{M}_Y, \nu)$ be two complete measure spaces. In order to define the product of μ and ν we first define an outer measure on $X \times Y$ in terms of μ and ν.

6.49. Definition. For each $S \subset X \times Y$ set

$$\varphi(S) = \inf\left\{\sum_{j=1}^{\infty} \mu(A_j)\nu(B_j)\right\}, \tag{6.30}$$

where the infimum is taken over all sequences $\{A_j \times B_j\}_{j=1}^{\infty}$ such that $A_j \in \mathcal{M}_X$, $B_j \in \mathcal{M}_Y$ for each j and $S \subset \bigcup_{j=1}^{\infty}(A_j \times B_j)$.

6.50. Theorem. *The set function φ is an outer measure on $X \times Y$.*

Proof. It is immediate from the definition that $\varphi \geq 0$ and $\varphi(\emptyset) = 0$. To see that φ is countably subadditive, suppose $S \subset \bigcup_{k=1}^{\infty} S_k$ and assume that $\varphi(S_k) < \infty$ for each k. Fix $\varepsilon > 0$. Then for each k there is a sequence $\{A_j^k \times B_j^k\}_{j=1}^{\infty}$ with $A_j^k \in \mathcal{M}_X$ and $B_j^k \in \mathcal{M}_Y$ for each j such that
$$S_k \subset \bigcup_{j=1}^{\infty}(A_j^k \times B_j^k)$$

and

$$\sum_{j=1}^{\infty} \mu(A_j^k)\nu(B_j^k) < \varphi(S_k) + \frac{\varepsilon}{2^k}.$$

Thus

$$\begin{aligned}
\varphi(S) &\leq \sum_{k=1}^{\infty}\sum_{j=1}^{\infty} \mu(A_j^k)\nu(B_j^k) \\
&\leq \sum_{k=1}^{\infty}(\varphi(S_k) + \frac{\varepsilon}{2^k}) \\
&\leq \sum_{k=1}^{\infty} \varphi(S_k) + \varepsilon
\end{aligned}$$

for any $\varepsilon > 0$. □

Since φ is an outer measure, we know that its measurable sets form a σ-algebra (See Corollary 4.11), which we denote by $\mathcal{M}_{X\times Y}$. Also, we denote by $\mu \times \nu$ the restriction of φ to $\mathcal{M}_{X\times Y}$. The main objective of this section is to show that $\mu \times \nu$ may appropriately be called the "product measure" corresponding to μ and ν, and that the integral of a function over $X \times Y$ with respect to $\mu \times \nu$ can be computed by iterated integration. This is the thrust of the next result.

6.51. THEOREM (Fubini's theorem). *Suppose $(X, \mathcal{M}_X, \mu)$ and $(Y, \mathcal{M}_Y, \nu)$ are complete measure spaces.*

(i) *If $A \in \mathcal{M}_X$ and $B \in \mathcal{M}_Y$, then $A \times B \in \mathcal{M}_{X\times Y}$ and*

$$(\mu \times \nu)(A \times B) = \mu(A)\nu(B).$$

(ii) *If $S \in \mathcal{M}_{X\times Y}$ and S is σ-finite with respect to $\mu \times \nu$, then*

$$\begin{aligned}
S_y &= \{x : (x,y) \in S\} \in \mathcal{M}_X, \quad \textit{for} \quad \nu\textit{-a.e. } y \in Y, \\
S_x &= \{y : (x,y) \in S\} \in \mathcal{M}_Y, \quad \textit{for} \quad \mu\textit{-a.e. } x \in X, \\
y &\mapsto \mu(S_y) \quad \textit{is } \mathcal{M}_Y\textit{-measurable}, \\
x &\mapsto \nu(S_x) \quad \textit{is } \mathcal{M}_X\textit{-measurable}, \\
(\mu \times \nu)(S) &= \int_X \nu(S_x)\, d\mu(x) = \int_X \left[\int_Y \chi_S(x,y) d\nu(y)\right] d\mu(x) \\
&= \int_Y \mu(S_y) d\nu(y) = \int_Y \left[\int_X \chi_S(x,y)\, d\mu(x)\right] d\nu(y).
\end{aligned}$$

(iii) *If* $f \in L^1(X \times Y, \mathcal{M}_{X\times Y}, \mu \times \nu)$, *then*

$$\begin{aligned} y &\mapsto f(x,y) \quad \textit{is } \nu\textit{-integrable for } \mu\textit{-a.e. } x \in X, \\ x &\mapsto f(x,y) \quad \textit{is } \mu\textit{-integrable for } \nu\textit{-a.e. } y \in Y, \\ x &\mapsto \int_Y f(x,y)d\nu(y) \quad \textit{is } \mu\textit{-integrable,} \\ y &\mapsto \int_X f(x,y)\, d\mu(x) \quad \textit{is } \nu\textit{-integrable,} \end{aligned}$$

$$\begin{aligned} \int_{X\times Y} f d(\mu \times \nu) &= \int_X \left[\int_Y f(x,y)d\nu(y) \right] d\mu(x) \\ &= \int_Y \left[\int_X f(x,y)\, d\mu(x) \right] d\nu(y). \end{aligned}$$

Proof. Let $\mathcal{F}$ denote the collection of all subsets S of $X \times Y$ such that

$$S_x := \{y : (x,y) \in S\} \in \mathcal{M}_Y \quad \text{for } \mu\text{-a.e.} x \in X$$

and the function

$$x \mapsto \nu(S_x) \quad \text{is } \mathcal{M}_X\text{-measurable.}$$

For $S \in \mathcal{F}$ set

$$\rho(S) = \int_X \nu(S_x)\, d\mu(x) = \int_X \left[\int_Y \chi_S(x,y)d\nu(y) \right] d\mu(x).$$

In other words, $\mathcal{F}$ is precisely the family of sets that makes is possible to define ρ. **Proof of (i).** First, note that ρ is monotone on $\mathcal{F}$. Next observe that if $\cup_{j=1}^{\infty} S_j$ is a countable union of disjoint $S_j \in \mathcal{F}$, then clearly $\cup_{j=1}^{\infty} S_j \in \mathcal{F}$, and the monotone convergence theorem implies

$$\sum_{j=1}^{\infty} \rho(S_j) = \rho(\bigcup_{j=1}^{\infty} S_j); \quad \text{hence } \bigcup_{i=1}^{\infty} S_i \in \mathcal{F}. \tag{6.31}$$

Finally, if $S_1 \supset S_2 \supset \cdots$ are members of $\mathcal{F}$, then

$$\bigcap_{j=1}^{\infty} S_j \in \mathcal{F}, \tag{6.32}$$

and if $\rho(S_1) < \infty$, then Lebesgue's dominated convergence theorem yields

$$\lim_{j\to\infty} \rho(S_j) = \rho\left(\bigcap_{j=1}^{\infty} S_j\right); \quad \text{hence } \bigcap_{j=1}^{\infty} S_j \in \mathcal{F}. \tag{6.33}$$

Set

$$P_0 = \{A \times B : A \in \mathcal{M}_X \quad \text{and} \quad B \in \mathcal{M}_Y\},$$
$$P_1 = \{\bigcup_{j=1}^{\infty} S_j : S_j \in P_0 \quad \text{for} \quad j = 1, 2, \ldots\},$$
$$P_2 = \{\bigcap_{j=1}^{\infty} S_j : S_j \in P_1 \quad \text{for} \quad j = 1, 2, \ldots\}.$$

Note that if $A \in \mathcal{M}_X$ and $B \in \mathcal{M}_Y$, then $A \times B \in \mathcal{F}$ and

$$\rho(A \times B) = \mu(A)\nu(B), \tag{6.34}$$

and thus $P_0 \subset \mathcal{F}$. If $A_1 \times B_1, A_2 \times B_2 \subset X \times Y$, then

$$(A_1 \times B_1) \cap (A_2 \times B_2) = (A_1 \cap A_2) \times (B_1 \cap B_2) \tag{6.35}$$

and

$$(A_1 \times B_1) \setminus (A_2 \times B_2) = ((A_1 \setminus A_2) \times B_1) \cup ((A_1 \cap A_2) \times (B_1 \setminus B_2)). \tag{6.36}$$

It follows from Lemma 4.7, (6.35), and (6.36) that each member of P_1 can be written as a countable disjoint union of members of P_0, and since $\mathcal{F}$ is closed under countable disjoint unions, we have $P_1 \subset \mathcal{F}$. It also follows from (6.35) that every finite intersection of members of P_1 is also a member of P_1. Therefore, from (6.32), we have $P_2 \subset \mathcal{F}$. In summary, we have

$$P_0, P_1, P_2 \subset \mathcal{F}. \tag{6.37}$$

Suppose $S \subset X \times Y$, $\{A_j\} \subset \mathcal{M}_X$, $\{B_j\} \subset \mathcal{M}_Y$, and $S \subset R = \cup_{j=1}^{\infty}(A_j \times B_j)$. Using (6.34) and that $R \in P_1 \subset \mathcal{F}$, we obtain

$$\rho(R) \leq \sum_{j=1}^{\infty} \rho(A_j \times B_j) = \sum_{j=1}^{\infty} \mu(A_j)\nu(B_j).$$

Thus, by the definition of σ, (6.30),

$$\inf\{\rho(R) : S \subset R \in P_1\} \leq \sigma(S). \tag{6.38}$$

To establish the opposite inequality, note that if $S \subset R = \cup_{j=1}^{\infty}(A_j \times B_j)$, where the sets $A_j \times B_j$ are disjoint, then referring to (6.31), we have

$$\sigma(S) \leq \sum_{j=1}^{\infty} \mu(A_j)\nu(B_j) = \rho(R),$$

and consequently, with (6.38), we have

$$\sigma(S) = \inf\{\rho(R) : S \subset R \in P_1\} \tag{6.39}$$

for each $S \subset X \times Y$. If $A \in \mathcal{M}_X$ and $B \in \mathcal{M}_Y$, then $A \times B \in P_0 \subset \mathcal{F}$, and hence for all $R \in P_1$ with $A \times B \subset R$,

$$\begin{aligned} \sigma(A \times B) &\leq \mu(A)\nu(B) && \text{by (6.30)} \\ &= \rho(A \times B) && \text{by (6.34)} \\ &\leq \rho(R) && \text{because } \rho \text{ is monotone.} \end{aligned}$$

Therefore, by (6.39) and (6.34),

$$\sigma(A \times B) = \rho(A \times B) = \mu(A)\nu(B). \tag{6.40}$$

Moreover, if $T \subset R \in P_1 \subset \mathcal{F}$, then using the additivity of ρ (see(6.31)), it follows that

$$\begin{aligned} &\sigma(T \setminus (A \times B)) + \sigma(T \cap (A \times B)) \\ &\leq \rho(R \setminus (A \times B)) + \rho(R \cap (A \times B)) && \text{by (6.39)} \\ &= \rho(R) && \text{since } \rho \text{ is additive.} \end{aligned}$$

In view of (6.39) we see that $\sigma(T \setminus (A \times B)) + \sigma(T \cap (A \times B))$ for all $T \subset X \times Y$, and thus (see Definition 4.3) that $A \times B$ is σ-measurable; that is, $A \times B \in \mathcal{M}_{X \times Y}$. Thus assertion (i) is proved.

Proof of (ii). Suppose $S \subset X \times Y$ and $\sigma(S) < \infty$. Then there is a sequence $\{R_j\} \subset P_1$ such that $S \subset R_j$ for each j and

$$\sigma(S) = \lim_{j \to \infty} \rho(R_j). \tag{6.41}$$

Set

$$R = \bigcap_{j=1}^{\infty} R_j \in P_2.$$

Since $P_2 \subset \mathcal{F}$ and $\sigma(S) < \infty$, the dominated convergence theorem implies

$$\rho(R) = \lim_{m \to \infty} \rho\left(\bigcap_{j=1}^{m} R_j\right). \tag{6.42}$$

Thus, since $S \subset R \subset \cap_{j=1}^{m} R_j \in P_2$ for each finite m, by (6.42) we have

$$\sigma(S) \leq \lim_{m \to \infty} \rho\left(\bigcap_{j=1}^{m} R_j\right) = \rho(R) \leq \lim_{m \to \infty} \rho(R_m) = \sigma(S),$$

which implies that

(6.43) for each $S \subset X \times Y$ there is $R \in P_2$ such that $S \subset R$ and $\sigma(S) = \rho(R)$.

We now are in a position to finish the proof of assertion (ii). First suppose $S \subset X \times Y$, $S \subset R \in P_2$, and $\rho(R) = 0$. Then $\nu(R_x) = 0$ for μ-a.e. $x \in X$ and $S_x \subset R_x$ for each $x \in X$. Since ν is complete, $S_x \in \mathcal{M}_Y$ for μ-a.e. $x \in X$ and $S \in \mathcal{F}$ with $\rho(S) = 0$. In particular, we see that if $S \subset X \times Y$ with $\sigma(S) = 0$, then $S \in \mathcal{F}$ and $\rho(S) = 0$.

Now suppose $S \in \mathcal{M}_{X\times Y}$ and $(\mu \times \nu)(S) < \infty$. Then from (6.43), there is an $R \in P_2$ such that $S \subset R$ and

$$(\mu \times \nu)(S) = \sigma(S) = \rho(R).$$

From assertion (i) we see that $R \in \mathcal{M}_{X\times Y}$, and since $(\mu \times \nu)(S) < \infty$, we have

$$(\mu \times \nu)(R \setminus S) = 0.$$

This in turn implies that $R \setminus S \in \mathcal{F}$ and $\rho(R \setminus S) = 0$. Since ν is complete and

$$R_x \setminus S_x \in \mathcal{M}_Y$$

for μ-a.e. $x \in X$, we see that $S_x \in \mathcal{M}_Y$ for μ-a.e. $x \in X$ and thus that $S \in \mathcal{F}$ with

$$(\mu \times \nu)(S) = \rho(S) = \int_X \nu(S_x)\, d\mu(x).$$

If $S \in \mathcal{M}_{X\times Y}$ is σ-finite with respect to the measure $\mu \times \nu$, then there exists a sequence $\{S_j\}$ of disjoint sets $S_j \in \mathcal{M}_{X\times Y}$ with $(\mu \times \nu)(S_j) < \infty$ for each j such that

$$S = \bigcup_{j=1}^{\infty} S_j.$$

Since the sets are disjoint and each S_j is in $\mathcal{F}$, we have $S \in \mathcal{F}$ and

$$(\mu \times \nu)(S) = \sum_{j=1}^{\infty} (\mu \times \nu)(S_j) = \sum_{j=1}^{\infty} \rho(S_j) = \rho(S).$$

Of course the above argument remains valid if the roles of μ and ν are interchanged, and thus we have proved assertion (ii).

Proof of (iii). Assume first that $f \in L^1(X \times Y, \mathcal{M}_{X\times Y}, \mu \times \nu)$, and $f \geq 0$. Fix $t > 1$ and set

$$E_k = \{(x, y) : t^k < f(x, y) \leq t^{k+1}\}$$

for each $k = 0, \pm 1, \pm 2, \ldots$. Then each E_k belongs to $\mathcal{M}_{X\times Y}$ with $(\mu \times \nu)$ $(E_k) < \infty$. In view of (ii), the function

$$f_t = \sum_{k=-\infty}^{\infty} t^k \chi_{E_k}$$

satisfies the first four assertions of (ii) and

$$\begin{aligned}
\int f_t d(\mu \times \nu) &= \sum_{k=-\infty}^{\infty} t^k (\mu \times \nu)(E_k) \\
&= \sum_{k=-\infty}^{\infty} t^k \rho(E_k) && \text{by (6.40)} \\
&= \sum_{k=-\infty}^{\infty} t^k \int_X \left[\int_Y \chi_{E_k}(x,y) d\nu(y) \right] d\mu(x) \\
&= \int_X \left[\sum_{k=-\infty}^{\infty} t^k \int_Y \chi_{E_k}(x,y) d\nu(y) \right] d\mu(x) \\
&= \int_X \left[\int_Y f_t(x,y) d\nu(y) \right] d\mu(x)
\end{aligned}$$

by the monotone convergence theorem. Similarly,

$$\int f_t d(\mu \times \nu) = \int_Y \left[\int_X f_t(x,y)\, d\mu(x) \right] d\nu(y).$$

Since

$$\frac{1}{t} f \leq f_t \leq f,$$

we see that $f_t(x,y) \to f(x,y)$ as $t \to 1^+$ for each $(x,y) \in X \times Y$. Thus the function

$$y \mapsto f(x,y)$$

is $\mathcal{M}_Y$-measurable for μ-a.e. $x \in X$. It follows that

$$\frac{1}{t} \int_Y f(x,y) d\nu(y) \leq \int_Y f_t(x,y) d\nu(y) \leq \int_Y f(x,y) d\nu(y)$$

for μ-a.e. $x \in X$, the function

$$x \mapsto \int_Y f(x,y) d\nu(y)$$

is $\mathcal{M}_X$-measurable and

$$\begin{aligned}
\frac{1}{t} \int_X \left[\int_Y f(x,y) d\nu(y) \right] d\mu(x) &\leq \int_X \left[\int_Y f_t(x,y) d\nu(y) \right] d\mu(x) \\
&\leq \int_X \left[\int_Y f(x,y) d\nu(y) \right] d\mu(x).
\end{aligned}$$

Thus we see that

$$\begin{aligned}\int f d(\mu\times\nu) &= \lim_{t\to 1^+}\int f_t(x,y)d(\mu\times\nu)\\ &= \lim_{t\to 1^+}\int_X\left[\int_Y f_t(x,y)d\nu(y)\right]d\mu(x)\\ &= \int_X\left[\int_Y f(x,y)d\nu(y)\right]d\mu(x).\end{aligned}$$

Since the first integral above is finite, we have established the first and third parts of assertion (iii) as well as the first half of the fifth part. The remainder of (iii) follows by an analogous argument.

To extend the proof to general $f\in L^1(X\times Y,\mathcal{M}_{X\times Y},\mu\times\nu)$ we need only recall that

$$f = f^+ - f^-,$$

where f^+ and f^- are nonnegative integrable functions. □

It is important to observe that the hypothesis (iii) in Fubini's theorem, namely that $f\in L^1(X\times Y,\mathcal{M}_{X\times Y},\mu\times\nu)$, is necessary. Indeed, consider the following example.

6.52. Example. Let Q denote the unit square $[0,1]\times[0,1]$ and consider a sequence of subsquares Q_k defined as follows: Let $Q_1 := [0,1/2]\times[0,1/2]$. Let Q_2 be a square with half the area of Q_1 and placed so that $Q_1\cap Q_2 = \{(1/2,1/2)\}$, that is, so that its "southwest" vertex is the same as the "northeast" vertex of Q_1. Similarly, let Q_3 be a square with half the area of Q_2, and as before, place it so that its "southwest" vertex is the same as the "northeast" vertex of Q_2. In this way, we obtain a sequence of squares $\{Q_k\}$ all of whose southwest–northeast diagonal vertices lie on the line $y=x$. Subdivide each subsquare Q_k into four equal squares $Q_k^{(1)}$, $Q_k^{(2)}$, $Q_k^{(3)}$, $Q_k^{(4)}$, where we will regard $Q_k^{(1)}$ as occupying the "first quadrant," $Q_k^{(2)}$ the "second quadrant," $Q_k^{(3)}$ the "third quadrant," and $Q_k^{(4)}$ the "fourth quadrant." In the next section it will be shown that the two-dimensional Lebesgue measure λ_2 is the same as the product $\lambda_1\times\lambda_1$, where λ_1 denotes the one-dimensional Lebesgue measure. Define a function f on Q such that $f=0$ on the complement of the Q_k, and otherwise, on each Q_k define $f=\frac{1}{\lambda_2(Q_k)}$ on subsquares in the first and third quadrants and $f=-\frac{1}{\lambda_2(Q_k)}$ on the subsquares in the second and fourth quadrants. Clearly,

$$\int_{Q_k}|f|\,d\lambda_2 = 1$$

and therefore

$$\int_Q |f|\,d\lambda_2 = \sum_k\int_{Q_k}|f| = \infty,$$

whereas

$$\int_0^1 f(x,y)\,d\lambda_1(x) = \int_0^1 f(x,y)\,d\lambda_1(y) = 0.$$

This is an example in which the iterated integral exists but Fubini's theorem does not hold because f is not integrable. However, this integrability hypothesis is not necessary if $f \geq 0$ and if f is measurable in each variable separately. The proof of this follows readily from the proof of Theorem 6.51.

6.53. COROLLARY (Tonelli). *If f is a nonnegative $\mathcal{M}_{X\times Y}$-measurable function and $\{(x,y) : f(x,y) \neq 0\}$ is σ-finite with respect to the measure $\mu \times \nu$, then the function*

$$y \mapsto f(x,y) \quad \text{is } \mathcal{M}_Y\text{-measurable for } \mu\text{-a.e. } x \in X,$$
$$x \mapsto f(x,y) \quad \text{is } \mathcal{M}_X\text{-measurable for } \nu\text{-a.e. } y \in Y,$$
$$x \mapsto \int_Y f(x,y)\,d\nu(y) \quad \text{is } \mathcal{M}_X\text{-measurable},$$
$$y \mapsto \int_X f(x,y)\,d\mu(x) \quad \text{is } \mathcal{M}_Y\text{-measurable},$$

and

$$\int_{X\times Y} f\,d(\mu\times\nu) = \int_X \left[\int_Y f(x,y)\,d\nu(y)\right] d\mu(x) = \int_Y \left[\int_X f(x,y)\,d\mu(x)\right] d\nu(y)$$

in the sense that either both expressions are infinite or both are finite and equal.

PROOF. Let $\{f_k\}$ be a sequence of nonnegative real-valued measurable functions with finite range such that $f_k \leq f_{k+1}$ and $\lim_{k\to\infty} f_k = f$. By assertion (ii) of Theorem 6.51 the conclusion of the corollary holds for each f_k. For each k let N_k be an $\mathcal{M}_X$-measurable subset of X such that $\mu(N_k) = 0$ and

$$y \mapsto f_k(x,y)$$

is $\mathcal{M}_Y$-measurable for each $x \in X - N_k$. Set $N = \cup_{k=1}^{\infty} N_k$. Then $\mu(N) = 0$ and for each $x \in X - N$,

$$y \mapsto f(x,y) = \lim_{k\to\infty} f_k(x,y)$$

is $\mathcal{M}_Y$-measurable, and by the monotone convergence theorem,

$$\int_Y f(x,y)\,d\nu(y) = \lim_{k\to\infty} \int_Y f_k(x,y)\,d\nu(y) \tag{6.44}$$

for $x \in X - N$.

Theorem 6.51 implies that $h_k(x) := \int_Y f_k(x,y)\,d\nu(y)$ is $\mathcal{M}_X$-measurable. Since $0 \le h_k \le h_{k+1}$, we can use again the monotone convergence theorem to obtain

$$\begin{aligned}
\int_{X\times Y} f\,d(\mu\times\nu) &= \lim_{k\to\infty}\int_{X\times Y} f_k(x,y)\,d(\mu\times\nu)\\
&= \lim_{k\to\infty}\int_X\left[\int_Y f_k(x,y)\,d\nu(y)\right]\,d\mu(x)\\
&= \lim_{k\to\infty}\int_X h_k(x)\,d\mu(x)\\
&= \int_X \lim_{k\to\infty} h_k(x)\,d\mu(x)\\
&= \int_X \lim_{k\to\infty}\left[\int_Y f_k(x,y)\,d\nu(y)\right]\,d\mu(x)\\
&= \int_X\left[\int_Y f(x,y)\,d\nu(y)\right]\,d\mu(x), \qquad \square
\end{aligned}$$

where the last line follows from (6.44). The reverse iteration can be obtained with a similar argument.

Exercises for Section 6.9

1. Let X be a well-ordered set (with ordering denoted by $<$) that is a representative of the ordinal number Ω and let $\mathcal{M}$ be the σ-algebra consisting of the sets E with the property that either E or its complement is at most countable. Let μ be the measure defined on $E \in \mathcal{M}$ as $\mu(E) = 0$ if E is at most countable and $\mu(E) = 1$ otherwise. Let $Y := X$ and $\nu := \mu$.
   (a) Show that if $A := \{(x,y) \in X \times Y : x < y\}$, then A_x and A_y are measurable for all x and y.
   (b) Show that both
   $$\int_Y\left(\int_X \chi_A(x,y)\,d\mu(x)\right)\,d\nu(y)$$
   and
   $$\int_X\left(\int_Y \chi_A(x,y)\,d\nu(y)\right)\,d\mu(x)$$
   exist.
   (c) Show that
   $$\int_Y\left(\int_X \chi_A(x,y)\,d\mu(x)\right)\,d\nu(y) \neq \int_X\left(\int_Y \chi_A(x,y)\,d\nu(y)\right)\,d\mu(x)$$
   (d) Why doesn't Fubini's theorem apply in this example?

2. Show that if $f(x,y) = \frac{x^2-y^2}{(x^2+y^2)^2}$, with $f(0,0) = 0$, then
$$\int_0^1 \left[\int_0^1 f(x,y)dx\right] dy = -\frac{\pi}{4} \text{ and } \int_0^1 \left[\int_0^1 f(x,y)dy\right] dx = \frac{\pi}{4}.$$
Why doesn't Fubini's theorem apply in this example?
3. Show that if $f(x,y) = ye^{-(1+x^2)y^2}$ for each x and y, then (see Section 6.4)
$$\int_0^\infty \left[\int_0^\infty f(x,y)dx\right] dy = \int_0^\infty \left[\int_0^\infty f(x,y)dy\right] dx.$$
Use this equality to show that
$$\int_0^\infty e^{-x^2} = \frac{\sqrt{\pi}}{2}.$$

6.10. Lebesgue Measure as a Product Measure

We will now show that n-dimensional Lebesgue measure on $\mathbb{R}^n$ is a product of lower-dimensional Lebesgue measures.

For each positive integer k let λ_k denote the Lebesgue measure on $\mathbb{R}^k$ and let $\mathcal{M}_k$ denote the σ-algebra of Lebesgue measurable subsets of $\mathbb{R}^k$.

6.54. THEOREM. *For each pair of positive integers n and m,*
$$\lambda_{n+m} = \lambda_n \times \lambda_m.$$

PROOF. Let φ denote the outer measure on $\mathbb{R}^n \times \mathbb{R}^m$ defined as in Definition 6.49 with $\mu = \lambda_n$ and $\nu = \lambda_m$. We will show that $\varphi = \lambda^*_{n+m}$.

If $A \in \mathcal{M}_n$ and $B \in \mathcal{M}_m$ are bounded sets and $\varepsilon > 0$, then there are open sets $U \supset A$, $V \supset B$ such that
$$\lambda_n(U - A) < \varepsilon,$$
$$\lambda_m(V - B) < \varepsilon,$$

and hence
$$\lambda_n(U)\lambda_m(V) \leq \lambda_n(A)\lambda_m(B) + \varepsilon(\lambda_n(A) + \lambda_m(B)) + \varepsilon^2. \tag{6.45}$$
Suppose E is a bounded subset of $\mathbb{R}^{n+m}$, and $\{A_k \times B_k\}_{k=1}^\infty$ is a sequence of subsets of $\mathbb{R}^n \times \mathbb{R}^m$ such that $A_k \in \mathcal{M}_n$, $B_k \in \mathcal{M}_m$, and $E \subset \cup_{k=1}^\infty A_k \times B_k$. Assume that the sequences $\{\lambda_n(A_k)\}_{k=1}^\infty$ and $\{\lambda_m(B_k)\}_{k=1}^\infty$ are bounded. Fix $\varepsilon > 0$. In view of (6.45), there exist open sets U_k, V_k such that $A_k \subset U_k$, $B_k \subset V_k$, and
$$\sum_{k=1}^\infty \lambda_n(A_k)\lambda_m(B_k) \geq \sum_{k=1}^\infty \lambda_n(U_k)\lambda_m(V_k) - \varepsilon.$$
It is not difficult to show that each of the open sets $U_k \times V_k$ can be written as a countable union of nonoverlapping closed intervals
$$U_k \times V_k = \bigcup_{l=1}^\infty I_l^k \times J_l^k,$$

where I_l^k, J_l^k are closed intervals in $\mathbb{R}^n$ and $\mathbb{R}^m$ respectively. Thus for each k,

$$\lambda_n(U_k)\lambda_m(V_k) = \sum_{l=1}^{\infty} \lambda_n(I_l^k)\lambda_m(J_l^k) = \sum_{l=1}^{\infty} \lambda_{n+m}(I_l^k \times J_l^k).$$

It follows that

$$\sum_{k=1}^{\infty} \lambda_n(A_k)\lambda_m(B_k) \geq \sum_{k=1}^{\infty} \lambda_n(U_k)\lambda_m(V_k) - \varepsilon \geq \lambda_{n+m}^*(E) - \varepsilon$$

and hence

$$\varphi(E) \geq \lambda_{n+m}^*(E)$$

whenever E is a bounded subset of $\mathbb{R}^{n+m}$.

If E is an unbounded subset of $\mathbb{R}^{n+m}$, we have

$$\varphi(E) \geq \lambda_{n+m}^*(E \cap B(0,j))$$

for each positive integer j. Since λ_{n+m}^* is a Borel regular outer measure (see Exercise 9, Section 4.3), there is a Borel set $A_j \supset E \cap B(0,j)$ such that $\lambda(A_j) = \lambda_{m+n}^*(E \cap B(0,j))$. With $A := \cup_{j=1}^{\infty} A_j$, we have

$$\lambda_{n+m}^*(E) \leq \lambda_{n+m}(A) = \lim_{j\to\infty} \lambda_{n+m}(A_j) = \lim_{j\to\infty} \lambda_{n+m}^*(E \cap B(0,j)) \leq \lambda_{n+m}^*(E)$$

and therefore

$$\lim_{j\to\infty} \lambda_{m+n}^*(E \cap B(0,j)) = \lambda_{m+n}^*(E).$$

This yields

$$\varphi(E) \geq \lambda_{m+n}^*(E)$$

for each $E \subset \mathbb{R}^{n+m}$.

On the other hand, it is immediate from the definitions of the two outer measures that

$$\varphi(E) \leq \lambda_{m+n}^*(E)$$

for each $E \subset \mathbb{R}^{n+m}$. □

6.11. Convolution

As an application of Fubini's theorem, we determine conditions on functions f and g that ensure the existence of the convolution $f * g$ and deduce the basic properties of convolution.

6.55. Definition. Given two Lebesgue measurable functions f and g on $\mathbb{R}^n$, we define the **convolution** $f * g$ of f and g to be the function defined for each $x \in \mathbb{R}^n$ by

$$(f * g)(x) = \int_{\mathbb{R}^n} f(y)g(x-y)\,dy.$$

Here and in the remainder of this section we will indicate integration with respect to Lebesgue measure by dx, dy, etc.

We first observe that if g is a nonnegative Lebesgue measurable function on $\mathbb{R}^n$, then

$$\int_{\mathbb{R}^n} g(x-y)\,dy = \int_{\mathbb{R}^n} g(y)\,dy$$

for all $x \in \mathbb{R}^n$. This follows readily from the definition of the integral and the fact that λ_n is invariant under translation.

To study the integrability properties of the convolution of two functions we will need the following lemma.

6.56. LEMMA. *If f is a Lebesgue measurable function on $\mathbb{R}^n$, then the function F defined on $\mathbb{R}^n \times \mathbb{R}^n = \mathbb{R}^{2n}$ by*

$$F(x,y) = f(x-y)$$

is λ_{2n}-measurable.

PROOF. First, define $F_1 : \mathbb{R}^{2n} \to \mathbb{R}$ by $F_1(x,y) := f(x)$ and observe that F_1 is λ_{2n}-measurable because for every Borel set $B \subset \mathbb{R}$, we have $F_1^{-1}(B) = f^{-1}(B) \times \mathbb{R}^n$. Then define $T : \mathbb{R}^{2n} \to \mathbb{R}^{2n}$ by $T(x,y) = (x-y, x+y)$ and note that $T^{-1}(x,y) = \left(\frac{x+y}{2}, \frac{y-x}{2}\right)$. The mean value theorem implies $|T(x_1,y_1) - T(x_2,y_2)| \le n^2|(x_1,y_1) - (x_2,y_2)|$ and $|T^{-1}(x_1,y_1) - T^{-1}(x_2,y_2)| \le \frac{1}{2}n^2|(x_1,y_1) - (x_2,y_2)|$ for every $(x_1,y_1),(x_2,y_2) \in \mathbb{R}^{2n}$. Therefore, T and T^{-1} are Lipschitz functions in $\mathbb{R}^{2n}$. Hence, it follows that $F_1 \circ T = F$ is λ_{2n}-measurable. Indeed, if $B \subset \mathbb{R}$ is a Borel set, then $E := F_1^{-1}(B)$ is λ_{2n}-measurable and thus can be expressed as $E = B_1 \cup N$, where $B_1 \subset \mathbb{R}^{2n}$ is a Borel set and $\lambda_{2n}(N) = 0$. Consequently, $T^{-1}(E) = T^{-1}(B_1) \cup T^{-1}(N)$, which is the union of a Borel set and a set of λ_{2n}-measure zero (see Exercise 4.11). □

We now prove a basic result concerning convolutions. Recall our notation $L^p(\mathbb{R}^n)$ for $L^p(\mathbb{R}^n, \mathcal{M}_n, \lambda_n)$ and $\|f\|_p$ for $\|f\|_{p,\mathbb{R}^n;\lambda_n}$.

6.57. THEOREM. *If $f \in L^p(\mathbb{R}^n)$, $1 \le p \le \infty$, and $g \in L^1(\mathbb{R}^n)$, then $f * g \in L^p(\mathbb{R}^n)$ and*

$$\|f * g\|_p \le \|f\|_p \|g\|_1 .$$

PROOF. Observe that $|f * g| \le |f| * |g|$, and thus it suffices to prove the assertion for $f, g \ge 0$. Then by Lemma 6.56 the function

$$(x,y) \mapsto f(y)g(x-y)$$

is nonnegative and $\mathcal{M}_{2n}$-measurable, and by Corollary 6.53,

$$\begin{aligned}\int (f*g)(x)\,dx &= \int\int f(y)g(x-y)\,dy\,dx \\ &= \int f(y)\left[\int g(x-y)\,dx\right]dy \\ &= \int f(y)\,dy \int g(x)\,dx.\end{aligned}$$

Thus the assertion holds if $p = 1$.

If $p = \infty$, we see that

$$(f * g)(x) \leq \|f\|_\infty \int g(x-y)\, dy = \|f\|_\infty \|g\|_1,$$

whence

$$\|f * g\|_\infty \leq \|f\|_\infty \|g\|_1 .$$

Finally, suppose $1 < p < \infty$. Then

$$\begin{aligned}(f * g)(x) &= \int f(y)(g(x-y))^{\frac{1}{p}}(g(x-y))^{1-\frac{1}{p}}\, dy \\ &\leq \left(\int f^p(y) g(x-y)\, dy\right)^{\frac{1}{p}} \left(\int g(x-y)\, dy\right)^{1-\frac{1}{p}} \\ &= (f^p * g)^{\frac{1}{p}}(x) \|g\|_1^{1-\frac{1}{p}} .\end{aligned}$$

Thus

$$\begin{aligned}\int (f * g)^p(x)\, dx &\leq \int (f^p * g)(x)\, dx \|g\|_1^{p-1} \\ &= \|f^p\|_1 \|g\|_1 \|g\|_1^{p-1} \\ &= \|f\|_p^p \|g\|_1^p ,\end{aligned}$$

and the assertion is proved. □

If we fix $g \in L^1(\mathbb{R}^n)$ and set

$$T(f) = f * g,$$

then we may interpret the theorem as saying that for all $1 \leq p \leq \infty$,

$$T : L^p(\mathbb{R}^n) \to L^p(\mathbb{R}^n)$$

is a bounded linear mapping. Such mappings induced by convolution will be further studied in Chapter 9.

Exercises for Section 6.11

1. (a) For $p > 1$ and $p' := p/(p-1)$, prove that if $f \in L^p(\mathbb{R}^n)$ and $g \in L^{p'}(\mathbb{R}^n)$, then

$$f * g(x) \leq \|f\|_p \|g\|_{p'}$$

for all $x \in \mathbb{R}^n$.

(b) Suppose that $f \in L^p(\mathbb{R}^n)$ and $g \in L^{p'}(\mathbb{R}^n)$. Prove that $f * g$ vanishes at infinity. That is, prove that for each $\varepsilon > 0$, there exists $R > 0$ such that

$$f * g(x) < \varepsilon \text{ for all } |x| > R.$$

2. Let ϕ be a nonnegative real-valued function in $C_0^\infty(\mathbb{R}^n)$ with the property that
$$\int_{\mathbb{R}^n} \phi(x)dx = 1, \quad \operatorname{spt}\phi \subset \overline{B}(0,1).$$
An example of such a function is given by
$$\phi(x) = \begin{cases} C\exp[-1/(1-|x|^2)] & \text{if } |x|<1, \\ 0 & \text{if } |x| \geq 1, \end{cases}$$
where C is chosen such that $\int_{R^n} \phi = 1$. For $\varepsilon > 0$, the function $\phi_\varepsilon(x) := \varepsilon^{-n}\phi(x/\varepsilon)$ belongs to $C_0^\infty(R^n)$ and $\operatorname{spt}\phi_\varepsilon \subset \overline{B}(0,\varepsilon)$. The function ϕ_ε is called a **regularizer** (or **mollifier**), and the convolution
$$u_\varepsilon(x) := \phi_\varepsilon * u(x) := \int_{\mathbb{R}^n} \phi_\varepsilon(x-y)u(y)dy$$
defined for functions $u \in L^1_{\text{loc}}(\mathbb{R}^n)$ is called the regularization (mollification) of u. As a consequence of Fubini's theorem, we have
$$\|u * v\|_p \leq \|u\|_p \|v\|_1$$
whenever $1 \leq p \leq \infty$, $u \in L^p(\mathbb{R}^n)$, and $v \in L^1(\mathbb{R}^n)$.

Prove the following (see Theorem 10.1):

(a) If $u \in L^1_{\text{loc}}(\mathbb{R}^n)$, then for every $\varepsilon > 0$, $u_\varepsilon \in C^\infty(\mathbb{R}^n)$.

(b) If u is continuous, then u_ε converges to u uniformly on compact subsets of $\mathbb{R}^n$.

3. If $u \in L^p(\mathbb{R}^n)$, $1 \leq p < \infty$, then $u_\varepsilon \in L^p(\mathbb{R}^n)$, $\|u_\varepsilon\|_p \leq \|u\|_p$, and $\lim_{\varepsilon\to 0} \|u_\varepsilon - u\|_p = 0$.

4. Let μ be a Radon measure on $\mathbb{R}^n$, $x \in \mathbb{R}^n$, and $0 < \alpha < n$. Then
$$\int_{\mathbb{R}^n} \frac{d\mu(y)}{|x-y|^{n-\alpha}} = (n-\alpha)\int_0^\infty r^{\alpha-n-1}\mu(B(x,r))\,dr,$$
provided that
$$\int_{\mathbb{R}^n} \frac{d\mu(y)}{|x-y|^{n-\alpha}} < \infty.$$

5. In this problem, we will consider $\mathbb{R}^2$ for simplicity, but everything carries over to $\mathbb{R}^n$. Let P be a polynomial in $\mathbb{R}^2$; that is, P has the form
$$P(x,y) = a_n x^n y^n + a_{n-1}x^n y^{n-1} + b_{n-1}x^{n-1}y^n + \cdots + a_1 x^1 y^0 + b_1 x y^1 + a_0,$$
where the a's and b's are real numbers and $n \in \mathbb{N}$. Let φ_ε denote the mollifying kernel discussed in the previous problem. Prove that $\varphi_\varepsilon * P$ is also a polynomial. In other words, it isn't possible to make a polynomial smoother than itself.

6. Consider $(X, \mathcal{M}, \mu)$, where μ is σ-finite and complete and suppose $f \in L^1(X)$ is nonnegative. Let
$$G_f := \{(x,y) \in X \times [0,\infty] : 0 \leq y \leq f(x)\}.$$
Prove the following:

(a) The set G_f is $\mu \times \lambda_1$-measurable.

(b) $\mu \times \lambda_1(G_f) = \int_X f \, d\mu.$

This shows that the "area under the graph is the integral of the function."

6.12. Distribution Functions

Here we will study an interesting and useful connection between abstract integration and Lebesgue integration.

Let $(X, \mathcal{M}, \mu)$ be a complete σ-finite measure space. Let f be a measurable function on X, and for $t \in \mathbb{R}$ set

$$E_t = \{x : |f(x)| > t\} \in \mathcal{M}.$$

We have the following definition.

6.58. Definition. The **distribution function** of f is the nonincreasing function defined as

$$A_f(t) := \mu(E_t).$$

An interesting relation between f and its distribution function can be deduced from Fubini's theorem.

6.59. Theorem. *If f is nonnegative and measurable, then*

$$\int_X f \ d\mu = \int_{[0,\infty)} A_f \, d\lambda = \int_{[0,\infty]} \mu(\{x : f(x) > t\}) d\lambda(t). \tag{6.46}$$

Proof. Let $\widetilde{\mathcal{M}}$ denote the σ-algebra of measurable subsets of $X \times \mathbb{R}$ corresponding to $\mu \times \lambda$. Set

$$W = \{(x, t) : 0 < t < f(x)\} \subset X \times \mathbb{R}.$$

Since f is measurable, there is a sequence $\{f_k\}$ of measurable simple functions such that $f_k \leq f_{k+1}$ and $\lim_{k\to\infty} f_k = f$ pointwise on X. If $f_k = \sum_{j=1}^{n_k} a_j^k \chi_{E_j^k}$, where for each k the sets $\{E_j^k\}$ are disjoint and measurable, then

$$W_k = \{(x, t) : 0 < t < f_k(x)\} = \bigcup_{j=1}^{n_k} E_j^k \times (0, a_j^k) \in \widetilde{\mathcal{M}}.$$

Since $\chi_W = \lim_{k\to\infty} \chi_{W_k}$, we see that $W \in \widetilde{\mathcal{M}}$. Thus by Corollary 6.53,

$$\begin{aligned}
\int_{[0,\infty)} A_f \, d\lambda &= \int_{\mathbb{R}} \int_X \chi_W(x, t) \ d\mu(x) \, d\lambda(t) \\
&= \int_X \int_{\mathbb{R}} \chi_W(x, t) \, d\lambda(t) \ d\mu(x) \\
&= \int_X \lambda(\{t : 0 < t < f(x)\}) \ d\mu(x) \\
&= \int_X f \, d\mu.
\end{aligned}$$

□

Thus a nonnegative measurable function f is integrable over X with respect to μ if and only if its distribution function A_f is integrable over $[0,\infty)$ with respect to the one-dimensional Lebesgue measure λ.

If $\mu(X) < \infty$, then A_f is a bounded monotone function and thus continuous λ-a.e. on $[0,\infty)$. In view of Theorem 6.19, this implies that A_f is Riemann integrable on every compact interval in $[0,\infty)$ and thus that the right-hand side of (6.46) can be interpreted as an improper Riemann integral.

The simple idea behind the proof of Theorem 6.59 can readily be extended as in the following theorem.

6.60. THEOREM. *If f is measurable and $1 \le p < \infty$, then*

$$\int_X |f|^p \, d\mu = p \int_{[0,\infty)} t^{p-1} \mu(\{x : |f(x)| > t\}) \, d\lambda(t).$$

PROOF. Set

$$W = \{(x,t) : 0 < t < |f(x)|\}$$

and note that the function

$$(x,t) \mapsto pt^{p-1}\chi_W(x,t)$$

is $\widetilde{\mathcal{M}}$-measurable. Thus by Corollary 6.53,

$$\begin{aligned}\int_X |f|^p \, d\mu &= \int_X \int_{(0,|f(x)|)} pt^{p-1} \, d\lambda(t) \, d\mu(x) \\ &= \int_X \int_{\mathbb{R}} pt^{p-1}\chi_W(x,t) \, d\lambda(t) \, d\mu(x) \\ &= \int_{\mathbb{R}} \int_X pt^{p-1}\chi_W(x,t) \, d\mu(x) \, d\lambda(t) \\ &= p \int_{[0,\infty)} t^{p-1}\mu(\{x : |f(x)| > t\}) \, d\lambda(t). \qquad \square\end{aligned}$$

6.61. REMARK. A useful mnemonic relating to the previous result is that if f is measurable and $1 \le p < \infty$, then

$$\int_X |f|^p \, d\mu = \int_0^\infty \mu(\{|f| > t\}) \, dt^p.$$

Exercise for Section 6.12

1. Suppose $f \in L^1(\mathbb{R}^n)$ and let $A_t := \{x : |f(x)| > t\}$. Prove that

$$\lim_{t\to\infty} \int_{A_t} |f| \, d\lambda = 0.$$

6.13. The Marcinkiewicz Interpolation Theorem

In the previous section, we employed Fubini's theorem extensively to investigate the properties of the distribution function. We close this chapter by pursuing this topic further to establish the Marcinkiewicz interpolation theorem, which has important applications in diverse areas of analysis, such as Fourier analysis and nonlinear potential theory. Later, in Chapter 7, we will see a beautiful interaction between this result and the Hardy–Littlewood maximal function, Definition 7.8.

In preparation for the main theorem of this section, we will need two preliminary results. The first, due to Hardy, gives two inequalities that are related to Jensen's inequality, Exercise 10, Section 6.5. If f is a nonnegative measurable function defined on the positive real numbers, let

$$F(x) = \frac{1}{x}\int_0^x f(t)dt, x > 0,$$
$$G(x) = \frac{1}{x}\int_x^\infty f(t)dt, x > 0.$$

Jensen's inequality states that for $p \geq 1$, one has $[F(x)]^p \leq \|f\|_{p;(0,x)}$ for each $x > 0$, and it thus provides an estimate of $F(x)^p$; Hardy's inequality (6.47) below gives an estimate of a weighted integral of F^p.

6.62. LEMMA (Hardy's inequalities). *If* $1 \leq p < \infty$, $r > 0$, *and* f *is a nonnegative measurable function on* $(0, \infty)$, *then with* F *and* G *defined as above,*

$$\int_0^\infty [F(x)]^p x^{p-r-1}dx \leq \left(\frac{p}{r}\right)^p \int_0^\infty [f(t)]^p t^{p-r-1}dt. \tag{6.47}$$

$$\int_0^\infty [G(x)]^p x^{p+r-1}dx \leq \left(\frac{p}{r}\right)^p \int_0^\infty [f(t)]^p t^{p+r-1}dt. \tag{6.48}$$

PROOF. To prove (6.47), we apply Jensen's inequality (Exercise 10, Section 6.5) with the measure $t^{(r/p)-1}dt$, obtaining

$$\left(\int_0^x f(t)dt\right)^p = \left(\int_0^x f(t)t^{1-(r/p)}t^{(r/p)-1}dt\right)^p \tag{6.49}$$

$$\leq \left(\frac{p}{r}\right)^{p-1} x^{r(1-1/p)} \int_0^x [f(t)]^p t^{p-r-1+r/p}dt. \tag{6.50}$$

Then by Fubini's theorem,

$$\begin{aligned}
&\int_0^\infty \left(\int_0^x f(t)dt\right)^p x^{p-r-1}dx \\
&\leq \left(\frac{p}{r}\right)^{p-1} \int_0^\infty x^{-1-(r/p)} \left(\int_0^x [f(t)]^p t^{p-r-1+(r/p)}dt\right) dx \\
&= \left(\frac{p}{r}\right)^{p-1} \int_0^\infty [f(t)]^p t^{p-r-1+(r/p)} \left(\int_t^\infty x^{-1-(r/p)}dx\right) dt \\
&= \left(\frac{p}{r}\right)^p \int_0^\infty [f(t)t]^p t^{-r-1}dt.
\end{aligned}$$

The proof of (6.48) proceeds in a similar way. □

6.63. Lemma. *If $f \geq 0$ is a nonincreasing function on $(0,\infty)$, $0<p\leq\infty$, and $p_1 \leq p_2 \leq \infty$, then*

$$\left(\int_0^\infty [x^{1/p}f(x)]^{p_2}\frac{d\lambda(x)}{x}\right)^{1/p_2} \leq C\left(\int_0^\infty [x^{1/p}f(x)]^{p_1}\frac{d\lambda(x)}{x}\right)^{1/p_1},$$

where $C = C(p, p_1, p_2)$.

Proof. Since f is nonincreasing, we have for all $x > 0$,

$$\begin{aligned} x^{1/p}f(x) &\leq C\left(\int_{x/2}^{x} [(x/2)^{1/p}f(x)]^{p_1}\frac{d\lambda(y)}{y}\right)^{1/p_1} \\ &\leq C\left(\int_{x/2}^{x} [y^{1/p}f(x)]^{p_1}\frac{d\lambda(y)}{y}\right)^{1/p_1} \\ &\leq C\left(\int_{x/2}^{x} [y^{1/p}f(y)]^{p_1}\frac{d\lambda(y)}{y}\right)^{1/p_1} \\ &\leq C\left(\int_{0}^{\infty} [y^{1/p}f(y)]^{p_1}\frac{d\lambda(y)}{y}\right)^{1/p_1}, \end{aligned}$$

which implies the desired result when $p_2 = \infty$. The general result follows by writing

$$\int_0^\infty [x^{1/p}f(x)]^{p_2}\frac{d\lambda(x)}{x} \leq \sup_{x>0}[x^{1/p}f(x)]^{p_2-p_1}\int_0^\infty [x^{1/p}f(x)]^{p_1}\frac{d\lambda(x)}{x}. \qquad \square$$

6.64. Definition. Let μ be a nonnegative Radon measure defined on $\mathbb{R}^n$ and suppose f is a μ-measurable function defined on $\mathbb{R}^n$. Its distribution function, $A_f(\cdot)$, is defined by

$$A_f(t) := \mu\big(\{x : |f(x)| > t\}\big).$$

The **nonincreasing rearrangement** of f, denoted by f^*, is defined as

$$f^*(t) = \inf\{\alpha : A_f(\alpha) \leq t\}. \tag{6.51}$$

For example, if μ is taken as Lebesgue measure, then f^* can be identified with that radial function F defined on $\mathbb{R}^n$ having the property that for all $t > 0$, $\{F > t\}$ is a ball centered at the origin whose Lebesgue measure is equal to $\mu\big(\{x : |f(x)| > t\}\big)$. Note that both f^* and A_f are nonincreasing and right-continuous. Since A_f is right-continuous, it follows that the infimum in (6.51) is attained. Therefore, if

$$f^*(t) = \alpha, \quad \text{then } A_f(\alpha') > t, \text{ where } \alpha' < \alpha. \tag{6.52}$$

Furthermore,

$$f^*(t) > \alpha \quad \text{if and only if} \quad t < A_f(\alpha).$$

Thus, it follows that $\{t : f^*(t) > \alpha\}$ is equal to the interval $\big(0, A_f(\alpha)\big)$. Hence, $A_f(\alpha) = \lambda(\{f^* > \alpha\})$, which implies that f and f^* have the same distribution function. Consequently, in view of Theorem 6.60, observe that for all $1 \le p \le \infty$,

$$\|f^*\|_p = \left(\int_0^\infty |f^*(t)|^p \, dt\right)^{1/p} = \left(\int_{\mathbb{R}^n} |f(x)|^p \, d\mu\right)^{1/p} = \|f\|_{p;\mu}. \tag{6.53}$$

6.65. REMARK. Observe that the L^p norm of f relative to the measure μ is thus expressed as the norm of f^* relative to Lebesgue measure.

Notice also that right continuity implies

$$A_f\big(f^*(t)\big) \le t \qquad \text{for all } t > 0. \tag{6.54}$$

6.66. LEMMA. *For all $t > 0$, $\sigma > 0$, suppose an arbitrary function $f \in L^p(\mathbb{R}^n)$ is decomposed as follows: $f = f^t + f_t$, where*

$$f^t(x) = \begin{cases} f(x) & \text{if } |f(x)| > f^*(t^\sigma), \\ 0 & \text{if } |f(x)| \le f^*(t^\sigma), \end{cases}$$

and $f_t := f - f^t$. Then

$$\begin{aligned} (f^t)^*(y) &\le f^*(y) && \text{if } 0 \le y \le t^\sigma, \\ (f^t)^*(y) &= 0 && \text{if } y > t^\sigma, \\ (f_t)^*(y) &\le f^*(y) && \text{if } y > t^\sigma, \\ (f_t)^*(y) &\le f^*(t^\sigma) && \text{if } 0 \le y \le t^\sigma. \end{aligned} \tag{6.55}$$

PROOF. We will prove only the first set, since the proof of the other set is similar.

For the first inequality, let $[f^t]^*(y) = \alpha$ as in (6.52), and similarly, let $f^*(y) = \alpha'$. If it were the case that $\alpha' < \alpha$, then we would have $A_{f^t}(\alpha') > y$. But by the definition of f^t,

$$\{|f^t| > \alpha'\} \subset \{|f| > \alpha'\},$$

which would imply

$$y < A_{f^t}(\alpha') \le A_f(\alpha') \le y,$$

a contradiction.

In the second inequality, assume $y > t^\sigma$. Now $f^t = f\chi_{\{|f|>f^*(t^\sigma)\}}$ and $f^*(t^\sigma) = \alpha$, where $A_f(\alpha) \le t^\sigma$ as in (6.52). Thus, $A_f[f^*(t^\sigma)] = A_f(\alpha) \le t^\sigma$, and therefore

$$\mu(\{|f^t| > \alpha'\}) = \mu(\{|f| > f^*(t^\sigma)\}) \le t^\sigma < y$$

for all $\alpha' > 0$. This implies $(f^t)^*(y) = 0$. $\square$

6.67. DEFINITION. Suppose $(X\mathcal{M}, \mu)$ is a measure space and let (p, q) be a pair of numbers such that $1 \le p,\, q < \infty$. Also, let μ be a Radon measure defined on X and suppose T is a subadditive operator defined on $L^p(X)$ whose values are μ-measurable functions. Thus, $T(f)$ is a μ-measurable function on

X, and we will write $Tf := T(f)$. The operator T is said to be of **weak type** (p, q) if there is a constant C such that for all $f \in L^p(X, \mu)$ and $\alpha > 0$,

$$\mu(\{x : |(Tf)(x)| > \alpha\}) \leq (\alpha^{-1}C\|f\|_{p;\mu})^q.$$

An operator T is said to be of **strong type** (p, q) if there is a constant C such that $\|Tf\|_{q;\mu} \leq C\,\|f\|_{p;\mu}$ for all $f \in L^p(X, \mu)$.

6.68. THEOREM (Marcinkiewicz interpolation theorem). *Let (p_0, q_0) and (p_1, q_1) be pairs of numbers such that $1 \leq p_i \leq q_i < \infty$, $i = 0, 1$, and $q_0 \neq q_1$. Let μ be a Radon measure defined on $\mathbb{R}^n$ and suppose T is a subadditive operator defined on $L^{p_0}(\mathbb{R}^n) + L^{p_1}(\mathbb{R}^n)$ whose values are μ-measurable functions. Suppose T is simultaneously of weak types (p_0, q_0) and (p_1, q_1). If $0 < \theta < 1$ and*

$$\begin{aligned} 1/p &= \frac{1-\theta}{p_0} + \frac{\theta}{p_1}, \\ 1/q &= \frac{1-\theta}{q_0} + \frac{\theta}{q_1}, \end{aligned} \tag{6.56}$$

then T is of strong type (p, q); that is,

$$\|Tf\|_{q;\mu} \leq C\,\|f\|_{p;\mu}, \qquad f \in L^p(\mathbb{R}^n),$$

where $C = C(p_0, q_0, p_1, q_1, \theta)$.

PROOF. The easiest case arises when $p_0 = p_1$; it is left as an exercise.

Henceforth, assume $p_0 < p_1$. Let $(Tf)^*(t) = \alpha$ as in (6.52). Then for $\alpha' < \alpha$, one has $A_{Tf}(\alpha') > t$. The weak type (p_0, q_0) assumption on T implies

$$\begin{aligned} \alpha' &\leq C_0\big(A_{Tf}(\alpha')\big)^{-1/q_0}\,\|f\|_{p_0;\mu} \\ &< C_0 t^{-1/q_0}\,\|f\|_{p_0;\mu} \end{aligned}$$

whenever $f \in L^{p_0}(\mathbb{R}^n)$. Since $\alpha' < C_0 t^{-1/q_0}\,\|f\|_{p_0;\mu}$ for all $\alpha' < \alpha = (Tf)^*(t)$, it follows that

$$(Tf)^*(t) \leq C_0 t^{-1/q_0}\,\|f\|_{p_0;\mu}\,. \tag{6.57}$$

A similar argument shows that if $f \in L^{p_1}(\mathbb{R}^n)$, then

$$(Tf)^*(t) \leq C_1 t^{-1/q_1}\,\|f\|_{p_1;\mu}\,. \tag{6.58}$$

We now appeal to Lemma 6.66, where σ is taken as

$$\sigma := \frac{1/q_0 - 1/q}{1/p_0 - 1/p} = \frac{1/q - 1/q_1}{1/p - 1/p_1}. \tag{6.59}$$

Recall the decomposition $f = f^t + f_t$; since $p_0 < p < p_1$, observe from (6.55) that $f^t \in L^{p_0}(\mathbb{R}^n)$ and $f_t \in L^{p_1}(\mathbb{R}^n)$. Also, we leave the following as Exercise 6.3:

$$(Tf)^*(t) \leq (Tf^t)^*(t/2) + (Tf_t)^*(t/2). \tag{6.60}$$

Since $p_i \le q_i$, $i = 0, 1$, by (6.56) we have $p \le q$. Thus, we obtain

$$\begin{aligned}
\|(Tf)^*\|_q &= \left(\int_0^\infty (t^{1/q}(Tf)^*(t))^q \frac{dt}{t}\right)^{1/q} \\
&\le C\left(\int_0^\infty (t^{1/q}(Tf)^*(t))^p \frac{dt}{t}\right)^{1/p} && \text{by Lemma 6.63} \\
&\le C\left(\int_0^\infty (t^{1/q}(Tf^t)^*(t))^p \frac{dt}{t}\right)^{1/p} \\
&\quad + C\left(\int_0^\infty (t^{1/q}(Tf_t)^*(t))^p \frac{dt}{t}\right)^{1/p} && \text{by (6.60)} \qquad (6.61) \\
&\le C\left(\int_0^1 (t^{1/q-1/q_0}\|f^t\|_{p_0})^p \frac{dt}{t}\right)^{1/p} && \text{by (6.57)} \qquad (6.62) \\
&\quad + C\left(\int_0^1 (t^{1/q-1/q_1}\|f_t\|_{p_1})^p \frac{dt}{t}\right)^{1/p} && \text{by (6.58)}. \qquad (6.63)
\end{aligned}$$

With σ defined by (6.59) we estimate the last two integrals with an appeal to Lemma 6.63 and write

$$\begin{aligned}
\|f^t\|_{p_0} &= \left(\int_0^\infty (y^{1/p_0}(f^t)^*(y))^{p_0} \frac{d\lambda(y)}{y}\right)^{1/p_0} \\
&\le C\int_0^{t^\sigma} y^{1/p_0}(f^t)^*(y)\frac{d\lambda(y)}{y} \\
&\le C\int_0^{t^\sigma} y^{1/p_0} f^*(y)\frac{d\lambda(y)}{y}.
\end{aligned}$$

Inserting this estimate for $\|f^t\|_p$ into (6.62), we obtain

$$\begin{aligned}
&\left(\int_0^1 \left(t^{1/q-1/q_0}\|f^t\|_{p_0}\right)^p \frac{dt}{t}\right)^{1/p} \\
&\le C\left(\int_0^1 \left((t^{1/q-1/q_0}\int_0^{t^\sigma} y^{1/p_0} f^*(y)\frac{d\lambda(y)}{y}\right)^p \frac{dt}{t}\right)^{1/p}.
\end{aligned}$$

Thus, to estimate (6.62) we analyze

$$\int_0^\infty \left(t^{1/q-1/q_0}\int_0^{t^\sigma} y^{1/p_0} f^*(y)\frac{d\lambda(y)}{y}\right)^p \frac{dt}{t},$$

which, under the change of variables $t^\sigma \mapsto s$, becomes

$$\frac{1}{\sigma}\int_0^\infty \left(s^{1/p-1/p_0}\int_0^s y^{1/p_0-1} f^*(y)\,d\lambda(y)\right)^p \frac{ds}{s},$$

which is equal to

$$\frac{1}{\sigma}\int_0^\infty \left(s^{1/p-1/p_0-1/p}\int_0^s y^{1/p_0-1} f^*(y)\,d\lambda(y)\right)^p ds.$$

Now apply Hardy's inequality (6.47) with $-r-1=-p/p_0$ and $f(y)=y^{1/p_0-1}f^*(y)$ to obtain

$$\left(\int_0^1 (t^{1/q-1/q_0}\left\|f^t\right\|_{p_0;\mu})^p\frac{dt}{t}\right)^{1/p} \le C(p,r)\left(\int_0^\infty f(y)^p y^{p-r-1}\right)^{1/p}$$
$$= C(p,r)\left\|f^*\right\|_p .$$

The estimate of

$$\int_0^\infty \left(t^{1/q-1/q_1}\int_0^{t^\sigma} y^{1/p_1}f^*(y)\frac{dy}{y}\right)^p\frac{dt}{t}$$

proceeds in a similar way, and thus our result is established. □

Exercises for Section 6.13

1. Let f be a measurable function on a measure space $(X,\mathcal{M})$. Define $A_f(s)=\mu(\{|f|>s\})$. The **nonincreasing rearrangement of** f on $(0,\infty)$ is defined as
$$f^*(t):=\inf\{s:A_f(s)\le t\}.$$
Prove the following:
(i) f^* is continuous from the right.
(ii) $A_{f^*}(s)=A_f(s)$ for all s if μ is Lebesgue measure on $\mathbb{R}^n$.
2. Prove the Marcinkiewicz interpolation theorem in the case that $p_0=p_1$.
3. If $f=f_1+f_2$, prove that

(6.64) $$(Tf)^*(t)\le(Tf_1)^*(t/2)+(Tf_2)^*(t/2).$$

CHAPTER 7

Differentiation

7.1. Covering Theorems

Certain covering theorems, such as the Vitali covering theorem, will be developed in this section. These covering theorems are of essential importance in the theory of differentiation of measures.

We depart from the theory of abstract measure spaces encountered in previous chapters and focus on certain aspects of functions defined in $\mathbb{R}$. A major result in elementary analysis is the fundamental theorem of calculus, which states that a C^1 function can be expressed as the integral of its derivative. One of the main objectives of this chapter is to show that this result still holds for a more general class of functions. In fact, we will determine precisely those functions for which the fundamental theorem holds. We will take a broader view of differentiation by developing a framework for differentiation of measures. This will include the usual notion of differentiability of a function. The following result, whose proof is left as an exercise, will serve to motivate our point of view.

7.1. REMARK. Suppose μ is a Borel measure on $\mathbb{R}$ and let

$$F(x) = \mu((-\infty, x]) \quad \text{for} \quad x \in \mathbb{R}.$$

Then the following two statements are equivalent:

(i) F is differentiable at x_0 and $F'(x_0) = c$.

(ii) For every $\varepsilon > 0$ there exists $\delta > 0$ such that

$$\left| \frac{\mu(I)}{\lambda(I)} - c \right| < \varepsilon$$

whenever I is a half-open interval whose left or right endpoint is x_0 and $\lambda(I) < \delta$.

Condition (ii) may be interpreted as the derivative of μ with respect to Lebesgue measure, λ. This concept will be developed more fully throughout this chapter. As an example in this framework, let $f \in L^1(\mathbb{R})$ be nonnegative and define a measure μ by

$$\mu(E) = \int_E f(y)\, d\lambda(y) \tag{7.1}$$

W.P. Ziemer, *Modern Real Analysis*, Graduate Texts in Mathematics 278,
https://doi.org/10.1007/978-3-319-64629-9_7

for every Lebesgue measurable set E. Then the function F introduced above can be expressed as

$$F(x) = \int_{-\infty}^{x} f(y)\, d\lambda(y).$$

Of course, the derivative of F at x_0 is the limit

$$\lim_{h\to 0} \frac{1}{h} \int_{x_0}^{x_0+h} f(y)\, d\lambda(y). \tag{7.2}$$

This, in turn, is equivalent to statement (ii) above. Given $f \in L^1(\mathbb{R}^n)$, we define μ as in (7.1) and we consider

$$\lim_{r\to 0} \frac{\mu[B(x_0,r)]}{\lambda[B(x_0,r)]} = \lim_{r\to 0} \frac{1}{\lambda[B(x_0,r)]} \int_{B(x_0,r)} f(y)\, d\lambda(y). \tag{7.3}$$

At this stage we know nothing about the existence of the limit.

7.2. Remark. Strictly speaking, (7.3) is not the precise analogue of (7.2), since we have considered only balls $B(x_0, r)$ centered at x_0. In $\mathbb{R}$ this would exclude the use of intervals whose left endpoint is x_0 as required by (7.2). Nevertheless, in our development we choose to use the family of open concentric balls for several reasons. First, they are slightly easier to employ than nonconcentric balls; second, we will see that it is immaterial to the main results of the theory whether or not concentric balls are used (see Theorem 7.17). Finally, in the development of the derivative of μ relative to an arbitrary measure ν, it is important that concentric balls be used. Thus, we formally introduce the notation

$$D_\lambda \mu(x_0) = \lim_{r\to 0} \frac{\mu[B(x_0,r)]}{\lambda[B(x_0,r)]}, \tag{7.4}$$

which is the **derivative of μ with respect to λ**.

One of the major objectives of the next section is to prove that the limit in (7.4) exists λ-almost everywhere and to see how it relates to the results surrounding the Radon–Nikodym theorem. In particular, in view of Theorems 4.32 and (7.1), it will follow from our development that a nondecreasing function is differentiable almost everywhere. In the following we will use the following notation: Given $B = B(x, r)$ we will call $B(x, 5r)$ the **enlargement** of B and denote it by $\widehat{B}$. The next lemma states that every collection of balls (which may be either open or closed) whose radii are bounded has a countable disjoint subcollection with the property that the union of their enlargements contains the union of the original collection. We emphasize here that the point of the lemma is that the subcollection consists of **disjoint** elements, a very important consideration, since countable additivity plays a central role in measure theory.

7.3. THEOREM. *Let $\mathcal{G}$ be a family of closed or open balls in $\mathbb{R}^n$ with*

$$R := \sup\{\operatorname{diam} B : B \in \mathcal{G}\} < \infty.$$

Then there is a countable subfamily $\mathcal{F} \subset \mathcal{G}$ of mutually disjoint elements such that

$$\bigcup\{B : B \in \mathcal{G}\} \subset \bigcup\{\widehat{B} : B \in \mathcal{F}\}.$$

In fact, for each $B \in \mathcal{G}$ there exists $B' \in \mathcal{F}$ such that $B \cap B' \neq 0$ and $B \subset \widehat{B}'$.

PROOF. Throughout this proof, we will adopt the following notation: if A is a set and $\mathcal{F}$ a family of sets, then we use the notation $A \cap \mathcal{F} \neq \emptyset$ to mean that $A \cap B \neq \emptyset$ for some $B \in \mathcal{F}$.

Let a be a number such that $1 < a < a^{1+2R} < 2$. For $j = 1, 2, \ldots$ let

$$\mathcal{G}_j = \left\{B := B(x,r) \in \mathcal{G} : a^{|x|-j} < \frac{r}{R} \leq a^{|x|-j+1}\right\},$$

and observe that $\mathcal{G} = \bigcup_{j=1}^{\infty} \mathcal{G}_j$. Since $r/R < 1$ and $a > 1$, observe that for every $B = B(x,r) \in \mathcal{G}_j$ we have

$$|x| < j \quad \text{and} \quad r > a^{-j}R.$$

Hence the elements of $\mathcal{G}_j$ are centered at points $x \in B(0,j)$, and their radii are bounded away from zero; this implies that there is a number $M_j > 0$ depending only on a, j, and R such that every disjoint subfamily of $\mathcal{G}_j$ has at most M_j elements.

The family $\mathcal{F}$ will be of the form

$$\mathcal{F} = \bigcup_{j=0}^{\infty} \mathcal{F}_j$$

where the $\mathcal{F}_j$ are finite disjoint families defined inductively as follows. We set $\mathcal{F}_0 = \emptyset$. Let $\mathcal{F}_1$ be the largest (in the sense of inclusion) disjoint subfamily of $\mathcal{G}_1$. Note that $\mathcal{F}_1$ can have no more than M_1 elements. Proceeding by induction, we assume that $\mathcal{F}_{j-1}$ has been determined, and then define $\mathcal{H}_j$ as the largest disjoint subfamily of $\mathcal{G}_j$ with the property that $B \cap \mathcal{F}_{j-1} = \emptyset$ for each $B \in \mathcal{H}_j$. Note that the number of elements in $\mathcal{H}_j$ could be 0 but no more than M_j. Define

$$\mathcal{F}_j := \mathcal{F}_{j-1} \cup \mathcal{H}_j.$$

We claim that the family $\mathcal{F} := \bigcup_{j=1}^{\infty} \mathcal{F}_j$ has the required properties: that is, we will show that

$$B \subset \bigcup\{\widehat{B} : B \in \mathcal{F}\} \quad \text{for each } B \in \mathcal{G}. \tag{7.5}$$

To verify this, first note that $\mathcal{F}$ is a disjoint family. Next, select $B := B(x,r) \in \mathcal{G}$, which implies $B \in \mathcal{G}_j$ for some j. If $B \cap \mathcal{H}_j \neq \emptyset$, then there exists $B' := B(x',r') \in \mathcal{H}_j$ such that $B \cap B' \neq \emptyset$, in which case

$$\frac{r'}{R} \geq a^{|x'|-j}. \tag{7.6}$$

On the other hand, if $B \cap \mathcal{H}_j = \emptyset$, then $B \cap \mathcal{F}_{j-1} \neq \emptyset$, for otherwise, the maximality of $\mathcal{H}_j$ would be violated. Thus, there exists $B' = B(x', r') \in \mathcal{F}_{j-1}$ such that $B \cap B' \neq \emptyset$, and in this case,

$$\frac{r'}{R} > a^{|x'|-j+1} > a^{|x'|-j}. \tag{7.7}$$

Since

$$\frac{r}{R} \leq a^{|x|-j+1},$$

it follows from (7.6) and (7.7) that

$$r \leq a^{|x|-j+1} R \leq a^{(|x|-|x'|+1)} r'.$$

Since a was chosen such that $1 < a < a^{1+2R} < 2$, we have

$$r \leq a^{1+|x|-|x'|} r' \leq a^{1+|x-x'|} r' \leq a^{1+2R} r' \leq 2r'.$$

This implies that $B \subset \widehat{B}'$, because if $z \in B(x, r)$ and $y \in B \cap B'$, then

$$\begin{aligned} |z - x'| &\leq |z - x| + |x - y| + |y - x'| \\ &\leq r + r + r' \\ &\leq 5r'. \end{aligned}$$

□

If we assume a bit more about $\mathcal{G}$, we can show that the union of elements in $\mathcal{F}$ contains almost all of the union $\bigcup\{B : B \in \mathcal{G}\}$. This requires the following definition.

7.4. DEFINITION. A collection $\mathcal{G}$ of balls is said to cover a set $E \subset \mathbb{R}^n$ **in the sense of Vitali** if for each $x \in E$ and each $\varepsilon > 0$, there exists $B \in \mathcal{G}$ containing x whose radius is positive and less than ε. We also say that $\mathcal{G}$ is a **Vitali covering** of E. Note that if $\mathcal{G}$ is a Vitali covering of a set $E \subset \mathbb{R}^n$ and $R > 0$ is arbitrary, then $\mathcal{G} \bigcap \{B : \text{diam} B < R\}$ is also a Vitali covering of E.

7.5. THEOREM. *Let $\mathcal{G}$ be a family of closed balls that covers a set $E \subset \mathbb{R}^n$ in the sense of Vitali. Then with $\mathcal{F}$ as in Theorem 7.3, we have*

$$E \setminus \bigcup\{B : B \in \mathcal{F}^*\} \subset \bigcup\{\widehat{B} : B \in \mathcal{F} \setminus \mathcal{F}^*\}$$

for each finite collection $\mathcal{F}^* \subset \mathcal{F}$.

PROOF. Since $\mathcal{G}$ is a Vitali covering of E, there is no loss of generality if we assume that the radius of each ball in $\mathcal{G}$ is less than some fixed number R. Let $\mathcal{F}$ be as in Theorem 7.3 and let $\mathcal{F}^*$ be any finite subfamily of $\mathcal{F}$. Since $R^n \setminus \cup\{B : B \in \mathcal{F}^*\}$ is open, for each $x \in E \setminus \cup\{B : B \in \mathcal{F}^*\}$ there exists $B \in \mathcal{G}$ such that $x \in B$ and $B \cap [\cup\{B : B \in \mathcal{F}^*\}] = \emptyset$. From Theorem 7.3, there is $B_1 \in \mathcal{F}$ such that $B \cap B_1 \neq \emptyset$ and $\widehat{B}_1 \supset B$. Since $\mathcal{F}^*$ is disjoint, it follows that $B_1 \notin \mathcal{F}^*$, since $B \cap B_1 \neq \emptyset$. Therefore,

$$x \in \widehat{B}_1 \subset \bigcup\{\widehat{B} : B \in \mathcal{F} \setminus \mathcal{F}^*\}.$$

□

7.6. Remark. The preceding result and the next one are not needed in the sequel, although they are needed in some of the exercises, such as Exercise 3, Section 7.9. We include them because they are frequently used in the analysis literature and because they follow so easily from the main result, Theorem 7.3.

Theorem 7.5 states that every finite family $\mathcal{F}^* \subset \mathcal{F}$ along with the enlargements of $\mathcal{F} \setminus \mathcal{F}^*$ provides a covering of E. But what covering properties does $\mathcal{F}$ itself have? The next result shows that $\mathcal{F}$ covers almost all of E.

7.7. Theorem. *Let $\mathcal{G}$ be a family of closed balls that covers a (possibly nonmeasurable) set $E \subset \mathbb{R}^n$ in the sense of Vitali. Then there exists a countable disjoint subfamily $\mathcal{F} \subset \mathcal{G}$ such that*

$$\lambda\left(E \setminus \bigcup\{B : B \in \mathcal{F}\}\right) = 0.$$

Proof. First, assume that E is a bounded set. Then we may as well assume that each ball in $\mathcal{G}$ is contained in some bounded open set $H \supset E$. Let $\mathcal{F}$ be the subfamily of disjoint balls provided by Theorem 7.3 and Theorem 7.5. Since all elements of $\mathcal{F}$ are disjoint and contained in the bounded set H, we have

$$\sum_{B \in \mathcal{F}} \lambda(B) \leq \lambda(H) < \infty. \tag{7.8}$$

Now, by Theorem 7.5, for any finite subfamily $\mathcal{F}^* \subset \mathcal{F}$, we obtain

$$\begin{aligned}
\lambda^*\left(E \setminus \cup\{B : B \in \mathcal{F}\}\right) &\leq \lambda^*\left(E \setminus \cup\{B : B \in \mathcal{F}^*\}\right) \\
&\leq \lambda^*\left(\cup\{\widehat{B} : B \in \mathcal{F} \setminus \mathcal{F}^*\right) \\
&\leq \sum_{B \in \mathcal{F}\setminus\mathcal{F}^*} \lambda(\widehat{B}) \\
&\leq 5^n \sum_{B \in \mathcal{F}\setminus\mathcal{F}^*} \lambda(B).
\end{aligned}$$

Referring to (7.8), we see that the last term can be made arbitrarily small by an appropriate choice of $\mathcal{F}^*$. This establishes our result in the case that E is bounded.

The general case can be handled by observing that there is a countable family $\{C_k\}_{k=1}^{\infty}$ of disjoint open cubes C_k such that

$$\lambda\left(\mathbb{R}^n \setminus \bigcup_{k=1}^{\infty} C_k\right) = 0.$$

The details are left to the reader. □

Exercises for Section 7.1

1. (a) Prove that for each open set $U \subset \mathbb{R}^n$, there exists a collection $\mathcal{F}$ of disjoint closed balls contained in U such that
$$\lambda\,(U \setminus \bigcup\{B : B \in \mathcal{F}\}) = 0.$$
   (b) Thus, $U = \bigcup_{B\in\mathcal{F}} B \cup N$ where $\lambda(N) = 0$. Prove that $N \neq \emptyset$. (Hint: To show that $N \neq \emptyset$, consider the proof in $\mathbb{R}^2$. Then consider the intersection of U with a line. This intersection is an open subset of the line, and note that the closed balls become closed intervals.)
2. Let μ be a finite Borel measure on $\mathbb{R}^n$ with the property that $\mu[B(x, 2r)] \leq c\mu[B(x, r)]$ for all $x \in \mathbb{R}^n$ and all $0 < r < \infty$, where c is a constant independent of x and r. Prove that the Vitali covering theorem, Theorem 7.7, is valid with λ replaced by μ.
3. Supply the proof of Theorem 7.1.
4. Prove the following alternative version of Theorem 7.7. Let $E \subset \mathbb{R}^n$ be an arbitrary set, possibly nonmeasurable. Suppose $\mathcal{G}$ is a family of closed cubes with the property that for each $x \in E$ and each $\varepsilon > 0$ there exists a cube $C \in \mathcal{G}$ containing x whose diameter is less than ε. Prove that there exists a countable disjoint subfamily $\mathcal{F} \subset \mathcal{G}$ such that
$$\lambda(E - \bigcup\{C : C \in \mathcal{F}\}) = 0.$$
5. With the help of the preceding exercise, prove that for each open set $U \subset \mathbb{R}^n$, there exists a countable family $\mathcal{F}$ of closed disjoint cubes, each contained in U, such that
$$\lambda(U \setminus \bigcup\{C : C \in \mathcal{F}\}) = 0.$$

7.2. Lebesgue Points

In integration theory, functions that differ only on a set of measure zero can be identified as one function. Consequently, with this identification a measurable function determines an equivalence class of functions. This raises the question whether it is possible to define a measurable function at almost all points in a way that is independent of any representative in the equivalence class. Our investigation of Lebesgue points provides a positive answer to this question.

7.8. DEFINITION. With each $f \in L^1(\mathbb{R}^n)$, we associate its **maximal function**, $\boldsymbol{M}f$, which is defined as
$$\boldsymbol{M}f(x) := \sup_{r>0} \fint_{B(x,r)} |f| \, d\lambda,$$

where

$$\fint_E |f|\, d\lambda := \frac{1}{\lambda[E]} \int_E |f|\, d\lambda$$

denotes the **integral average** of $|f|$ over an arbitrary measurable set E. In other words, $\boldsymbol{M}f(x)$ is the upper envelope of integral averages of $|f|$ over balls centered at x.

Clearly, $\boldsymbol{M}f\colon \mathbb{R}^n \to \overline{\mathbb{R}}$ is a nonnegative function. Furthermore, it is Lebesgue measurable. To see this, note that for each fixed $r > 0$,

$$x \mapsto \fint_{B(x,r)} |f|\, d\lambda$$

is a continuous function of x (see Exercise 4, Section 7.2). Therefore, we see that $\{\boldsymbol{M}f > t\}$ is an open set for each real number t, thus showing that $\boldsymbol{M}f$ is lower semicontinuous and therefore measurable.

The next question is whether $\boldsymbol{M}f$ is integrable over $\mathbb{R}^n$. In order for this to be true, it follows from Theorem 6.59 that it would be necessary that

$$\int_0^\infty \lambda(\{\boldsymbol{M}f > t\})\, d\lambda(t) < \infty. \tag{7.9}$$

It turns out that $\boldsymbol{M}f$ is never integrable unless f is identically zero (see Exercise 3, Section 7.2). However, the next result provides an estimate of how the measure of the set $\{\boldsymbol{M}f > t\}$ becomes small as t increases. It also shows that inequality (7.9) fails to be true by only a small margin.

7.9. THEOREM (Hardy–Littlewood). *If $f \in L^1(\mathbb{R}^n)$, then*

$$\lambda[\{\boldsymbol{M}f > t\}] \leq \frac{5^n}{t} \int_{\mathbb{R}^n} |f|\, d\lambda$$

for every $t > 0$.

PROOF. For fixed $t > 0$, the definition implies that for each $x \in \{\boldsymbol{M}f > t\}$ there exists a ball B_x centered at x such that

$$\fint_{B_x} |f|\, d\lambda > t,$$

or equivalently,

$$\frac{1}{t} \int_{B_x} |f|\, d\lambda > \lambda(B_x). \tag{7.10}$$

Since f is integrable and t is fixed, the radii of all balls satisfying (7.10) are bounded. Thus, with $\mathcal{G}$ denoting the family of these balls, we may appeal to Theorem 7.3 to obtain a countable subfamily $\mathcal{F} \subset \mathcal{G}$ of disjoint balls such that

$$\{\boldsymbol{M}f > t\} \subset \bigcup\{\widehat{B} : B \in \mathcal{F}\}.$$

Therefore,

$$\begin{aligned}\lambda(\{\boldsymbol{M}f > t\}) &\le \lambda\Big(\bigcup_{B\in\mathcal{F}} \widehat{B}\Big)\\ &\le \sum_{B\in\mathcal{F}} \lambda(\widehat{B})\\ &= 5^n \sum_{B\in\mathcal{F}} \lambda(B)\\ &< \frac{5^n}{t} \sum_{B\in\mathcal{F}} \int_B |f|\, d\lambda\\ &\le \frac{5^n}{t} \int_{\mathbb{R}^n} |f|\, d\lambda\end{aligned}$$

which establishes the desired result. □

We now appeal to the results of Section 6.13 concerning the Marcinkiewicz interpolation theorem. Clearly, the operator $\boldsymbol{M}$ is subadditive, and our previous result shows that it is of weak type $(1,1)$ (see Definition 6.67). Also, it is clear that

$$\|\boldsymbol{M}f\|_\infty \le \|f\|_\infty$$

for all $f \in L^\infty$. Therefore, we appeal to the Marcinkiewicz interpolation theorem to conclude that M is of strong type (p,p). That is, we have the following corollary.

7.10. Corollary. *There exists a constant $C > 0$ such that*

$$\|\boldsymbol{M}f\|_p \le Cp(p-1)^{-1} \|f\|_p$$

whenever $1 < p < \infty$ and $f \in L^p(\mathbb{R}^n)$.

If $f \in L^1_{\text{loc}}(\mathbb{R}^n)$ is continuous, then it follows from elementary considerations that

$$\lim_{r\to 0} \fint_{B(x,r)} f(y)\, d\lambda(y) = f(x) \quad \text{for} \quad x \in \mathbb{R}^n. \tag{7.11}$$

Since Lusin's theorem tells us that a measurable function is almost continuous, one might suspect that (7.11) is true in some sense for an integrable function. Indeed, we have the following.

7.11. Theorem. *If $f \in L^1_{loc}(\mathbb{R}^n)$, then*

$$\lim_{r\to 0} \fint_{B(x,r)} f(y)\, d\lambda(y) = f(x) \tag{7.12}$$

for a.e. $x \in \mathbb{R}^n$.

Proof. Since the limit in (7.12) depends only on the values of f in an arbitrarily small neighborhood of x, and since $\mathbb{R}^n$ is a countable union of bounded measurable sets, we may assume without loss of generality that f

vanishes on the complement of a bounded set. Choose $\varepsilon > 0$. From Exercise 12, Section 6.5, we can find a continuous function $g \in L^1(\mathbb{R}^n)$ such that

$$\int_{\mathbb{R}^n} |f(y) - g(y)| \, d\lambda(y) < \varepsilon.$$

For each such g we have

$$\lim_{r \to 0} \fint_{B(x,r)} g(y) \, d\lambda(y) = g(x)$$

for every $x \in \mathbb{R}^n$. This implies

$$\begin{aligned}
&\limsup_{r \to 0} \left| \fint_{B(x,r)} f(y) \, d\lambda(y) - f(x) \right| \\
&= \limsup_{r \to 0} \left| \fint_{B(x,r)} [f(y) - g(y)] \, d\lambda(y) \right. \\
&\qquad \left. + \left(\fint_{B(x,r)} g(y) \, d\lambda(y) - g(x) \right) + [g(x) - f(x)] \right| \\
&\leq M(f-g)(x) + 0 + |f(x) - g(x)| \, .
\end{aligned} \tag{7.13}$$

For each positive number t let

$$E_t = \{x : \limsup_{r \to 0} \left| \fint_{B(x,r)} f(y) \, d\lambda(y) - f(x) \right| > t\},$$
$$F_t = \{x : |f(x) - g(x)| > t\},$$

and

$$H_t = \{x : M(f-g)(x) > t\}.$$

Then by (7.13), $E_t \subset F_{t/2} \cup H_{t/2}$. Furthermore,

$$t\lambda(F_t) \leq \int_{F_t} |f(y) - g(y)| \, d\lambda(y) < \varepsilon,$$

and Theorem 7.9 implies

$$\lambda(H_t) \leq \frac{5^n \varepsilon}{t}.$$

Hence

$$\lambda(E_t) \leq 2\frac{\varepsilon}{t} + 2\frac{5^n \varepsilon}{t}.$$

Since ε is arbitrary, we conclude that $\lambda(E_t) = 0$ for all $t > 0$, thus establishing the conclusion. □

The theorem states that

$$\lim_{r \to 0} \fint_{B(x,r)} f(y) \, d\lambda(y) \tag{7.14}$$

exists for a.e. x and that the limit defines a function that is equal to f almost everywhere. The limit in (7.14) provides a way to define the value of f at x

that is independent of the choice of representative in the equivalence class of f. Observe that (7.12) can be written as

$$\lim_{r\to 0} \fint_{B(x,r)} [f(y) - f(x)]\, d\lambda(y) = 0.$$

It is rather surprising that Theorem 7.11 implies the following apparently stronger result.

7.12. THEOREM. *If $f \in L^1_{loc}(\mathbb{R}^n)$, then*

$$\lim_{r\to 0} \fint_{B(x,r)} |f(y) - f(x)|\, d\lambda(y) = 0 \tag{7.15}$$

for a.e. $x \in \mathbb{R}^n$.

PROOF. For each rational number ρ apply Theorem 7.11 to conclude that there is a set E_ρ of measure zero such that

$$\lim_{r\to 0} \fint_{B(x,r)} |f(y) - \rho|\, d\lambda(y) = |f(x) - \rho| \tag{7.16}$$

for all $x \notin E_\rho$. Thus, with

$$E := \bigcup_{\rho\in\mathbb{Q}} E_\rho,$$

we have $\lambda(E) = 0$. Moreover, for $x \notin E$ and $\rho \in \mathbb{Q}$, since $|f(y) - f(x)| < |f(y) - \rho| + |f(x) - \rho|$, (7.16) implies

$$\limsup_{r\to 0} \fint_{B(x,r)} |f(y) - f(x)|\, d\lambda(y) \leq 2\,|f(x) - \rho|\,.$$

Since

$$\inf\{|f(x) - \rho| \ :\ \rho \in \mathbb{Q}\} = 0,$$

the proof is complete. □

A point x for which (7.15) holds is called a **Lebesgue point** of f. Thus, almost all points are Lebesgue points for all $f \in L^1_{\mathrm{loc}}(\mathbb{R}^n)$.

An important special case of Theorem 7.11 occurs when f is taken as the characteristic function of a set. For $E \subset \mathbb{R}^n$ a Lebesgue measurable set, let

$$\overline{D}(E,x) = \limsup_{r\to 0} \frac{\lambda(E \cap B(x,r))}{\lambda(B(x,r))}, \tag{7.17}$$

$$\underline{D}(E,x) = \liminf_{r\to 0} \frac{\lambda(E \cap B(x,r))}{\lambda(B(x,r))}. \tag{7.18}$$

7.13. THEOREM (Lebesgue density theorem). *If $E \subset \mathbb{R}^n$ is a Lebesgue measurable set, then*

$$D(E,x) = 1 \quad \textit{for } \lambda\textit{-almost all } x \in E,$$

and

$$D(E,x) = 0 \quad \textit{for } \lambda\textit{-almost all } x \in \widetilde{E}.$$

PROOF. For the first part, let $B(r)$ denote the open ball centered at the origin of radius r and let $f = \chi_{E\cap B(r)}$. Since $E \cap B(r)$ is bounded, it follows that f is integrable, and then Theorem 7.11 implies that $D(E \cap B(r), x) = 1$ for λ-almost all $x \in E \cap B(r)$. Since r is arbitrary, the result follows.

For the second part, take $f = \chi_{\widetilde{E}\cap B(r)}$ and conclude, as above, that $D(\widetilde{E} \cap B(r), x) = 1$ for λ-almost all $x \in \widetilde{E} \cap B(r)$. Observe that $D(E, x) = 0$ for all such x, and thus the result follows, since r is arbitrary. □

Exercises for Section 7.2

1. Let f be a measurable function defined on $\mathbb{R}^n$ with the property that for some constant C, $f(x) \geq C\,|x|^{-n}$ for $|x| \geq 1$. Prove that f is not integrable on $\mathbb{R}^n$. Hint: One way to proceed is to use Theorem 6.59.
2. Let $f \in L^1(\mathbb{R}^n)$ be a function that does not vanish identically on $B(0, 1)$. Show that $Mf \notin L^1(\mathbb{R}^n)$ by establishing the following inequality for all x with $|x| > 1$:

$$
\begin{aligned}
Mf(x) \geq \fint_{B(x,|x|+1)} |f|\, d\lambda &\geq \frac{1}{C(|x|+1)^n} \int_{B(0,1)} |f|\, d\lambda \\
&> \frac{1}{2^n C\,|x|^n} \int_{B(0,1)} |f|\, d\lambda,
\end{aligned}
$$

   where $C = \lambda[B(0, 1)]$.
3. Prove that the maximal function Mf is not integrable on $\mathbb{R}^n$ unless f is identically 0 (cf. the previous exercise).
4. Let $f \in L^1_{\text{loc}}(\mathbb{R}^n)$. Prove for each fixed $r > 0$ that

$$
\fint_{B(x,r)} |f|\, d\lambda
$$

   is a continuous function of x.

7.3. The Radon–Nikodym Derivative: Another View

We return to the concept of the Radon–Nikodym derivative in the setting of Lebesgue measure on $\mathbb{R}^n$. In this section it is shown that the Radon–Nikodym derivative can be interpreted as a classical limiting process, very similar to that of the derivative of a function.

We now turn to the question of relating the derivative in the sense of (7.4) to the Radon–Nikodym derivative. Consider a σ-finite measure μ on $\mathbb{R}^n$ that is absolutely continuous with respect to Lebesgue measure. The Radon–Nikodym theorem asserts the existence of a measurable function f (the Radon–Nikodym derivative) such that μ can be represented as

$$
\mu(E) = \int_E f(y)\, d\lambda(y)
$$

for every Lebesgue measurable set $E \subset \mathbb{R}^n$. Theorem 7.11 implies that

$$D_\lambda \mu(x) = \lim_{r \to 0} \frac{\mu[B(x,r)]}{\lambda[B(x,r)]} = f(x) \tag{7.19}$$

for λ-a.e. $x \in \mathbb{R}^n$. Thus, the Radon–Nikodym derivatives of μ with respect to λ and $D_\lambda \mu$ agree almost everywhere. Now we turn to measures that are singular with respect to Lebesgue measure.

7.14. THEOREM. *Let σ be a Radon measure that is singular with respect to λ. Then*

$$D_\lambda \sigma(x) = 0$$

for λ-almost all $x \in \mathbb{R}^n$.

PROOF. Since $\sigma \perp \lambda$, we know that σ is concentrated on a Borel set A with $\sigma(\widetilde{A}) = \lambda(A) = 0$. For each positive integer k, let

$$E_k = \widetilde{A} \cap \left\{ x : \limsup_{r \to 0} \frac{\sigma[B(x,r)]}{\lambda[B(x,r)]} > \frac{1}{k} \right\}.$$

In view of Exercise 4.8, we see for fixed r that $\sigma[B(x,r)]$ is lower semicontinuous and therefore that E_k is a Borel set. It suffices to show that

$$\lambda(E_k) = 0 \quad \text{for all} \quad k,$$

because $D_\lambda \sigma(x) = 0$ for all $x \in \tilde{A} - \cup_{k=1}^{\infty} E_k$ and $\lambda(A) = 0$. Referring to Theorems 4.63 and 4.52, it follows that for every $\varepsilon > 0$ there exists an open set $U_\varepsilon \supset E_k$ such that $\sigma(U_\varepsilon) < \varepsilon$. For each $x \in E_k$ there exists a ball $B(x,r)$ with $0 < r < 1$ such that $B(x,r) \subset U_\varepsilon$ and $\lambda[B(x,r)] < k\sigma[B(x,r)]$. The collection of all such balls $B(x,r)$ provides a covering of E_k. Now employ Theorem 7.3 with $R = 1$ to obtain a disjoint collection of balls, $\mathcal{F}$, such that

$$E_k \subset \bigcup_{B \in \mathcal{F}} \widehat{B}.$$

Then

$$\begin{aligned}
\lambda(E_k) &\leq \lambda\left\{ \bigcup_{B \in \mathcal{F}} \widehat{B} \right\} \\
&\leq 5^n \sum_{B \in \mathcal{F}} \lambda(B) \\
&< 5^n k \sum_{B \in \mathcal{F}} \sigma(B) \\
&\leq 5^n k \sigma(U_\varepsilon) \\
&\leq 5^n k \varepsilon.
\end{aligned}$$

Since ε is arbitrary, this shows that $\lambda(E_k) = 0$. □

This result together with (7.19) establishes the following theorem.

7.15. THEOREM. *Suppose ν is a Radon measure on $\mathbb{R}^n$. Let $\nu = \mu + \sigma$ be its Lebesgue decomposition with $\mu \ll \lambda$ and $\sigma \perp \lambda$. Finally, let f denote the Radon–Nikodym derivative of μ with respect to λ. Then*

$$\lim_{r\to 0} \frac{\nu[B(x,r)]}{\lambda[B(x,r)]} = f(x)$$

for λ-a.e. $x \in \mathbb{R}^n$.

7.16. DEFINITION. Now we address the issue raised in Remark 7.2 concerning the use of concentric balls in the definition of (7.4). It can easily be shown that nonconcentric balls or even a more general class of sets could be used. For $x \in \mathbb{R}^n$, a sequence of Borel sets $\{E_k(x)\}$ is called a **regular differentiation basis at** x if there is a number $\alpha_x > 0$ with the following property: there is a sequence of balls $B(x, r_k)$ with $r_k \to 0$ such that $E_k(x) \subset B(x, r_k)$ and

$$\lambda(E_k(x)) \geq \alpha_x \lambda[B(x, r_k)].$$

The sets $E_k(x)$ are in no way related to x except for the condition $E_k \subset B(x, r_k)$. In particular, the sets are not required to contain x.

The next result shows that Theorem 7.15 can be generalized to include regular differentiation bases.

7.17. THEOREM. *Suppose the hypotheses and notation of Theorem 7.15 are in force. Then for λ almost every $x \in \mathbb{R}^n$, we have*

$$\lim_{k\to\infty} \frac{\sigma[E_k(x)]}{\lambda[E_k(x)]} = 0$$

and

$$\lim_{k\to\infty} \frac{\mu[E_k(x)]}{\lambda[E_k(x)]} = f(x)$$

whenever $\{E_k(x)\}$ is a regular differentiation basis at x.

PROOF. In view of the inequalities

$$\frac{\alpha_x \sigma[E_k(x)]}{\lambda[E_k(x)]} \leq \frac{\sigma[E_k(x)]}{\lambda[B(x,r_k)]} \leq \frac{\sigma[B(x,r_k)]}{\lambda[B(x,r_k)]}, \tag{7.20}$$

the first conclusion of the theorem follows from Theorem 7.14.

Concerning the second conclusion, Theorem 7.12 implies

$$\lim_{r_k\to 0} \fint_{B(x,r_k)} |f(y) - f(x)|\, d\lambda(y) = 0$$

for almost all x, and consequently, by the same reasoning as in (7.20),

$$\lim_{k\to\infty} \fint_{E_k(x)} |f(y) - f(x)|\, d\lambda(y) = 0$$

for almost all x. Hence, for almost all x it follows that

$$\lim_{k\to\infty}\frac{\mu[E_k(x)]}{\lambda[E_k(x)]}=\lim_{k\to\infty}\fint_{E_k(x)} f(y)\,d\lambda(y)=f(x). \qquad \square$$

This leads immediately to the following theorem, which is fundamental to the theory of functions of a single variable. For the companion result for functions of several variables, see Theorem 11.1. Also, see Exercise 2, Section 7.3, for a completely different proof.

7.18. THEOREM. *Let $f\colon \mathbb{R}\to\mathbb{R}$ be a nondecreasing function. Then $f'(x)$ exists at λ-a.e. $x\in\mathbb{R}$.*

PROOF. Since f is nondecreasing, Theorem 3.61 implies that f is continuous except possibly on the countable set $D=\{x_1,x_2,\dots\}$. Indeed, from Theorem 3.61 we have

$$f(x_i-)<f(x_i+),$$

for each $x_i\in D$. Define $g:\mathbb{R}\to\mathbb{R}$ as $g(x)=f(x)$ if $x\notin D$ and $g(x)=f(x_i+)$ if $x_i\in D$. Then g is a right-continuous nondecreasing function that agrees with f except on the countable set D (see Exercise 1, Section 7.3). Now refer to Theorems 4.31 and 4.32 to obtain a Borel measure μ such that

$$\mu((a,b])=g(b)-g(a) \tag{7.21}$$

whenever $a<b$. For $x\in\mathbb{R}$ take as a regular differentiation basis an arbitrary sequence of half-open intervals $\{I_k(x)\}$ with $I_k=(x,x+h_k]$, $h_k>0$, $h_k\to 0$. Indeed, note that $\lambda(I_k(x))=\frac{1}{2}\lambda[(x-h_k,x+h_k)]$. From Theorem 7.15 we have the decomposition $\mu=\tilde\mu+\sigma$, where $\tilde\mu\ll\lambda$ and $\sigma\ll\lambda$. Theorem 7.17 and Theorem 7.14 state that $D_\lambda\mu(x)=D_\lambda\tilde\mu(x)(x)+D_\lambda\sigma(x)=D_\lambda\tilde\mu(x)(x)$, for λ-almost every x. Hence, for λ-a.e. x, there exists c_x such that

$$\lim_{k\to\infty}\frac{\mu(I_k(x))}{\lambda(I_k(x))}=\lim_{k\to\infty}\frac{\mu((x,x+h_k])}{\lambda((x,x+h_k])}=c_x.$$

Clearly, the limit above holds for every sequence $h_k\to 0$, $h_k>0$, and thus

$$\lim_{h\to 0+}\frac{\mu((x,x+h])}{\lambda((x,x+h])}=c_x \text{ for } \lambda\text{-a.e. } x. \tag{7.22}$$

In a similar way we see that

$$\lim_{h\to 0+}\frac{\mu((x-h,x])}{\lambda((x-h,x])}=c_x, \text{ for } \lambda\text{-a.e. } x. \tag{7.23}$$

From (7.22), (7.23), and (7.21) we have, for λ-a.e. x,

$$\lim_{h\to 0+}\frac{g(x+h)-g(x)}{h}=c_x=\lim_{h\to 0+}\frac{g(x)-g(x-h)}{h}, \tag{7.24}$$

which means that $g'(x)$ exists for λ-a.e. x and $g'(x)=c_x$ for such x. Let $G:=\{x\in\mathbb{R}: g'(x) \text{ exists and } g(x)=f(x)\}$. Clearly, $\mathbb{R}\setminus G$ is a set of λ-measure zero. We now show that f is differentiable at each $x\in G$ and

$f'(x) = g'(x)$. Consider the sequence $h_k \to 0$, $h_k > 0$. For each h_k choose $0 < h'_k < h_k < h''_k$ such that $\frac{h''_k}{h_k} \to 1$ and $\frac{h'_k}{h_k} = 1$. Then

$$\frac{h'_k}{h_k} \cdot \frac{f(x+h'_k) - f(x)}{h'_k} \leq \frac{f(x+h_k) - f(x)}{h_k} \leq \frac{f(x+h''_k) - f(x)}{h''_k} \cdot \frac{h''_k}{h_k}.$$

Note that h'_k and h''_k can be chosen so that $f(x+h'_k) = g(x+h'_k)$ and $f(x+h''_k) = g(x+h''_k)$. Since $g(x) = f(x)$ for $\in G$, we obtain

$$\frac{h'_k}{h_k} \cdot \frac{g(x+h'_k) - g(x)}{h'_k} \leq \frac{f(x+h_k) - f(x)}{h_k} \leq \frac{g(x+h''_k) - g(x)}{h''_k} \cdot \frac{h''_k}{h_k}.$$

Letting $h_k \to 0$, we obtain

$$\lim_{h_k \to 0+} \frac{f(x+h_k) - f(x)}{h_k} = g'(x). \tag{7.25}$$

The same argument shows that

$$\lim_{h_k \to 0+} \frac{f(x) - f(x-h_k)}{h_k} = g'(x). \tag{7.26}$$

Since (7.25) and (7.26) hold for all $h_k \to 0$, $h_k > 0$, we conclude that $f'(x) = g'(x) = c_x$ for each $x \in G$. □

Another consequence of the above results is the following theorem concerning the derivative of the indefinite integral.

7.19. THEOREM. *Suppose f is a Lebesgue integrable function defined on $[a, b]$. For each $x \in [a, b]$ let*

$$F(x) = \int_a^x f(t)\, d\lambda(t).$$

Then $F' = f$ almost everywhere on $[a, b]$.

PROOF. The derivative $F'(x)$ is given by

$$F'(x) = \lim_{h \to 0} \frac{1}{h} \int_x^{x+h} f(t)\, d\lambda(t).$$

Let μ be the measure defined by

$$\mu(E) = \int_E f\, d\lambda$$

for every measurable set E. Using intervals of the form $I_h(x) = [x, x+h]$ as a regular differentiation basis, it follows from Theorem 7.17 that

$$\lim_{h \to 0} \frac{1}{h} \int_x^{x+h} f(t)\, d\lambda(t) = \lim_{h \to 0} \frac{\mu[I_h(x)]}{\lambda[I_h(x)]} = f(x)$$

for almost all $x \in [a, b]$. □

Exercises for Section 7.3

1. Show that $g\colon \mathbb{R} \to \mathbb{R}$ defined in Theorem 7.18 is a nondecreasing and right-continuous function.
2. Here is an outline of an alternative proof of Theorem 7.18, which states that a nondecreasing function f defined on (a, b) is differentiable (Lebesgue) almost everywhere. For this, we introduce the **Dini derivatives**:

$$D^+ f(x_0) = \limsup_{h\to 0^+} \frac{f(x_0+h)-f(x_0)}{h},$$
$$D_+ f(x_0) = \liminf_{h\to 0^+} \frac{f(x_0+h)-f(x_0)}{h},$$
$$D^- f(x_0) = \limsup_{h\to 0^-} \frac{f(x_0+h)-f(x_0)}{h},$$
$$D_- f(x_0) = \liminf_{h\to 0^-} \frac{f(x_0+h)-f(x_0)}{h}.$$

If all four Dini derivatives are finite and equal, then $f'(x_0)$ exists and is equal to the common value. Clearly, $D_+ f(x_0) \le D^+ f(x_0)$ and $D_- f(x_0) \le D^- f(x_0)$. To prove that f' exists almost everywhere, it suffices to show that the set

$$\{x : D^+ f(x) > D_- f(x)\}$$

has measure zero. A similar argument would apply to any two Dini derivatives. For two rational numbers $r, s > 0$, let

$$E_{r,s} = \{x : D^+ f(x) > r > s > D_- f(x)\}.$$

The proof reduces to showing that $E_{r,s}$ has measure zero. If $x \in E_{r,s}$, then there exist arbitrarily small positive numbers h such that

$$\frac{f(x-h)-f(x)}{-h} < s.$$

Fix $\varepsilon > 0$ and use the Vitali covering theorem to find a countable family of closed, disjoint intervals $[x_k - h_k, x_k]$, $k = 1, 2, \ldots$ such that

$$f(x_k) - f(x_k - h_k) < s h_k,$$
$$\lambda(E_{r,s} \cap \left(\bigcup_{k=1}^{\infty} [x_k - h_k, x_k]\right) = \lambda(E_{r,s}),$$
$$\sum_{k=1}^{m} h_k < (1+\varepsilon)\lambda(E_{r,s}).$$

From this it follows that

$$\sum_{k=1}^{\infty} [f(x_k) - f(x_k - h_k)] < s(1+\varepsilon)\lambda(E_{r,s}).$$

For each point

$$y \in A := E_{r,s} \bigcap \left(\bigcup_{k=1}^{\infty} (x_k - h_k, x_k) \right),$$

there exists an arbitrarily small $h > 0$ such that

$$\frac{f(y+h) - f(y)}{h} > r.$$

Employ the Vitali covering theorem again to obtain a countable family of disjoint closed intervals $[y_j, y_j + h_j]$ such that

$$\begin{gathered} \text{each } [y_j, y_j + h_j] \text{ lies in some } [x_k - h_k, x_k], \\ f(y_j + h_j) - f(y_j) > r h_j \; j = 1, 2, \ldots, \\ \sum_{j=1}^{\infty} h_j \geq \lambda(A). \end{gathered}$$

Since f is nondecreasing, it follows that

$$\sum_{j=1}^{\infty} [f(y_j + h_j) - f(y_j)] \leq \sum_{k=1}^{\infty} [f(x_k) - f(x_k - h_k)],$$

and from this a contradiction is readily reached.

7.4. Functions of Bounded Variation

The main objective of this and the next section is to completely determine the conditions under which the following equation holds on an interval $[a, b]$:

$$f(x) - f(a) = \int_a^x f'(t)\, dt \quad \text{for} \quad a \leq x \leq b.$$

This formula is well known in the context of Riemann integration, and our purpose is to investigate its validity via the Lebesgue integral. It will be shown that the formula is valid precisely for the class of absolutely continuous functions. In this section we begin by introducing functions of bounded variation.

In the elementary version of the fundamental theorem of calculus, it is assumed that f' exists at every point of $[a, b]$ and that f' is continuous. Since the Lebesgue integral is more general than the Riemann integral, one would expect a more general version of the fundamental theorem in the Lebesgue theory. What then would be the necessary assumptions? Perhaps it would be sufficient to assume that f' exists almost everywhere on $[a, b]$ and that $f' \in L^1$. But this is obviously not true in view of the Cantor–Lebesgue function, f; see Example 5.7. We have seen that it is continuous, nondecreasing on $[0, 1]$, and constant on each interval in the complement of the Cantor set. Consequently, $f' = 0$ at each point of the complement, and thus

$$1 = f(1) - f(0) > \int_0^1 f'(t)\, dt = 0.$$

The quantity $f(1) - f(0)$ indicates how much the function varies on $[0, 1]$. Intuitively, one might have guessed that the quantity

$$\int_0^1 |f'| \, d\lambda$$

provides a measurement of the variation of f. Although this is false in general, for what class of functions is it true? We will begin to investigate the ideas surrounding these questions by introducing functions of bounded variation.

7.20. DEFINITIONS. Suppose a function f is defined on $I = [a, b]$. The **total variation** of f from a to x, $x \leq b$, is defined by

$$V_f(a; x) = \sup \sum_{i=1}^{k} |f(t_i) - f(t_{i-1})|,$$

where the supremum is taken over all finite sequences $a = t_0 < t_1 < \cdots < t_k = x$. A function f is said to be of **bounded variation** (abbreviated BV) on $[a, b]$ if $V_f(a; b) < \infty$. If there is no danger of confusion, we will sometimes write $V_f(x)$ in place of $V_f(a; x)$.

Note that if f is of bounded variation on $[a, b]$ and $x \in [a, b]$, then

$$|f(x) - f(a)| \leq V_f(a; x) \leq V_f(a; b),$$

from which we see that f is bounded.

It is easy to see that a bounded function that is either nonincreasing or nondecreasing is of bounded variation. Also, the sum (or difference) of two functions of bounded variation is again of bounded variation. The converse, which is not so immediate, is also true.

7.21. THEOREM. *Suppose f is of bounded variation on $[a, b]$. Then f can be written as*

$$f = f_1 - f_2,$$

where both f_1 and f_2 are nondecreasing.

PROOF. Let $x_1 < x_2 \leq b$ and let $a = t_0 < t_1 < \cdots < t_k = x_1$. Then

$$V_f(x_2) \geq |f(x_2) - f(x_1)| + \sum_{i=1}^{k} |f(t_i) - f(t_{i-1})| . \tag{7.27}$$

Now,

$$V_f(x_1) = \sup \sum_{i=1}^{k} |f(t_i) - f(t_{i-1})|$$

over all sequences $a = t_0 < t_1 < \cdots < t_k = x_1$. Hence,

$$V_f(x_2) \geq |f(x_2) - f(x_1)| + V_f(x_1). \tag{7.28}$$

In particular,

$$V_f(x_2) - f(x_2) \geq V_f(x_1) - f(x_1) \quad \text{and} \quad V_f(x_2) + f(x_2) \geq V_f(x_1) + f(x_1).$$

This shows that $V_f - f$ and $V_f + f$ are nondecreasing functions. The assertions thus follow by taking

$$f_1 = \tfrac{1}{2}(V_f + f) \quad \text{and} \quad f_2 = \tfrac{1}{2}(V_f - f). \qquad \square$$

7.22. THEOREM. *Suppose f is of bounded variation on $[a,b]$. Then f is Borel measurable and has at most a countable number of discontinuities. Furthermore, f' exists almost everywhere on $[a,b]$, f' is Lebesgue measurable,*

$$|f'(x)| = V'(x) \tag{7.29}$$

for a.e. $x \in [a,b]$, *and*

$$\int_a^b |f'(x)| \, d\lambda(t) \leq V_f(b). \tag{7.30}$$

In particular, if f is nondecreasing on $[a,b]$, then

$$\int_a^b f'(x) \, d\lambda(x) \leq f(b) - f(a). \tag{7.31}$$

PROOF. We will first prove (7.31). Assume that f is nondecreasing and extend f by defining $f(x) = f(b)$ for $x > b$ and for each positive integer i, let g_i be defined by

$$g_i(x) = i\,[f(x + 1/i) - f(x)].$$

Since f in nondecreasing, it follows that g_i is a Borel function. Consequently, the functions u and v defined by

$$\begin{aligned} u(x) &= \limsup_{i\to\infty} g_i(x), \\ v(x) &= \liminf_{i\to\infty} g_i(x), \end{aligned} \tag{7.32}$$

are also Borel functions. We know from Theorem 7.18 that f' exists a.e. Hence, it follows that $f' = u$ a.e. and is therefore Lebesgue measurable.

Now, each g_i is nonnegative because f is nondecreasing, and therefore we may employ Fatou's lemma to conclude that

$$\begin{aligned}
\int_a^b f'(x)\,d\lambda(x) &\le \liminf_{i\to\infty} \int_a^b g_i(x)\,d\lambda(x) \\
&= \liminf_{i\to\infty} i \int_a^b [f(x+1/i)-f(x)]\,d\lambda(x) \\
&= \liminf_{i\to\infty} i \left[\int_{a+1/i}^{b+1/i} f(x)\,d\lambda(x) - \int_a^b f(x)\,d\lambda(x)\right] \\
&= \liminf_{i\to\infty} i \left[\int_b^{b+1/i} f(x)\,d\lambda(x) - \int_a^{a+1/i} f(x)\,d\lambda(x)\right] \\
&\le \liminf_{i\to\infty} i \left[\frac{f(b+1/i)}{i} - \frac{f(a)}{i}\right] \\
&= \liminf_{i\to\infty} i \left[\frac{f(b)}{i} - \frac{f(a)}{i}\right] = f(b)-f(a).
\end{aligned}$$

In establishing the last inequality, we have used the fact that f is nondecreasing.

Now suppose that f is an arbitrary function of bounded variation. Since f can be written as the difference of two nondecreasing functions, Theorem 7.21, it follows from Theorem 3.61 that the set D of discontinuities of f is countable. For each real number t let $A_t := \{f > t\}$. Then

$$(a,b)\cap A_t = ((a,b)\cap(A_t - D)) \cup ((a,b)\cap A_t \cap D).$$

The first set on the right is open, since f is continuous at each point of $(a,b)-D$. Since D is countable, the second set is a Borel set; therefore, so is $(a,b)\cap A_t$, which implies that f is a Borel function.

The statements in the theorem referring to the almost everywhere differentiability of f follow from Theorem 7.18; the measurability of f' is addressed in (7.32).

Similarly, since V_f is a nondecreasing function, we have that V_f' exists almost everywhere. Furthermore, with $f = f_1 - f_2$ as in the previous theorem and recalling that $f_1', f_2' \ge 0$ almost everywhere, it follows that

$$|f'| = |f_1' - f_2'| \le |f_1'| + |f_2'| = f_1' + f_2' = V_f' \quad \text{almost everywhere on } [a,b].$$

To prove (7.29) we will show that

$$E := [a,b] \bigcap \{t \,:\, V_f'(t) > |f'(t)|\}$$

has measure zero. For each positive integer m let E_m be the set of all $t \in E$ such that $\tau_1 \le t \le \tau_2$ with $0 < \tau_2 - \tau_1 < \frac{1}{m}$ implies

$$\frac{V_f(\tau_2)-V_f(\tau_1)}{\tau_2-\tau_1} > \frac{|f(\tau_2)-f(\tau_1)|}{\tau_2-\tau_1} + \frac{1}{m}. \tag{7.33}$$

Since each $t \in E$ belongs to E_m for sufficiently large m, we see that

$$E = \bigcup_{m=1}^{\infty} E_m$$

and thus it suffices to show that $\lambda(E_m) = 0$ for each m. Fix $\varepsilon > 0$ and let $a = t_0 < t_1 < \cdots < t_k = b$ be a partition of $[a, b]$ such that $|t_i - t_{i-1}| < \frac{1}{m}$ for each i and

$$\sum_{i=1}^{k} |f(t_i) - f(t_{i-1})| > V_f(b) - \frac{\varepsilon}{m}. \tag{7.34}$$

For each interval in the partition, (7.28) states that

$$V_f(t_i) - V_f(t_{i-1}) \geq |f(t_i) - f(t_{i-1})|, \tag{7.35}$$

while (7.33) implies

$$V_f(t_i) - V_f(t_{i-1}) \geq |f(t_i) - f(t_{i-1})| + \frac{t_i - t_{i-1}}{m} \tag{7.36}$$

if the interval contains a point of E_m. Let $\mathcal{F}_1$ denote the intervals of the partition that do not contain any points of E_m and let $\mathcal{F}_2$ denote the intervals that do contain points of E_m. Then, since $\lambda(E_m) \leq \sum_{I \in \mathcal{F}_2} b_I - a_I$,

$$\begin{aligned}
V_f(b) &= \sum_{i=1}^{k} V_f(t_i) - V_f(t_{i-1}) \\
&= \sum_{I \in \mathcal{F}_1} V_f(b_I) - V_f(a_I) + \sum_{I \in \mathcal{F}_2} V_f(b_I) - V_f(a_I) \\
&= \sum_{I \in \mathcal{F}_1} f(b_I) - f(a_I) + \sum_{I \in \mathcal{F}_2} f(b_I) - f(a_I) + \frac{b_I - a_I}{m} && \text{by (7.35) and (7.36)} \\
&\geq \sum_{i=1}^{k} |f(t_i) - f(t_{i-1})| + \frac{\lambda(E_m)}{m} \\
&\geq V_f(b) - \frac{\varepsilon}{m} + \frac{\lambda(E_m)}{m}, && \text{by (7.34)}
\end{aligned}$$

and therefore $\lambda(E_m) \leq \varepsilon$, from which we conclude that $\lambda(E_m) = 0$, since ε is arbitrary. Thus (7.29) is established.

Finally we apply (7.29) and (7.31) to obtain

$$\int_a^b |f'| \, d\lambda = \int_a^b V' \, d\lambda \leq V_f(b) - V_f(a) = V_f(b),$$

and the proof is complete. □

Exercises for Section 7.4

1. Prove that a function of bounded variation is a Borel measurable function.
2. If f is of bounded variation on $[a,b]$, Theorem 7.21 states that $f = f_1 - f_2$, where both f_1 and f_2 are nondecreasing. Prove that

$$f_1(x) = \sup\left\{\sum_{i=1}^{k}(f(t_i) - f(t_{i-1}))^+\right\},$$

where the supremum is taken over all partitions $a = t_0 < t_1 < \cdots < t_k = x$.

7.5. The Fundamental Theorem of Calculus

We introduce absolutely continuous functions and show that they are precisely those functions for which the fundamental theorem of calculus is valid.

7.23. Definition. A function f defined on an interval $I = [a,b]$ is said to be **absolutely continuous** on I (briefly, AC on I) if for every $\varepsilon > 0$ there exists $\delta > 0$ such that

$$\sum_{i=1}^{k} |f(b_i) - f(a_i)| < \varepsilon$$

for every finite collection of nonoverlapping intervals $[a_1,b_1], [a_2,b_2], \ldots, [a_k,b_k]$ in I with

$$\sum_{i=1}^{k} |b_i - a_i| < \delta.$$

Observe that if f is AC, then it is easy to show that

$$\sum_{i=1}^{\infty} |f(b_i) - f(a_i)| \leq \varepsilon$$

for every countable collection of nonoverlapping intervals with

$$\sum_{i=1}^{\infty} |b_i - a_i| < \delta. \tag{7.37}$$

Indeed, if f is AC, then (7.37) holds for every partial sum and therefore for the limit of the partial sums. Thus, it holds for the whole series.

From the definition it follows that an absolutely continuous function is uniformly continuous. The converse is not true, as we shall see illustrated later by the Cantor–Lebesgue function. (Of course, it can be shown directly that the Cantor–Lebesgue function is not absolutely continuous; see Exercise 1, Section 7.5.) The reader can easily verify that every Lipschitz function is absolutely continuous. Another example of an AC function is given by the indefinite integral; let f be an integrable function on $[a,b]$ and set

$$F(x) = \int_a^x f(t)\,d\lambda(t). \tag{7.38}$$

For every nonoverlapping collection of intervals in $[a,b]$ we have

$$\sum_{i=1}^{k} |F(b_i) - F(a_i)| \leq \int_{\cup[a_i,b_i]} |f| \, d\lambda.$$

With the help of Exercise 1, Section 6.6, we know that the set function μ defined by

$$\mu(E) = \int_E |f| \, d\lambda$$

is a measure, and clearly it is absolutely continuous with respect to Lebesgue measure. Referring to Theorem 6.40, we see that F is absolutely continuous.

7.24. Notation. The following notation will be used frequently throughout. If $I \subset \mathbb{R}$ is an interval, we will denote its endpoints by a_I, b_I; thus, if I is closed, then $I = [a_I, b_I]$.

7.25. Theorem. *An absolutely continuous function on* $[a,b]$ *is of bounded variation.*

Proof. Let f be absolutely continuous. Choose $\varepsilon = 1$ and let $\delta > 0$ be the corresponding number provided by the definition of absolute continuity. Subdivide $[a,b]$ into a finite collection $\mathcal{F}$ of nonoverlapping subintervals $I = [a_I, b_I]$ each of which has length less than δ. Then

$$|f(b_I) - f(a_I)| < 1$$

for each $I \in \mathcal{F}$. Consequently, if $\mathcal{F}$ consists of M elements, we have

$$\sum_{I \in \mathcal{F}} |f(b_I) - f(a_I)| < M.$$

To show that f is of bounded variation on $[a,b]$, consider an arbitrary partition $a = t_0 < t_1 < \cdots < t_k = b$. Since the sum

$$\sum_{i=1}^{k} |f(t_i) - f(t_{i-1})|$$

is not decreased by adding more points to this partition, we may assume that each interval of this partition is a subset of some $I \in \mathcal{F}$. But then,

$$\sum_{i=1}^{k} \chi_I(t_i)\chi_I(t_{i-1}) \, |f(t_i) - f(t_{i-1})| < 1,$$

for each $I \in \mathcal{F}$ (this is simply saying that the sum is taken over only those intervals $[t_{i-1}, t_i]$ that are contained in I), and hence

$$\sum_{i=1}^{k} |f(t_i) - f(t_{i-1})| < M,$$

thus proving that the total variation of f on $[a,b]$ is no more than M. □

Next, we introduce a property that is of great importance concerning absolutely continuous functions. Later we will see that this property is one among three that characterize absolutely continuous functions (see Corollary 7.36). This concept is due to Lusin, now called "condition N."

7.26. DEFINITION. A function f defined on $[a, b]$ is said to satisfy **condition** N if f preserves sets of Lebesgue measure zero; that is, $\lambda[f(E)] = 0$ whenever $E \subset [a, b]$ with $\lambda(E) = 0$.

7.27. THEOREM. *If f is an absolutely continuous function on $[a, b]$, then f satisfies condition N.*

PROOF. Choose $\varepsilon > 0$ and let $\delta > 0$ be the corresponding number provided by the definition of absolute continuity. Let E be a set of measure zero. Then there is an open set $U \supset E$ with $\lambda(U) < \delta$. Since U is the union of a countable collection $\mathcal{F}$ of disjoint open intervals, we have

$$\sum_{I \in \mathcal{F}} \lambda(I) < \delta.$$

The closure of each interval I contains an interval $I' = [a_{I'}, b_{I'}]$ at whose endpoints f assumes its maximum and minimum on the closure of I. Then

$$\lambda[f(I)] = |f(b_{I'}) - f(a_{I'})|$$

and the absolute continuity of f along with (7.37) imply

$$\lambda[f(E)] \leq \sum_{I \in \mathcal{F}} \lambda[f(I)] = \sum_{I \in \mathcal{F}} |f(b_{I'}) - f(a_{I'})| < \varepsilon.$$

Since ε is arbitrary, this shows that $f(E)$ has measure zero. □

7.28. REMARK. This result shows that the Cantor–Lebesgue function is not absolutely continuous, since it maps the Cantor set (of Lebesgue measure zero) onto $[0, 1]$. Thus, there are continuous functions of bounded variation that are not absolutely continuous.

7.29. THEOREM. *Suppose f is an arbitrary function defined on $[a, b]$. Let*

$$E_f := (a, b) \cap \{x : f'(x) \text{ exists and } f'(x) = 0\}.$$

Then $\lambda[f(E_f)] = 0$.

PROOF. **Step 1:**

Initially, we will assume that f is bounded; let $M := \sup\{f(x) : x \in [a, b]\}$ and $m := \inf\{f(x) : x \in [a, b]\}$. Choose $\varepsilon > 0$. For each $x \in E_f$ there exists $\delta = \delta(x) > 0$ such that

$$|f(x + h) - f(x)| < \varepsilon h$$

and

$$|f(x-h)-f(x)|<\varepsilon h$$

whenever $0<h<\delta(x)$. Thus, for each $x\in E_f$ we have a collection of intervals of the form $[x-h,x+h]$, $0<h<\delta(x)$, with the property that for arbitrary points $a',b'\in I=[x-h,x+h]$, one has

$$\begin{aligned}|f(b')-f(a')| &\leq |f(b')-f(x)|+|f(a')-f(x)| \\ &< \varepsilon\,|b'-x|+\varepsilon\,|a'-x| \\ &\leq \varepsilon\lambda(I).\end{aligned}\tag{7.39}$$

We will adopt the following notation: For each interval I let I' be an interval such that $I'\supset I$ has the same center as I but is 5 times as long. Using the definition of Lebesgue measure, we can find an open set $U\supset E_f$ with the property $\lambda(U)-\lambda(E_f)<\varepsilon$ and let $\mathcal{G}$ be the collection of intervals I such that $I'\subset U$ and I' satisfies (7.39). We then appeal to Theorem 7.3 to find a disjoint subfamily $\mathcal{F}\subset\mathcal{G}$ such that

$$E_f\subset\bigcup_{I\in\mathcal{F}}I'.$$

Since f is bounded, each interval I' that is associated with $I\in\mathcal{F}$ contains points $a_{I'},b_{I'}$ such that $f(b_{I'})>M_{I'}-\varepsilon\lambda(I')$ and $f(a_{I'})<m_{I'}+\varepsilon\lambda(I')$, where $\varepsilon>0$ is chosen as above. Thus

$$f(I')\subset[m_{I'},M_{I'}]\subset[f(a_{I'})+\varepsilon\lambda(I'),f(b_{I'})-\varepsilon\lambda(I')],$$

and thus by (7.39),

$$\lambda[f(I')]\leq|M_{I'}-m_{I'}|\leq|f(b_{I'})-f(a_{I'})|+2\varepsilon\lambda(I')<3\varepsilon\lambda(I').$$

Then, using the fact that $\mathcal{F}$ is a disjoint family, we obtain

$$\begin{aligned}\lambda\Big(f(E_f)\Big)\subset\lambda\left[f\left(\bigcup_{I\in\mathcal{F}}I'\right)\right] &\leq \bigcup_{I\in\mathcal{F}}\lambda(f(I'))<\sum_{I\in\mathcal{F}}\varepsilon\lambda(I) \\ &=5\sum_{I\in\mathcal{F}}\varepsilon\lambda(I) \\ &\leq 5\varepsilon\lambda(U)<5\varepsilon(\lambda(E_f)+\varepsilon).\end{aligned}$$

Since $\varepsilon>0$ is arbitrary, we conclude that $\lambda(f(E_f))=0$, as desired.

Step 2:

Now assume that f is an arbitrary function on $[a,b]$. Let $H\colon\mathbb{R}\to[0,1]$ be a smooth, strictly increasing function with $H'>0$ on $\mathbb{R}$. Note that both H and H^{-1} are absolutely continuous. Define $g:=H\circ f$. Then g is bounded and $g'(x)=H'(f(x))f'(x)$ holds whenever either g or f is differentiable at x. Then $f'(x)=0$ iff $g'(x)=0$, and therefore $E_f=E_g$. From Step 1, we know that $\lambda[g(E_g)]=0=\lambda[g(E_f)]$. But $f(A)=H^{-1}[g(A)]$ for all $A\subset\mathbb{R}$ and we have $f(E_f)=H^{-1}[g(E_f)]$, and therefore $\lambda[f(E_f)]=0$, since H^{-1} preserves sets of measure zero. □

7.30. THEOREM. *If f is absolutely continuous on $[a,b]$ with the property that $f'=0$ almost everywhere, then f is constant.*

PROOF. Let

$$E = (a,b) \cap \{x : f'(x) = 0\},$$

so that $[a,b] = E \cup N$, where N is of measure zero. Then $\lambda[f(E)] = \lambda[f(N)] = 0$ by the previous result and Theorem 7.27. Thus, since $f([a,b])$ is an interval and also of measure zero, f must be constant. □

We now have reached our main objective.

7.31. THEOREM (The fundamental theorem of calculus). $f\colon [a,b] \to \mathbb{R}$ *is absolutely continuous if and only if f' exists a.e. on (a,b), f' is integrable on (a,b), and*

$$f(x) - f(a) = \int_a^x f'(t)\, d\lambda(t) \quad \textit{for} \quad x \in [a,b].$$

PROOF. The sufficiency follows from integration theory as discussed in (7.38).

As for necessity, recall that f is BV (Theorem 7.25), and therefore by Theorem 7.22 that f' exists almost everywhere and is integrable. Hence, it is meaningful to define

$$F(x) = \int_a^x f'(t)\, d\lambda(t).$$

Then F is absolutely continuous (as in (7.38)). By Theorem 7.19 we have that $F' = f'$ almost everywhere on $[a,b]$. Thus, $F - f$ is an absolutely continuous function whose derivative is zero almost everywhere. Therefore, by Theorem 7.30, $F - f$ is constant on $[a,b]$, so that $[F(x) - f(x)] = [F(a) - f(a)]$ for all $x \in [a,b]$. Since $F(a) = 0$, we have $F = f - f(a)$. □

7.32. COROLLARY. *If f is an absolutely continuous function on $[a,b]$, then the total variation function V_f of f is also absolutely continuous on $[a,b]$ and*

$$V_f(x) = \int_a^x |f'|\, d\lambda$$

for each $x \in [a,b]$.

PROOF. We know that V_f is a bounded, nondecreasing function and thus of bounded variation. From Theorem 7.22 we know that

$$\int_a^x |f'|\, d\lambda = \int_a^x V_f'\, d\lambda \le V_f(x)$$

for each $x \in [a,b]$.

Fix $x \in [a, b]$ and let $\varepsilon > 0$. Choose a partition $\{a = t_0 < t_1 < \cdots < t_k = x\}$ such that

$$V_f(x) \leq \sum_{i=1}^{k} |f(t_i) - f(t_{i-1})| + \varepsilon.$$

In view of the Theorem 7.31,

$$|f(t_i) - f(t_{i-1})| = \left| \int_{t_{i-1}}^{t_i} f' \, d\lambda \right| \leq \int_{t_{i-1}}^{t_i} |f'| \, d\lambda$$

for each i. Thus

$$V_f(x) \leq \sum_{i=1}^{k} \int_{t_{i-1}}^{t_i} |f'| \, d\lambda + \varepsilon \leq \int_a^x |f'| \, d\lambda + \varepsilon.$$

Since ε is arbitrary, we conclude that

$$V_f(x) = \int_a^x |f'| \, d\lambda$$

for each $x \in [a, b]$. □

Exercises for Section 7.5

1. Prove directly from the definition that the Cantor–Lebesgue function is not absolutely continuous.
2. Prove that a Lipschitz function on $[a, b]$ is absolutely continuous.
3. Prove directly from definitions that a Lipschitz function satisfies condition N.
4. Suppose f is a Lipschitz function defined on $\mathbb{R}$ having Lipschitz constant C. Prove that $\lambda[f(A)] \leq C\lambda(A)$ whenever $A \subset \mathbb{R}$ is Lebesgue measurable.
5. Let $\{f_k\}$ be a uniformly bounded sequence of absolutely continuous functions on $[0, 1]$. Suppose that $f_k \to f$ is in $L^1[0, 1]$ and that $\{f'_k\}$ is Cauchy in $L^1[0, 1]$. Prove that $f = g$ almost everywhere, where g is absolutely continuous on $[0, 1]$.
6. Let f be a strictly increasing continuous function defined on $(0, 1)$. Prove that $f' > 0$ almost everywhere if and only if f^{-1} is absolutely continuous.
7. Prove that the sum and product of absolutely continuous functions are absolutely continuous.
8. Prove that the composition of BV functions is not necessarily BV. (Hint: Consider the composition of $f(x) = \sqrt{x}$ and $g(x) = x^2 \cos^2\left(\frac{\pi}{2x}\right)$, $x \neq 0$, $g(0) = 0$, both defined on $[0, 1]$.)
9. Find two absolutely continuous functions $f, g : [0, 1] \to [0, 1]$ such that their composition is not absolutely continuous. However, show that if $g : [a, b] \to [c, d]$ is absolutely continuous and $f : [c, d] \to \mathbb{R}$ is Lipschitz, then $f \circ g$ is absolutely continuous.

10. Establish the integration by parts formula: If f and g are absolutely continuous functions defined on $[a, b]$, then

$$\int_a^b f'g\,dx = f(b)g(b) - f(a)g(a) - \int_a^b fg'\,dx.$$

11. Give an example of a function $f\colon (0,1) \to \mathbb{R}$ that is differentiable everywhere but is not absolutely continuous. Compare this with Theorem 7.50.
12. Given a Lebesgue measurable set E, prove that $\{x : D(E,x) = 1\}$ is a Borel set.
13. Let $f : \mathbb{R} \to \mathbb{R}$. Prove that f satisfies the Lipschitz condition

$$|f(x) - f(y)| \leq M|x - y|,$$

for some M and all $x, y \in \mathbb{R}$, if and only if f satisfies the following two properties:
(a) f is absolutely continuous.
(b) $|f'(x)| \leq M$ for a.e. x.

7.6. Variation of Continuous Functions

One possibility of determining the variation of a function is the following. Consider the graph of f in the (x, y)-plane and for each y, let $N(y)$ denote the number of times the horizontal line passing through $(0, y)$ intersects the graph of f. It seems plausible that

$$\int_{\mathbb{R}} N(y)\,d\lambda(y)$$

should equal the variation of f on $[a, b]$. If f is a continuous nondecreasing function, this is easily seen to be true. The next theorem provides the general result. First, we introduce some notation: if $f\colon \mathbb{R} \to \mathbb{R}$ and $E \subset \mathbb{R}$, then

$$N(f, E, y) \tag{7.40}$$

denotes the (possibly infinite) number of points in the set $E \cap f^{-1}\{y\}$. Thus, $N(f, E, y)$ is the number of points in E that are mapped onto y.

7.33. THEOREM. *Let f be a continuous function defined on $[a, b]$. Then $N(f, [a, b], y)$ is a Borel measurable function (of y) and*

$$V_f(b) = \int_{R^1} N(f, [a, b], y)\,d\lambda(y).$$

PROOF. For brevity throughout the proof, we will simply write $N(y)$ for $N(f, [a, b], y)$.

Let $m \leq N(y)$ be a nonnegative integer and let $x_1, x_2, \ldots, x_m$ be points that are mapped into y. Thus, $\{x_1, x_2, \ldots, x_m\} \subset f^{-1}\{y\}$. For each positive integer i, consider a partition $\mathcal{P}_i = \{a = t_0 < t_1 < \cdots < t_k = b\}$ of $[a, b]$ such that the length of each interval I is less than $1/i$. Choose i so large that each interval of $\mathcal{P}_i$ contains at most one x_j, $j = 1, 2, \ldots, m$. Then

$$m \leq \sum_{I \in \mathcal{P}_i} \chi_{f(I)}(y).$$

Consequently,

$$m \leq \liminf_{i \to \infty} \left\{ \sum_{I \in \mathcal{P}_i} \chi_{f(I)}(y) \right\}. \tag{7.41}$$

Since m is an arbitrary positive integer with $m \leq N(y)$, we obtain

$$N(y) \leq \liminf_{i \to \infty} \left\{ \sum_{I \in \mathcal{P}_i} \chi_{f(I)}(y) \right\}. \tag{7.42}$$

On the other hand, for every partition $\mathcal{P}_i$ we obviously have

$$N(y) \geq \sum_{I \in \mathcal{P}_i} \chi_{f(I)}(y), \tag{7.43}$$

provided that each point of $f^{-1}(y)$ is contained in the interior of some interval $I \in \mathcal{P}_i$. Thus (7.43) holds for all but finitely many y and therefore

$$N(y) \geq \limsup_{i \to \infty} \left\{ \sum_{I \in \mathcal{P}_i} \chi_{f(I)}(y) \right\}$$

holds for all but countably many y. Hence, with (7.42), we have

$$N(y) = \lim_{i \to \infty} \left\{ \sum_{I \in \mathcal{P}_i} \chi_{f(I)}(y) \right\} \quad \text{for all but countably many } y. \tag{7.44}$$

Since $\sum_{I \in \mathcal{P}_i} \chi_{f(I)}$ is a Borel measurable function, it follows that N is also Borel measurable. For every interval $I \in P_i$,

$$\lambda[f(I)] = \int_{R^1} \chi_{f(I)}(y)\, d\lambda(y),$$

and therefore by (7.43),

$$\begin{aligned} \sum_{I \in \mathcal{P}_i} \lambda[f(I)] &= \sum_{I \in \mathcal{P}_i} \int_{R^1} \chi_{f(I)}(y)\, d\lambda(y) \\ &= \int_{R^1} \sum_{I \in \mathcal{P}_i} \chi_{f(I)}(y)\, d\lambda(y) \\ &\leq \int_{R^1} N(y)\, d\lambda(y), \end{aligned}$$

which implies

$$\limsup_{i \to \infty} \left\{ \sum_{I \in \mathcal{P}_i} \lambda[f(I)] \right\} \leq \int_{R^1} N(y)\, d\lambda(y). \tag{7.45}$$

For the opposite inequality, observe that Fatou's lemma and (7.44) yield

$$\begin{aligned}\liminf_{i\to\infty}\left\{\sum_{I\in\mathcal{P}_i}\lambda[f(I)]\right\} &= \liminf_{i\to\infty}\left\{\sum_{I\in\mathcal{P}_i}\int_{R^1}\chi_{f(I)}(y)\,d\lambda(y)\right\}\\ &= \liminf_{i\to\infty}\left\{\int_{R^1}\sum_{I\in\mathcal{P}_i}\chi_{f(I)}(y)\,d\lambda(y)\right\}\\ &\geq \int_{R^1}\left\{\liminf_{i\to\infty}\sum_{I\in\mathcal{P}_i}\chi_{f(I)}(y)\right\}\,d\lambda(y)\\ &= \int_{R^1}N(y)\,d\lambda(y).\end{aligned}$$

Thus, we have

$$\lim_{i\to\infty}\left\{\sum_{I\in\mathcal{P}_i}\lambda[f(I)]\right\} = \int_{R^1}N(y)\,d\lambda(y). \tag{7.46}$$

We will conclude the proof by showing that the limit on the left-hand side is equal to $V_f(b)$. First, recall the notation introduced in Notation 7.24: If I is an interval belonging to a partition $\mathcal{P}_i$, we will denote the endpoints of this interval by a_I, b_I. Thus, $I = [a_I, b_I]$. We now proceed with the proof by selecting a sequence of partitions $\mathcal{P}_i$ with the property that each subinterval I in $\mathcal{P}_i$ has length less than $\frac{1}{i}$ and

$$\lim_{i\to\infty}\sum_{I\in\mathcal{P}_i}|f(b_i)-f(a_i)| = V_f(b).$$

Then,

$$\begin{aligned}V_f(b) &= \lim_{i\to\infty}\left\{\sum_{I\in\mathcal{P}_i}|f(b_I)-f(a_I|)\right\}\\ &\leq \liminf_{i\to\infty}\left\{\sum_{I\in\mathcal{P}_i}\lambda[f(I)]\right\}.\end{aligned}$$

We now show that

$$\limsup_{i\to\infty}\left\{\sum_{I\in\mathcal{P}_i}\lambda[f(I)]\right\} \leq V_f(b), \tag{7.47}$$

which will conclude the proof. For this, let $I' = [a_{I'}, b_{I'}]$ be an interval contained in $I = [a_i, b_i]$ such that f assumes its maximum and minimum on

I at the endpoints of I'. Let $\mathcal{Q}_i$ denote the partition formed by the endpoints of $I \in \mathcal{P}_i$ along with the endpoints of the intervals I'. Then

$$\begin{aligned}\sum_{I\in\mathcal{P}_i} \lambda[f(I)] &= \sum_{I\in\mathcal{P}_i} |f(b_{I'}) - f(a_{I'})| \\ &\leq \sum_{I\in\mathcal{Q}_i} |f(b_i) - f(a_i)| \\ &\leq V_f(b),\end{aligned}$$

thereby establishing (7.47). □

7.34. COROLLARY. *Suppose f is a continuous function of bounded variation on $[a,b]$. Then the total variation function $V_f(\cdot)$ is continuous on $[a,b]$. In addition, if f also satisfies condition N, then so does $V_f(\cdot)$.*

PROOF. Fix $x_0 \in [a,b]$. By the previous result,

$$V_f(x_0) = \int_{\mathbb{R}} N(f,[a,x_0],y)\,d\lambda(y).$$

For $a \leq x_0 < x \leq b$,

$$N(f,[a,x],y) - N(f,[a,x_0],y) = N(f,(x_0,x],y)$$

for each y such that $N(f,[a,b],y) < \infty$, i.e., for a.e. $y \in \mathbb{R}$ because $N(f,[a,b],\cdot)$ is integrable. Thus

$$\begin{aligned}0 \leq V_f(x) - V_f(x_0) &= \int_{\mathbb{R}} N(f,[a,x],y)\,d\lambda(y) \\ &\quad - \int_{\mathbb{R}} N(f,[a,x_0],y)\,d\lambda(y) \\ &= \int_{\mathbb{R}} N(f,(x_0,x],y)\,d\lambda(y) \\ &\leq \int_{f((x_0,x])} N(f,[a,b],y)\,d\lambda(y).\end{aligned} \tag{7.48}$$

Since f is continuous at x_0, we have $\lambda[f((x_0,x])] \to 0$ as $x \to x_0$, and thus

$$\lim_{x\to x_0^+} V_f(x) = V_f(x_0).$$

A similar argument shows that

$$\lim_{x\to x_0^-} V_f(x) = V_f(x_0)$$

and thus that V_f is continuous at x_0.

Now assume that f also satisfies condition N (see Definition 7.26) and let A be a set with $\lambda(A) = 0$. Hence $\lambda(f(A)) = 0$. Observe that for every measurable set $E \subset \mathbb{R}$, the set function

$$\mu(E) := \int_E N(f,[a,b],y)\,dy$$

is a measure that is absolutely continuous with respect to Lebesgue measure. Consequently, for each $\varepsilon > 0$ there exists $\delta > 0$ such that $\mu(E) < \varepsilon$ whenever $\lambda(E) < \delta$. Hence if $U \supset f(A)$ is an open set with $\lambda(f(A)) < \delta$, we have

$$\int_U N(f, [a,b], y)\, dy < \varepsilon. \tag{7.49}$$

The open set $f^{-1}(U)$ can be expressed as the countable disjoint union of intervals $\bigcup_{i=1}^{\infty} I_i \supset A$. Thus, with the notation $I_i := [a_i, b_i]$ we have

$$\begin{aligned}
\lambda(V_f(A)) &\leq \lambda\left(V_f(\bigcup_i^{\infty} I_i)\right) \\
&\leq \sum_{i=1}^{\infty} \lambda(V_f(I_i)) \\
&= \sum_{i=1}^{\infty} V_f(b_i) - V_f(a_i) \\
&= \sum_{i=1}^{\infty} \int_{\mathbb{R}} N(f, (a_i, b_i], y)\, dy && \text{by (7.48)} \\
&= \int_{\cup_{i=1}^{\infty} f((a_i,b_i])} N(f, [a,b], y)\, dy && \text{since the } I_i's \text{ are disjoint} \\
&= \int_U N(f, [a,b], y), \\
&< \varepsilon && \text{by (7.49)}
\end{aligned}$$

which implies that $\lambda(V_f(A)) = 0$. □

7.35. THEOREM. *If f is a nondecreasing function defined on $[a, b]$, then it is absolutely continuous if and only if the following two conditions are satisfied:*

(i) *f is continuous,*
(ii) *f satisfies condition N.*

PROOF. Clearly condition (i) is necessary for absolute continuity, and Theorem 7.27 shows that condition (ii) is also necessary.

To prove that the two conditions are sufficient, let $g(x) := x + f(x)$, so that g is a strictly increasing continuous function. Note that g satisfies condition N since f does and $\lambda(g(I)) = \lambda(I) + \lambda(f(I))$ for every interval I. Also, if E is a measurable set, then so is $g(E)$, because $E = F \cup N$, where F is an F_σ set and N has measure zero, by Theorem 4.25. Since g is continuous and since F can be expressed as the countable union of compact sets, it follows that $g(E) = g(F) \cup g(N)$ is the union of a countable number of compact sets and a set of measure zero and is therefore a measurable set. Now define a measure μ by

$$\mu(E) = \lambda(g(E))$$

for each measurable set E. Observe that μ is, in fact, a measure, since g is injective. Furthermore, $\mu \ll \lambda$, because g satisfies condition N. Consequently, the Radon–Nikodym theorem applies (Theorem 6.43), and we obtain a function $h \in L^1(\lambda)$ such that

$$\mu(E) = \int_E h\,d\lambda \quad \text{for every measurable set } E.$$

In particular, taking $E = [a, x]$, we obtain

$$g(x) - g(a) = \lambda(g(E)) = \mu(E) = \int_E h\,d\lambda = \int_a^x h\,d\lambda.$$

Thus, as in (7.38), we conclude that g is absolutely continuous, and therefore so is f. □

7.36. Corollary. *A function f defined on $[a, b]$ is absolutely continuous if and only if f satisfies the following three conditions on $[a, b]$:*

(i) *f is continuous,*
(ii) *f is of bounded variation,*
(iii) *f satisfies condition N.*

Proof. The necessity of the three conditions is established by Theorems 7.25 and 7.27.

To prove sufficiency suppose f satisfies conditions (i)–(iii). It follows from Corollary 7.34 that $V_f(\cdot)$ is continuous, satisfies condition N, and therefore is absolutely continuous by Theorem 7.35. Since $f = f_1 - f_2$, where $f_1 = \frac{1}{2}(V_f + f)$ and $f_1 = \frac{1}{2}(V_f - f)$, it follows that both f_1 and f_2 are absolutely continuous, and therefore so is f. □

7.7. Curve Length

Adapting the methods of the previous section, the notion of length is developed and shown to be closely related to 1-dimensional Hausdorff measure.

7.37. Definitions. A **curve in** $\mathbb{R}^n$ is a continuous mapping γ: $[a, b] \to \mathbb{R}^n$, and its **length** is defined as

$$L_\gamma = \sup \sum_{i=1}^{k} |\gamma(t_i) - \gamma(t_{i-1})|, \tag{7.50}$$

where the supremum is taken over all finite sequences $a = t_0 < t_1 < \cdots < t_k = b$. Note that $\gamma(x)$ is a vector in $\mathbb{R}^n$ for each $x \in [a, b]$; writing $\gamma(x)$ in terms of its component functions, we have

$$\gamma(x) = (\gamma_1(x), \gamma_2(x), \ldots, \gamma_n(x)).$$

Thus in (7.50), $\gamma(t_i) - \gamma(t_{i-1})$ is a vector in $\mathbb{R}^n$ and $|\gamma(t_i) - \gamma(t_{i-1})|$ is its length. For $x \in [a, b]$, we will use the notation $L_\gamma(x)$ to denote the length of γ restricted to the interval $[a, x]$; γ is said to have **finite length** or to be **rectifiable** if $L_\gamma(b) < \infty$.

We will show that there is a strong parallel between the notions of length and bounded variation. If γ is a curve in $\mathbb{R}$, i.e., if $\gamma\colon [a, b] \to \mathbb{R}$, the two notions coincide. More generally, we have the following.

7.38. THEOREM. *A continuous curve $\gamma\colon [a, b] \to \mathbb{R}^n$ is rectifiable if and only if each component function, γ_i, is of bounded variation on $[a, b]$.*

PROOF. Suppose each component function is of bounded variation. Then there are numbers $M_1, M_2, \ldots, M_n$ such that for every finite partition $\mathcal{P}$ of $[a, b]$ into nonoverlapping intervals $I = [a_I, b_I]$,

$$\sum_{I \in \mathcal{P}} |\gamma_j(b_I) - \gamma_j(a_I)| \le M_j, \; j = 1, 2, \ldots, n.$$

Thus, with $M = M_1 + M_2 + \cdots + M_n$ we have

$$\begin{aligned}
&\sum_{I \in \mathcal{P}} |\gamma(b_I) - \gamma(a_I)| \\
&\quad = \sum_{I \in \mathcal{P}} \left[(\gamma_1(b_I) - \gamma_1(a_I))^2 + (\gamma_2(b_I) - \gamma_2(a_I))^2 \right. \\
&\quad\quad \left. + \cdots + (\gamma_n(b_I) - \gamma_n(a_I))^2\right]^{1/2} \\
&\quad \le \sum_{I \in \mathcal{P}} |\gamma_1(b_I) - \gamma_1(a_I)| + \sum_{I \in \mathcal{P}} |\gamma_2(b_I) - \gamma_2(a_I)| \\
&\quad\quad + \cdots + \sum_{I \in \mathcal{P}} |\gamma_n(b_I) - \gamma_n(a_I)| \\
&\quad \le M.
\end{aligned}$$

Since the partition $\mathcal{P}$ is arbitrary, we conclude that the length of γ is less than or equal to M.

Now assume that γ is rectifiable. Then for every partition $\mathcal{P}$ of $[a, b]$ and integer $j \in [1, n]$,

$$\sum_{I \in \mathcal{P}} |\gamma(b_I) - \gamma(a_I)| \ge \sum_{I \in \mathcal{P}} |\gamma_j(b_I) - \gamma_j(a_I)|,$$

thus showing that the total variation of γ_j is no more than the length of γ. □

In elementary calculus, we know that the formula for the length of a curve $\gamma = (\gamma_1, \gamma_2)$ defined on $[a, b]$ is given by

$$L_\gamma = \int_a^b \sqrt{(\gamma_1'(t))^2 + (\gamma_2'(t))^2}\, dt.$$

We will proceed to investigate the conditions under which this formula holds using our definition of length. We will consider a curve γ in $\mathbb{R}^n$; thus we have $\gamma\colon [a,b] \to \mathbb{R}^n$ with $\gamma(t) = (\gamma_1(t), \gamma_2(t), \dots, \gamma_n(t))$, and we recall the notation $L_\gamma(t)$ introduced earlier that denotes the length of γ from a to t.

7.39. THEOREM. *If γ is rectifiable, then*

$$L'_\gamma(t) = \sqrt{(\gamma'_1(t))^2 + (\gamma'_2(t))^2 + \cdots + (\gamma'_n(t))^2}$$

for almost all $t \in [a,b]$.

PROOF. The number $|L_\gamma(t+h) - L_\gamma(t)|$ denotes the length along the curve between the points $\gamma(t+h)$ and $\gamma(t)$, which is clearly not less than the straight-line distance. Therefore, it is intuitively clear that

$$|L_\gamma(t+h) - L_\gamma(t)| \geq |\gamma(t+h) - \gamma(t)|. \tag{7.51}$$

The rigorous argument to establish this is very similar to the proof of (7.27). Thus, for $h > 0$, consider an arbitrary partition $a = t_0 < t_1 < \cdots < t_k = x$. Then from the definition of L_γ,

$$L_\gamma(x+h) \geq |\gamma(x+h) - \gamma(x)| + \sum_{i=1}^{k} |\gamma(t_i) - \gamma(t_{i-1})|.$$

Since

$$L_\gamma(x) = \sup\left\{\sum_{i=1}^{k} |\gamma(t_i) - \gamma(t_{i-1})|\right\}$$

over all partitions $a = t_0 < t_1 < \cdots < t_k = x$, it follows that

$$L_\gamma(x+h) \geq |\gamma(x+h) - \gamma(x)| + L_\gamma(x).$$

A similar inequality holds for $h < 0$, and therefore we obtain

$$|L_\gamma(t+h) - L_\gamma(t)| \geq |\gamma(t+h) - \gamma(t)|. \tag{7.52}$$

consequently,

$$\begin{aligned} |L_\gamma(t+h) - L_\gamma(t)| &\geq |\gamma(t+h) - \gamma(t)| \\ &= \big[(\gamma_1(t+h) - \gamma_1(t))^2 + (\gamma_2(t+h) - \gamma_2(t))^2 \\ &\quad + \cdots + (\gamma_n(t+h) - \gamma_n(t))^2\big]^{1/2} \end{aligned} \tag{7.53}$$

whenever $t+h, t \in [a,b]$. Consequently,

$$\begin{aligned} \frac{|L_\gamma(t+h) - L_\gamma(t)|}{h} \geq \Bigg[\left(\frac{\gamma_1(t+h) - \gamma_1(t)}{h}\right)^2 \\ + \cdots + \left(\frac{\gamma_n(t+h) - \gamma_n(t)}{h}\right)^2\Bigg]^{1/2}. \end{aligned}$$

Taking the limit as $h \to 0$, we obtain

$$L'_\gamma(t) \geq \sqrt{(\gamma'_1(t))^2 + (\gamma'_2(t))^2 + \cdots + (\gamma'_n(t))^2} \tag{7.54}$$

whenever all derivatives exist, which is almost everywhere in view of Theorem 7.38. The remainder of the proof is completely analogous to the proof of (7.29) in Theorem 7.22 and is left as an exercise. □

The proof of the following result is completely analogous to the proof of Corollary 7.32 and is also left as an exercise.

7.40. THEOREM. *If each component function γ_i of the curve $\gamma : [a,b] \to \mathbb{R}^n$ is absolutely continuous, then the function $L_\gamma(\cdot)$ is absolutely continuous, and*

$$L_\gamma(x) = \int_a^x L'_\gamma \, d\lambda = \int_a^x \sqrt{(\gamma'_1)^2 + (\gamma'_2)^2 + \cdots + (\gamma'_n)^2} \, d\lambda$$

for each $x \in [a,b]$.

7.41. EXAMPLE. Intuitively, one might expect that the trace of a continuous curve would resemble a piece of string, perhaps badly crumpled, but still like a piece of string. However, Peano discovered that the situation could be far worse. He was the first to demonstrate the existence of a continuous mapping $\gamma : [0,1] \to \mathbb{R}^2$ such that $\gamma[0,1]$ occupies the unit square, Q. In other words, he showed the existence of an "area-filling curve." In the figure below, we show the first three stages of the construction of such a curve. This construction is due to Hilbert.

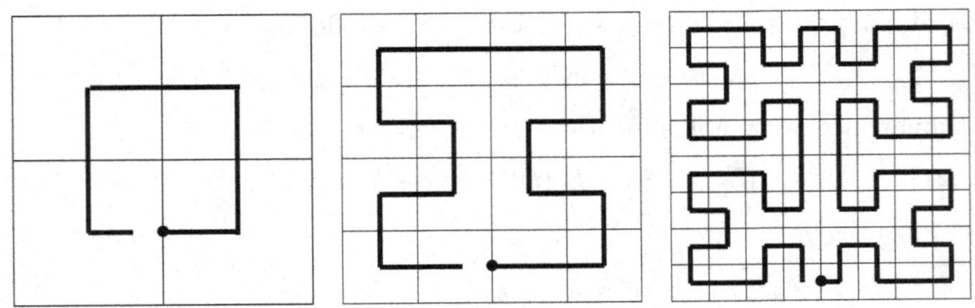

Each stage represents the graph of a continuous (piecewise linear) mapping $\gamma_k : [0,1] \to \mathbb{R}^2$. From the way the construction is made, we find that

$$\sup_{t\in[0,1]} |\gamma_k(t) - \gamma_l(t)| \leq \frac{\sqrt{2}}{2^k},$$

where $k \leq l$. Hence, since the space of continuous functions with the topology of uniform convergence is complete, there exists a continuous mapping γ that is the uniform limit of $\{\gamma_k\}$. To see that $\gamma[0,1]$ is the unit square Q, first observe that each point x_0 in the unit square belongs to some square of each of the partitions of Q. Denoting by $\mathcal{P}_k$ the kth partition of Q into 4^k subsquares of side length $(1/2)^k$, we see that each point $x_0 \in Q$ belongs to some square, Q_k, of $\mathcal{P}_k$ for $k = 1, 2, \ldots$. For each k the curve γ_k passes through Q_k, and so it is clear that there exist points $t_k \in [0,1]$ such that

$$x_0 = \lim_{k\to\infty} \gamma_k(t_k). \tag{7.55}$$

For $\varepsilon > 0$, choose K_1 such that $|x_0 - \gamma_k(t_k)| < \varepsilon/2$ for $k \geq K_1$. Since $\gamma_k \to \gamma$ uniformly, we see by Theorem 3.56 that there exists $\delta = \delta(\varepsilon) > 0$ such that $|\gamma_k(t) - \gamma_k(s)| < \varepsilon$ for all k whenever $|s - t| < \delta$. Choose K_2 such that $|t - t_k| < \delta$ for $k \geq K_2$. Since $[0,1]$ is compact, there is a point $t_0 \in [0,1]$ such that (for a subsequence) $t_k \to t_0$ as $k \to \infty$. Then $x_0 = \gamma(t_0)$, because

$$|x_0 - \gamma(t_0)| \leq |x_0 - \gamma_k(t_k)| + |\gamma_k(t_k) - \gamma_k(t_0)| < \varepsilon$$

for $k \geq \max K_1, K_2$. One must be careful to distinguish a curve in $\mathbb{R}^n$ from the point set described by its trace. For example, compare the curve in $\mathbb{R}^2$ given by

$$\gamma(x) = (\cos x, \sin x),\ x \in [0, 2\pi]$$

to the curve

$$\eta(x) = (\cos 2x, \sin 2x),\ x \in [0, 2\pi].$$

Their traces occupy the same point set, namely, the unit circle. However, the length of γ is 2π, whereas the length of η is 4π. This simple example serves as a model for the relationship between the length of a curve and the 1-dimensional Hausdorff measure of its trace. Roughly speaking, we will show that they are the same if one takes into account the number of times each point in the trace is covered. In particular, if γ is injective, then they are the same.

7.42. THEOREM. *Let $\gamma\colon [a,b] \to \mathbb{R}^n$ be a continuous curve. Then*

$$L_\gamma(b) = \int N(\gamma, [a,b], y)\, dH^1(y), \tag{7.56}$$

where $N(\gamma, [a,b], y)$ denotes the (possibly infinite) number of points in the set $\gamma^{-1}\{y\} \cap [a,b]$. Equality in (7.56) is understood in the sense that either both sides are finite and equal or both sides are infinite.

PROOF. Let M denote the length of γ on $[a,b]$. We will first prove

$$M \geq \int N(\gamma, [a,b], y)\, dH^1(y), \tag{7.57}$$

and therefore, we may as well assume $M < \infty$. The function $L_\gamma(\cdot)$ is nondecreasing, and its range is the interval $[0, M]$. As in (7.53), we have

$$|L_\gamma(x) - L_\gamma(y)| \geq |\gamma(x) - \gamma(y)| \tag{7.58}$$

for all $x, y \in [a,b]$. Since L_γ is nondecreasing, it follows that $L_\gamma^{-1}\{s\}$ is an interval (possibly degenerate) for all $s \in [0, M]$. In fact, there are only countably many s for which $L_\gamma^{-1}\{s\}$ is a nondegenerate interval. Now define $g\colon [0, M] \to \mathbb{R}^n$ as follows:

$$g(s) = \gamma(x), \tag{7.59}$$

where x is any point in $L_\gamma^{-1}(s)$. Observe that if $L_\gamma^{-1}\{s\}$ is a nondegenerate interval and if $x_1, x \in L_\gamma^{-1}\{s\}$, then $\gamma(x_1) = \gamma(x)$, thus ensuring that $g(s)$ is well defined. Also, for $x \in L_\gamma^{-1}\{s\}$, $y \in L_\gamma^{-1}\{t\}$, notice from (7.58) that

$$|g(s) - g(t)| = |\gamma(x) - \gamma(y)| \leq |L_\gamma(x) - L_\gamma(y)| = |s - t|, \tag{7.60}$$

so that g is a Lipschitz function with Lipschitz constant 1. From the definition of g, we clearly have $\gamma = g \circ L_\gamma$. Let S be the set of points in $[0, M]$ such that $L_\gamma^{-1}\{s\}$ is a nondegenerate interval. Then it follows that

$$N(\gamma, [a, b], y) = N(g, [0, M], y)$$

for all $y \notin g(S)$. Since S is countable and therefore $g(S)$ is countable as well, we have

$$\int N(\gamma, [a, b], y)\, dH^1(y) = \int N(g, [0, M], y)\, dH^1(y). \tag{7.61}$$

We will appeal to the proof of Theorem 7.33 to show that

$$M \geq \int N(g, [0, M], y)\, dH^1(y).$$

Let $\mathcal{P}_i$ be a sequence of partitions of $[0, M]$ each having the property that its intervals have length less than $1/i$. Then, using the argument that established (7.44), we have

$$N(g, [0, M], y) = \lim_{i\to\infty} \left\{ \sum_{I \in \mathcal{P}_i} \chi_{g(I)}(y) \right\}.$$

Since

$$H^1(g(I)) = \int \chi_{g(I)}(y)\, dH^1(y)$$

for each interval I, we adapt the proof of (7.46) to obtain

$$\lim_{i\to\infty} \left\{ \sum_{I \in \mathcal{P}_i} H^1[g(I)] \right\} = \int N(g, [0, M], y)\, dH^1(y). \tag{7.62}$$

Now use (7.60) and Exercise 2, Section 7.7, to conclude that $H^1[g(I)] \leq \lambda(I)$, so that (7.62) yields

$$\liminf_{i\to\infty} \left\{ \sum_{I \in \mathcal{P}_i} \lambda(I) \right\} \geq \int_0^\infty N(g, [0, M], y)\, dH^1(y). \tag{7.63}$$

It is necessary to use the lim inf here because we don't know that the limit exists. However, since

$$\sum_{I \in \mathcal{P}_i} \lambda(I) = \lambda([0, M]) = M,$$

we see that the limit does exist and that the left-hand side of (7.63) equals M. Thus, we obtain

$$M \geq \int_0^\infty N(g, [0, M], y)\, dH^1(y),$$

which, along with (7.61), establishes (7.57).

We now will prove

$$M \leq \int_0^\infty N(\gamma, [a, b], y)\, dH^1(y) \tag{7.64}$$

to conclude the proof of the theorem. Again, we adapt the reasoning leading to (7.62) to obtain

$$\lim_{i \to \infty} \left\{ \sum_{I \in \mathcal{P}_i} H^1[\gamma(I)] \right\} = \int N(\gamma, [a, b], y)\, dH^1(y). \tag{7.65}$$

In this context, $\mathcal{P}_i$ is a sequence of partitions of $[a, b]$, each of whose intervals has maximum length $1/i$. Each term on the left-hand side involves $H^1[\gamma(I)]$. Now $\gamma(I)$ is the trace of a curve defined on the interval $I = [a_I, b_I]$. By Exercise 2, Section 7.7, we know that $H^1[\gamma(I)]$ is greater than or equal to the H^1 measure of the orthogonal projection of $\gamma(I)$ onto any straight line, l. That is, if $p \colon \mathbb{R}^n \to l$ is an orthogonal projection, then

$$H^1[p(\gamma(I))] \leq H^1[\gamma(I)].$$

In particular, consider the straight line l that passes through the points $\gamma(a_I)$ and $\gamma(b_I)$. Since I is connected and γ is continuous, $\gamma(I)$ is connected, and therefore, its projection, $p[\gamma(I)]$, onto l is also connected. Thus, $p[\gamma(I)]$ must contain the interval with endpoints $\gamma(a_I)$ and $\gamma(b_I)$. Hence $|\gamma(b_I) - \gamma(a_I)| \leq H^1[\gamma(I)]$. Thus, from (7.65), we have

$$\lim_{i \to \infty} \left\{ \sum_{I \in \mathcal{P}_i} |\gamma(b_I) - \gamma(a_I)| \right\} \leq \int N(\gamma, [a, b], y)\, dH^1(y).$$

By Exercise 7.3, the expression on the left is the length of γ on $[a, b]$, which is M, thus proving (7.64). □

7.43. Remark. The function g defined in (7.59) is called a **parametrization of γ with respect to arc length.** The purpose of g is to give an alternative and equivalent description of the curve γ. It is equivalent in the sense that the trace and length of g are the same as those of γ. If γ is not constant on any interval, then L_γ is a homeomorphism, and thus g and γ are related by a homeomorphic change of variables.

Exercises for Section 7.7

1. In the proof of Theorem 7.48 we established (7.69) by means of Theorem 7.47. Another way to obtain (7.69) is the following. Prove that if f is Lipschitz on a set E with Lipschitz constant C, then $\lambda[f(A)] \leq C\lambda(A)$ whenever $A \subset E$ is Lebesgue measurable. This is the same as Exercise 4, Section 7.5, except that f is defined only on E, not necessarily on $\mathbb{R}$. First prove that Lebesgue measure can be defined as follows:
$$\lambda(A) = \inf\left\{\sum_{i=1}^{\infty} \text{diam } E_i : A \subset \bigcup_{i=1}^{\infty} E_i\right\}$$
where the E_i are arbitrary sets.
2. Suppose $f\colon \mathbb{R}^n \to \mathbb{R}^m$ is a Lipschitz mapping with Lipschitz constant C. Prove that
$$H^k[f(E)] \leq C^k H^k(E)$$
for $E \subset \mathbb{R}^n$.
3. Prove that if $\gamma : [a,b] \to \mathbb{R}^n$ is a continuous curve and $\{\mathcal{P}_i\}$ is a sequence of partitions of $[a,b]$ such that
$$\lim_{i\to\infty} \max_{I\in\mathcal{P}_i} |b_I - a_I| = 0,$$
then
$$\lim_{i\to\infty} \sum_{I\in\mathcal{P}_i} |\gamma(b_I) - \gamma(a_I)| = L_\gamma(b).$$
4. Prove that the example of an area-filling curve in Section 7.7 actually has the unit square as its trace.
5. It follows from Theorem 7.42 that the area-filling curve is not rectifiable. Prove this directly from the construction of the curve.
6. Give an example of a continuous curve that fills the unit cube in $\mathbb{R}^3$.
7. Give a proof of Theorem 7.40.
8. Let $\gamma : [0,1] \to \mathbb{R}^2$ be defined by $\gamma(x) = (x, f(x))$, where f is the Cantor–Lebesgue function described in Example 5.7. Thus, γ describes the graph of the Cantor function. Find the length of γ.

7.8. The Critical Set of a Function

During the course of our development of the fundamental theorem of calculus in Section 7.5, we found that absolutely continuous functions are continuous functions of bounded variation that satisfy condition N. We will show here that these properties characterize AC functions. This will be done by carefully analyzing the behavior of a function on the set where its derivative is 0.

Recall that a function f defined on $[a,b]$ is said to satisfy **condition N** if $\lambda[f(E)] = 0$ whenever $\lambda(E) = 0$ for $E \subset [a,b]$.

An example of a function that does not satisfy condition N is the Cantor–Lebesgue function. Indeed, it maps the Cantor set (of Lebesgue measure zero) onto the unit interval $[0,1]$. On the other hand, a Lipschitz function is an example of a function that does satisfy condition N (see Exercise 3, Section 7.5). Recall Definition 3.10, which states that f satisfies a **Lipschitz** condition on $[a,b]$ if there exists a constant $C = C_f$ such that

$$|f(x) - f(y)| \leq C\,|x-y| \quad \text{whenever} \quad x,y \in [a,b].$$

One of the important aspects of a function is its behavior on the **critical set**, the set where its derivative is zero. One would expect that the critical set of a function f would be mapped onto a set of measure zero, since f is neither increasing nor decreasing at points where $f' = 0$. For convenience, we state this result, which was proved earlier, Theorem 7.29.

7.44. THEOREM. *Suppose f is defined on $[a,b]$. Let*

$$E = (a,b) \cap \{x : f'(x) = 0\}.$$

Then $\lambda[f(E)] = 0$.

Now we will investigate the behavior of a function on the complement of its critical set and show that good things happen there. We will prove that the set on which a continuous function has a nonzero derivative can be decomposed into a countable collection of disjoint sets on each of which the function is bi-Lipschitz. That is, on each of these sets the function is Lipschitz and injective; furthermore, its inverse is Lipschitz on the image of each such set.

7.45. THEOREM. *Suppose f is defined on $[a,b]$ and let A be defined by*

$$A := [a,b] \cap \{x : f'(x) \text{ exists and } f'(x) \neq 0\}.$$

Then for each $\theta > 1$, there is a countable collection $\{E_k\}$ of disjoint Borel sets such that

(i) $A = \bigcup_{k=1}^{\infty} E_k$

(ii) *For each positive integer k there is a positive rational number r_k such that*

$$\frac{r_k}{\theta} \leq |f'(x)| \leq \theta r_k \quad \text{for} \quad x \in E_k,$$

$$|f(y) - f(x)| \leq \theta r_k\,|x-y| \quad \text{for} \quad x,y \in E_k, \tag{7.66}$$

$$|f(y) - f(x)| \geq \frac{r_k}{\theta}(y-x) \quad \text{for} \quad x,y \in E_k. \tag{7.67}$$

PROOF. Since $\theta > 1$, there exists $\varepsilon > 0$ such that

$$\frac{1}{\theta} + \varepsilon < 1 < \theta - \varepsilon.$$

For each positive integer k and each positive rational number r let $A(k,r)$ be the set of all points $x \in A$ such that

$$\left(\frac{1}{\theta} + \varepsilon\right) r \leq |f'(x)| \leq (\theta - \varepsilon)r.$$

With the help of the triangle inequality, observe that if $x, y \in [a, b]$ with $x \in A(k, r)$ and $|y - x| < 1/k$, then

$$|f(y) - f(x)| \leq \varepsilon r \, |(y - x)| + |f'(x)(y - x)| \leq \theta r \, |y - x|$$

and

$$|f(y) - f(x)| \geq -\varepsilon r \, |y - x| + |f'(x)(y - x)| \geq \frac{r}{\theta} \, |y - x| \, .$$

Clearly,

$$A = \bigcup A(k, r),$$

where the union is taken as k and r range through the positive integers and positive rationals, respectively. To ensure that (7.66) and (7.67) hold for all $a, b \in E_k$ (defined below), we express $A(k, r)$ as the countable union of sets each having diameter $1/k$ by writing

$$A(k, r) = \bigcup_{s \in \mathbb{Q}} [A(k, r) \cap I(s, 1/k)]$$

where $I(s, 1/k)$ denotes the open interval of length $1/k$ centered at the rational number s. The sets $[A(k, r) \cap I(s, 1/k)]$ constitute a countable collection as k, r, and s range through their respective sets. Relabel the sets $[A(k, r) \cap I(s, 1/k)]$ as E_k, $k = 1, 2, \ldots$. We may assume that the E_k are disjoint by appealing to Lemma 4.7, thus obtaining the desired result. □

The next result differs from the preceding one only in that the hypothesis now allows the set A to include critical points of f. There is no essential difference in the proof.

7.46. THEOREM. *Suppose f is defined on $[a, b]$ and let A be defined by*

$$A = [a, b] \cap \{x : f'(x) \textit{ exists}\}.$$

Then for each $\theta > 1$, there is a countable collection $\{E_k\}$ of disjoint sets such that

(i) $A = \cup_{k=1}^{\infty} E_k$.

(ii) *For each positive integer k there is a positive rational number r such that*

$$|f'(x)| \leq \theta r \quad \textit{for} \quad x \in E_k$$

and

$$|f(y) - f(x)| \leq \theta r \, |y - x| \quad \textit{for} \quad x, y \in E_k.$$

Before giving the proof of the next main result, we take a slight diversion that is concerned with the extension of Lipschitz functions. We will give a proof of a special case of **Kirzbraun's theorem**, whose general formulation we do not require and is more difficult to prove.

7.47. THEOREM. *Let $A \subset \mathbb{R}$ be an arbitrary set and suppose $f \colon A \to \mathbb{R}$ is a Lipschitz function with Lipschitz constant C. Then there exists a Lipschitz function $\bar{f} \colon \mathbb{R} \to \mathbb{R}$ with the same Lipschitz constant C such that $\bar{f} = f$ on A.*

PROOF. Define

$$\bar{f}(x) = \inf\{f(a) + C\,|x-a| : a \in A\}.$$

Clearly, $\bar{f} = f$ on A, because if $b \in A$, then

$$f(b) - f(a) \le C\,|b-a|$$

for all $a \in A$. This shows that $f(b) \le \bar{f}(b)$. On the other hand, $\bar{f}(b) \le f(b)$ follows immediately from the definition of $\bar{f}$. Finally, to show that $\bar{f}$ has Lipschitz constant C, let $x, y \in \mathbb{R}$. Then

$$\begin{aligned}\bar{f}(x) &\le \inf\{f(a) + C(|y-a| + |x-y|) : a \in A\}\\ &= \bar{f}(y) + C\,|x-y|,\end{aligned}$$

which proves $\bar{f}(x) - \bar{f}(y) \le C\,|x-y|$. The proof with x and y interchanged is similar. □

7.48. THEOREM. *Suppose f is a continuous function on $[a,b]$ and let*

$$A = [a,b] \cap \{x : f'(x) \text{ exists and } f'(x) \ne 0\}.$$

Then for every Lebesgue measurable set $E \subset A$,

$$\int_E |f'|\, d\lambda = \int_{-\infty}^{\infty} N(f, E, y)\, d\lambda(y). \tag{7.68}$$

Equality is understood in the sense that either both sides are finite and equal or both sides are infinite.

PROOF. We apply Theorem 7.45 with A replaced by E. Thus, for each $k \in \mathbb{N}$ there is a positive rational number r such that

$$\frac{r}{\theta}\lambda(E_k) \le \int_{E_k} |f'|\, d\lambda \le \theta r \lambda(E_k)$$

and f restricted to E_k satisfies a Lipschitz condition with constant θr. Therefore by Theorem 7.47, f has a Lipschitz extension to $\mathbb{R}$ with the same Lipschitz constant. From Exercise 4, Section 7.5, we obtain

$$\lambda[f(E_k)] \le \theta r \lambda(E_k). \tag{7.69}$$

Theorem 7.45 states that f restricted to E_k is univalent and that its inverse function is Lipschitz with constant θ/r. Thus, with the same reasoning as before,

$$\lambda(E_k) \le \frac{\theta}{r}\lambda[f(E_k)].$$

Hence, we obtain

$$\frac{1}{\theta^2}\lambda(f[E_k]) \le \int_{E_k} |f'|\, d\lambda \le \theta^2 \lambda[f(E_k)]. \tag{7.70}$$

Each E_k can be expressed as the countable union of compact sets and a set of Lebesgue measure zero. Since f restricted to E_k satisfies a Lipschitz condition, f maps the set of measure zero into a set of measure zero, and each compact set is mapped into a compact set. Consequently, $f(E_k)$ is the countable union of compact sets and a set of measure zero and iş therefore Lebesgue measurable. Let

$$g(y) = \sum_{k=1}^{\infty} \chi_{f(E_k)}(y),$$

so that $g(y)$ is the number of sets $\{f(E_k)\}$ that contain y. Observe that g is Lebesgue measurable. Since the sets $\{E_k\}$ are disjoint, their union is E, and f restricted to each E_k is univalent, we have

$$g(y) = N(f, E, y).$$

Finally,

$$\frac{1}{\theta^2} \sum_{k=1}^{\infty} \lambda(f[E_k]) \leq \int_E |f'| \, d\lambda \leq \theta^2 \sum_{k=1}^{\infty} \lambda[f(E_k)]$$

and, with the aid of Corollary 6.15,

$$\begin{aligned} \int N(f, E, y) \, d\lambda(y) &= \int g(y) \, d\lambda(y) \\ &= \int \sum_{k=1}^{\infty} \chi_{f(E_k)}(y) \, d\lambda(y) \\ &= \sum_{k=1}^{\infty} \int \chi_{f(E_k)}(y) \, d\lambda(y) \\ &= \sum_{k=1}^{\infty} \lambda[f(E_k)]. \end{aligned}$$

The result now follows from (7.70), since $\theta > 1$ is arbitrary. □

7.49. COROLLARY. *If f satisfies* (7.68), *then f satisfies condition N on the set A.*

We conclude this section with a another result concerning absolute continuity. Theorem 7.46 states that a function possesses some regularity properties on the set on which it is differentiable. Therefore, it seems reasonable to expect that if f is differentiable everywhere in its domain of definition, then it will have to be a "nice" function. This is the thrust of the next result.

7.50. THEOREM. *Suppose $f\colon (a, b) \to \mathbb{R}$ has the property that f' exists everywhere and f' is integrable. Then f is absolutely continuous.*

PROOF. Referring to Theorem 7.45, we find that (a, b) can be written as the union of a countable collection $\{E_k,\ k = 1, 2, \ldots\}$ of disjoint Borel sets such that the restriction of f to each E_k is Lipschitz. Hence, it follows that

f satisfies condition N on (a, b). Since f is continuous on (a, b), it remains to show that f is of bounded variation.

For this, let $E_0 = (a, b) \cap \{x : f'(x) = 0\}$. According to Theorem 7.44, $\lambda[f(E_0)] = 0$. Therefore, Theorem 7.48 implies

$$\int_{E_k} |f'| \, d\lambda = \int_{-\infty}^{\infty} N(f, E_k, y) \, d\lambda(y)$$

for $k = 1, 2, \ldots$. Hence,

$$\int_a^b |f'| \, d\lambda = \int_{-\infty}^{\infty} N(f, (a, b), y) \, d\lambda(y),$$

and since f' is integrable by assumption, it follows that f is of bounded variation by Theorem 7.33. □

Exercises for Section 7.8

1. Show that the conclusion of Theorem 7.50 still holds if the following assumptions are satisfied: f' exists everywhere on (a, b) except for a countable set, f' is integrable, and f is continuous.
2. Prove that the sets $A(k, r)$ defined in the proof of Theorem 7.45 are Borel sets. Hints:
   (a) For each positive rational number r, let $A_1(r)$ denote all points $x \in A$ such that
   $$\left(\frac{1}{\theta} + \varepsilon\right) r \leq |f'(x)| \leq (\theta - \varepsilon) r.$$
   Show that $A_1(r)$ is a Borel set.
   (b) Let f be a continuous function on $[a, b]$. Let
   $$F_1(x, y) := \frac{f(y) - f(x)}{x - y} \quad \text{for all } x, y \in [a, b] \text{ with } a \neq b.$$
   Prove that F_1 is a Borel function on $[a, b] \times [a, b] \setminus \{(x, y) : x = y\}$.
   (c) Let $F_2(x, y) := f'(x)$ for $x \in [a, b]$. Show that F_2 is a Borel function on $A \times \mathbb{R}$.
   (d) For each positive integer k, let $A(k) = \{(x, y) : |y - x| < 1/k\}$. Note that $A(k)$ is open.
   (e) $A(k, r)$ is thus a Borel set, since
   $$A(k, r) = \{x : |F_1(x, y) - F_2(x, y)| \leq \varepsilon r\} \cap A_1(r) \cap \{x : (x, y) \in A(k)\}.$$

7.9. Approximate Continuity

In Section 7.2 the notion of Lebesgue point allowed us to define an integrable function, f, at almost all points in a way that does not depend on the choice of function in the equivalence class determined by f. In the development below, the concept of approximate continuity will permit us to carry through a similar program for functions that are merely measurable.

A key ingredient in the development of Section 7.2 occurred in the proof of Theorem 7.11, where a continuous function was used to approximate an integrable function in the L^1-norm. A slightly disquieting feature of that development is that it does not allow the approximation of measurable functions, only integrable ones. In this section this objection is addressed by introducing the concept of **approximate continuity**.

Throughout this section, we use the following notation. Recall that some of it was introduced in (7.17) and (7.18):

$$A_t = \{x : f(x) > t\},$$

$$B_t = \{x : f(x) < t\},$$

$$\overline{D}(E, x) = \limsup_{r \to 0} \frac{\lambda(E \cap B(x,r))}{\lambda(B(x,r))},$$

and

$$\underline{D}(E, x) = \liminf_{r \to 0} \frac{\lambda(E \cap B(x,r))}{\lambda(B(x,r))}.$$

If the upper and lower limits are equal, we denote their common value by $D(E, x)$. Note that the sets A_t and B_t are defined up to sets of Lebesgue measure zero.

7.51. DEFINITION. Before giving the next definition, let us first review the definition of limit superior of a function that we discussed earlier on page p. 64. Recall that

$$\limsup_{x \to x_0} f(x) := \lim_{r \to 0} M(x_0, r),$$

where $M(x_0, r) = \sup\{f(x) : 0 < |x - x_0| < r\}$. Since $M(x_0, r)$ is a nondecreasing function of r, the limit of the right-hand side exists. We define $L(x_0) := \limsup_{x \to x_0} f(x)$ and let

$$T := \{t : B(x_0, r) \cap A_t = \emptyset \quad \text{for all small } r\}. \tag{7.71}$$

If $t \in T$, then there exists $r_0 > 0$ such that for all $0 < r < r_0$, $M(x_0, r) \leq t$. Since $M(x_0, r) \downarrow L(x_0)$, it follows that $L(x_0) \leq t$, and therefore, $L(x_0)$ is a lower bound for T. On the other hand, if $L(x_0) < t < t'$, then the definition of $M(x_0, r)$ implies that $M(x_0, r) < t'$ for all small $r > 0$, which implies $B(x_0, r) \cap A_{t'} = \emptyset$ for all small r. Since this is true for each $t' > L(x_0)$, we conclude that t is not a lower bound for T and therefore that $L(x_0)$ is the greatest lower bound; that is,

$$\limsup_{x \to x_0} f(x) = L(x_0) = \inf T. \tag{7.72}$$

With the above serving as motivation, we proceed with the measure-theoretic counterpart of (7.72): If f is a Lebesgue measurable function defined on $\mathbb{R}^n$, the **upper (lower) approximate limit** of f at a point x_0 is defined by

$$\operatorname{ap}\limsup_{x\to x_0} f(x) = \inf\{t : D(A_t, x_0) = 0\},$$
$$\operatorname{ap}\liminf_{x\to x_0} f(x) = \sup\{t : D(B_t, x_0) = 0\}.$$

We speak of the **approximate limit** of f at x_0 when

$$\operatorname{ap}\limsup_{x\to x_0} f(x) = \operatorname{ap}\liminf_{x\to x_0} f(x)$$

and f is said to be **approximately continuous** at x_0 if

$$\operatorname{ap}\lim_{x\to x_0} f(x) = f(x_0).$$

Note that if $g = f$ a.e., then the sets A_t and B_t corresponding to g differ from those for f by at most a set of Lebesgue measure zero, and thus the upper and lower approximate limits of g coincide with those of f everywhere.

In topology, a point x is interior to a set E if there is a ball $B(x,r)$ that is a subset of E. In other words, x is interior to E if it is completely surrounded by other points in E. In measure theory, it would be natural to say that x is interior to E (in the measure-theoretic sense) if $D(E,x) = 1$. See Exercise 1, Section 7.9.

The following is a direct consequence of Theorem 7.11, which implies that almost every point of a measurable set is interior to it (in the measure-theoretic sense).

7.52. THEOREM. *If $E \subset \mathbb{R}^n$ is a Lebesgue measurable set, then*

$$D(E,x) = 1 \quad \text{for } \lambda\text{-almost all } x \in E,$$
$$D(E,x) = 0 \quad \text{for } \lambda\text{-almost all } x \in \widetilde{E}.$$

Recall that a function f is continuous at x if for every open interval I containing $f(x)$, x is interior to $f^{-1}(I)$. This remains true in the measure-theoretic context.

7.53. THEOREM. *Suppose $f \colon \mathbb{R}^n \to \mathbb{R}$ is a Lebesgue measurable function. Then f is approximately continuous at x if and only if for every open interval I containing $f(x)$, $D[f^{-1}(I), x] = 1$.*

PROOF. Assume that f is approximately continuous at x and let I be an arbitrary open interval containing $f(x)$. We will show that $D[f^{-1}(I), x] = 1$. Let $J = (t_1, t_2)$ be an interval containing $f(x)$ whose closure is contained in I. From the definition of approximate continuity, we have

$$D(A_{t_2}, x) = D(B_{t_1}, x) = 0$$

and therefore

$$D(A_{t_2} \cup B_{t_1}, x) = 0.$$

Since

$$\mathbb{R}^n - f^{-1}(I) \subset A_{t_2} \cup B_{t_1},$$

it follows that $D(\mathbb{R}^n - f^{-1}(I)) = 0$ and therefore that $D[f^{-1}(I), x] = 1$, as desired.

For the proof in the opposite direction, assume that $D(f^{-1}(I), x) = 1$ whenever I is an open interval containing $f(x)$. Let t_1 and t_2 be any numbers $t_1 < f(x) < t_2$. With $I = (t_1, t_2)$ we have $D(f^{-1}(I), x) = 1$. Hence $D[\mathbb{R}^n - f^{-1}(I), x] = 0$. This implies $D(A_{t_2}, x) = D(B_{t_1}, x) = 0$, which implies that the approximate limit of f at x is $f(x)$. □

The next result shows the great similarity between continuity and approximate continuity.

7.54. THEOREM. *A function f is approximately continuous at x if and only if there exists a Lebesgue measurable set E containing x such that $D(E, x) = 1$ and the restriction of f to E is continuous at x.*

PROOF. We will prove only the difficult direction. The other direction is left to the reader. Thus, assume that f is approximately continuous at x. The definition of approximate continuity implies that there are positive numbers $r_1 > r_2 > r_3 > \dots$ tending to zero such that

$$\lambda\left[B(x,r) \cap \left\{y : |f(y) - f(x)| > \frac{1}{k}\right\}\right] < \frac{\lambda[B(x,r)]}{2^k}, \quad \text{for } r \le r_k.$$

Define

$$E = \mathbb{R}^n \setminus \bigcup_{k=1}^{\infty} \left[\{B(x,r_k) \setminus B(x,r_{k+1})\} \cap \left\{y : |f(y) - f(x)| > \frac{1}{k}\right\}\right].$$

From the definition of E, it follows that the restriction of f to E is continuous at x. In order to complete the assertion, we will show that $D(\widetilde{E}, x) = 0$. For this purpose, choose $\varepsilon > 0$ and let J be such that $\sum_{k=J}^{\infty} \frac{1}{2^k} < \varepsilon$. Furthermore, choose r such that $0 < r < r_J$ and let $K \ge J$ be the integer such that $r_{K+1} \le r < r_K$. Then,

$$\begin{aligned}
\lambda[(\mathbb{R}^n \setminus E) \cap B(x,r)] &\le \lambda\left[B(x,r) \cap \{y \,:\, |f(y) - f(x)| > \frac{1}{K}\}\right] \\
&\quad + \sum_{k=K+1}^{\infty} \lambda\left[\{B(x,r_k) \setminus B(x,r_{k+1})\}\right. \\
&\quad \left. \cap \left\{y : |f(y) - f(x)| > \frac{1}{k}\right\}\right] \\
&\le \frac{\lambda[B(x,r)]}{2^K} + \sum_{k=K+1}^{\infty} \frac{\lambda[B(x,r_k)]}{2^k}
\end{aligned}$$

$$\leq \frac{\lambda[B(x,r)]}{2^K} + \sum_{k=K+1}^{\infty} \frac{\lambda[B(x,r)]}{2^k}$$
$$\leq \lambda[B(x,r)] \sum_{k=K}^{\infty} \frac{1}{2^k}$$
$$\leq \lambda[B(x,r)] \cdot \varepsilon,$$

which yields the desired result, since ε is arbitrary. □

7.55. THEOREM. *Assume that $f\colon \mathbb{R}^n \to \mathbb{R}$ is Lebesgue measurable. Then f is approximately continuous λ-almost everywhere.*

PROOF. First, we will prove that there exist disjoint compact sets $K_i \subset \mathbb{R}^n$ such that

$$\lambda\left[\mathbb{R}^n - \bigcup_{i=1}^{\infty} K_i\right] = 0$$

and f restricted to each K_i is continuous. To this end, set $B_i = B(0,i)$ for each positive integer i. By Lusin's theorem, there exists a compact set $K_1 \subset B_1$ with $\lambda(B_1 - K_1) \leq 1$ such that f restricted to K_1 is continuous. Assuming that $K_1, K_2, \ldots, K_j$ have been constructed, we appeal to Lusin's theorem again to obtain a compact set K_{j+1} such that

$$K_{j+1} \subset B_{j+1} - \bigcup_{i=1}^{j} K_i, \qquad \lambda\left[B_{j+1} - \bigcup_{i=1}^{j+1} K_i\right] \leq \tfrac{1}{j+1},$$

and f restricted to K_{j+1} is continuous. Let

$$E_i = K_i \cap \{x : D(K_i, x) = 1\}$$

and recall from Theorem 7.13 that $\lambda(K_i - E_i) = 0$. Thus, E_i has the property that $D(E_i, x) = 1$ for each $x \in E_i$. Furthermore, f restricted to E_i is continuous, since $E_i \subset K_i$. Hence, by Theorem 7.54 we have that f is approximately continuous at each point of E_i and therefore at each point of

$$\bigcup_{i=1}^{\infty} E_i.$$

Since

$$\lambda\left[\mathbb{R}^n - \bigcup_{i=1}^{\infty} E_i\right] = \lambda\left[\mathbb{R}^n - \bigcup_{i=1}^{\infty} K_i\right] = 0,$$

we obtain the conclusion of the theorem. □

Exercises for Section 7.9

1. Define a set E to be "density open" if E is Lebesgue measurable and $D(E, x) = 1$ for all $x \in E$. Prove that the density open sets form a topology. The issue here is the following: In order to show that the density

open sets form a topology, let $\{E_\alpha\}$ denote an arbitrary (possibly uncountable) collection of density open sets. It must be shown that $E := \cup_\alpha E_\alpha$ is density open. In particular, it must be shown that E is measurable.

2. Prove that a function $f\colon \mathbb{R}^n \to \mathbb{R}$ is approximately continuous at every point if and only if f is continuous in the density open topology of Exercise 1, Section 7.9.
3. Let f be an arbitrary function with the property that for each $x \in \mathbb{R}^n$ there exists a measurable set E such that $D(E, x) = 1$ and f restricted to E is continuous at x. Prove that f is a measurable function.
4. Suppose f is a bounded measurable function on $\mathbb{R}^n$ that is approximately continuous at x_0. Prove that x_0 is a Lebesgue point for f. Hint: Use Definition 7.51.
5. Show that if f has a Lebesgue point at x_0, then f is approximately continuous at x_0.

CHAPTER 8

Elements of Functional Analysis

8.1. Normed Linear Spaces

We have already encountered examples of normed linear spaces, namely the L^p spaces. Here we introduce the notion of abstract normed linear spaces and begin the investigation of the structure of such spaces.

8.1. DEFINITION. A **linear space** (or vector space) is a set X that is endowed with two operations, **addition** and **scalar multiplication**, that satisfy the following conditions: for every $x, y, z \in X$ and $\alpha, \beta \in \mathbb{R}$:

(i) $x + y = y + x \in X$.
(ii) $x + (y + z) = (x + y) + z$.
(iii) There is an element $0 \in X$ such that $x + 0 = x$ for each $x \in X$.
(iv) For each $x \in X$ there is an element $w \in X$ such that $x + w = 0$.
(v) $\alpha x \in X$.
(vi) $\alpha(\beta x) = (\alpha\beta)x$.
(vii) $\alpha(x + y) = \alpha x + \alpha y$.
(viii) $(\alpha + \beta)x = \alpha x + \beta y$.
(xi) $1x = x$.

We note here some immediate consequences of the definition of a linear space. If $x, y, z \in X$ and

$$x + y = x + z,$$

then by conditions (i) and (iv), there is a $w \in X$ such that $w + x = 0$ and hence

$$y = 0 + y = (w + x) + y = w + (x + y) = w + (x + z) = (w + x) + z = 0 + z = z.$$

Thus, in particular, for each $x \in X$ there is exactly one element $w \in X$ such that $x + w = 0$. We will denote that element by $-x$, and we will write $y - x$ for $y + (-x)$.

If $\alpha \in \mathbb{R}$ and $x \in X$, then $\alpha x = \alpha(x + 0) = \alpha x + \alpha 0$, from which we can conclude that $\alpha 0 = 0$. Similarly, $\alpha x = (\alpha + 0)x = \alpha x + 0x$, from which we conclude that $0x = 0$.

W.P. Ziemer, *Modern Real Analysis*, Graduate Texts in Mathematics 278,
https://doi.org/10.1007/978-3-319-64629-9_8

If $\lambda \neq 0$ and $\lambda x = 0$, then

$$x = (\frac{\lambda}{\lambda})x = \frac{1}{\lambda}(\lambda x) = \frac{1}{\lambda}0 = 0.$$

8.2. Definition. A subset Y of a linear space X is a **subspace** of X if $\alpha x + \beta y \in Y$ for all $x, y \in Y$ and $\alpha, \beta \in \mathbb{R}$.

Thus if Y is a subspace of X, then Y is itself a linear space with respect to the addition and scalar multiplication it inherits from X. The notion of subspace that we have defined above might more properly be called **linear** subspace to distinguish it from the notion of topological subspace in case X is also a topological space. In this chapter we will use the term "subspace" in the sense of the definition above. If we have occasion to refer to a topological subspace we will mention it explicitly.

If S is a nonempty subspace of a linear space X, then the set Y of all elements of X of the form

$$\alpha_1 x_1 + \alpha_2 x_2 + \cdots + \alpha_m x_m,$$

where m is any positive integer and $\alpha_j \in \mathbb{R}$, $x_j \in S$ for $1 \leq j \leq m$, is easily seen to be a subspace of X. The subspace Y will be called the subspace **spanned** by S. It is the smallest subspace of X that contains S.

8.3. Definition. If $S = \{x_1, x_2, \ldots, x_m\}$ is a finite subset of a linear space X, then S is **linearly independent** if

$$\alpha_1 x_1 + \alpha_2 x_2 + \cdots + \alpha_m x_m = 0$$

implies $\alpha_1 = \alpha_2 = \cdots = \alpha_m = 0$. In general, a subset S of X is linearly independent if every finite subset of S is linearly independent.

Suppose S is a subset of a linear space X. If S is linearly independent and X is spanned by S, then for every $x \in X$ there exist a finite subset $\{x_i\}_{i=1}^m$ of S and a finite sequence of real numbers $\{\alpha_i\}_{i=1}^m$ such that

$$x = \sum_{i=1}^{m} \alpha_i x_i,$$

where $\alpha_i \neq 0$ for each i.

Suppose that for some other choice $\{y_j\}_{j=1}^k$ of elements in S and real numbers $\{\beta_j\}_{j=1}^k$ one has

$$x = \sum_{j=1}^{k} \beta_j y_j,$$

where $\beta_j \neq 0$ for each j. Then

$$\sum_{i=1}^{m} \alpha_i x_i - \sum_{j=1}^{k} \beta_j y_j = 0.$$

If for some $i \in \{1, 2, \ldots, m\}$ there were no $j \in \{1, 2, \ldots, k\}$ such that $x_i = y_j$, then since S is linearly independent, we would have $\alpha_i = 0$. Similarly, for each $j \in \{1, 2, \ldots, k\}$ there is an $i \in \{1, 2, \ldots, m\}$ such that $x_i = y_j$. Thus the two sequences $\{x_i\}_{i=1}^m$ and $\{y_j\}_{j=1}^k$ must contain exactly the same elements. Renumbering the β_j if necessary, we see that

$$\sum_{i=1}^{m} (\alpha_i - \beta_i) x_i = 0.$$

Since S is linearly independent, we have $\alpha_i = \beta_i$ for $1 \leq i \leq k$. Thus each $x \in X$ has a unique representation as a finite linear combination of elements of S.

8.4. DEFINITION. A subset of a linear space X that is linearly independent and spans X is a **basis** for X. A linear space is **finite-dimensional** if it has a finite basis.

The proof of the following result is a consequence of the Hausdorff maximal principle; see Exercise 1, Section 8.1.

8.5. THEOREM. *Every linear space has a basis.*

8.6. EXAMPLES. (i) The set $\mathbb{R}^n$ is a linear space with respect to the addition and scalar multiplication defined by

$$\begin{aligned} (x_1, x_2, \ldots, x_n) + (y_1, y_2, \ldots, y_n) &= (x_1 + y_1, \ldots, x_n + y_n), \\ \alpha(x_1, x_2, \ldots, x_n) &= (\alpha x_1, \alpha x_2, \ldots, \alpha x_n). \end{aligned}$$

(ii) If A is a set, then the set of all real-valued functions on A is a linear space with respect to the addition and scalar multiplication

$$\begin{aligned} (f + g)(x) &= f(x) + g(x), \\ (\alpha f)(x) &= \alpha f(x). \end{aligned}$$

(iii) If S is a topological space, then the set $C(S)$ of all real-valued continuous functions on S is a linear space with respect to the addition and scalar multiplication defined in (ii).

(iv) If (X, μ) is a measure space, then by Theorem 6.23, $L^p(X, \mu)$ is a linear space for $1 \leq p \leq \infty$ with respect to the addition and scalar multiplication defined in (ii).

(v) For $1 \leq p < \infty$ let l^p denote the set of all sequences $\{a_k\}_{k=1}^{\infty}$ of real numbers such that $\sum_{k=1}^{\infty} |a_k|^p$ converges. Each such sequence may be viewed as a real-valued function on the set of positive integers. If addition and scalar are defined as in example (ii), then each l^p is a linear space.

(vi) Let l^∞ denote the set of all bounded sequences of real numbers. Then with respect to the addition and scalar multiplication of (v), l^∞ is a linear space.

8.7. DEFINITION. Let X be a linear space. A function $\|\cdot\| : X \to \mathbb{R}$ is a **norm** on X if

(i) $\|x+y\| \le \|x\| + \|y\|$ for all $x, y \in X$,
(ii) $\|\alpha x\| = |\alpha| \, \|x\|$ for all $x \in X$ and $\alpha \in \mathbb{R}$,
(iii) $\|x\| \ge 0$ for each $x \in X$,
(iv) $\|x\| = 0$ only if $x = 0$.

A real-valued function on X satisfying conditions (i), (ii), and (iii) is a **seminorm** on X. A linear space X equipped with a norm $\|\cdot\|$ is a **normed linear space.**

Suppose X is a normed linear space with norm $\|\cdot\|$. For $x, y \in X$ set

$$\rho(x,y) = \|x - y\| \,.$$

Then ρ is a nonnegative real-valued function on $X \times X$, and from the properties of $\|\cdot\|$ we see that

(i) $\rho(x,y) = 0$ if and only if $x = y$,
(ii) $\rho(x,y) = \rho(y,x)$ for all $x, y \in X$,
(iii) $\rho(x,z) \le \rho(x,y) + \rho(y,z)$ for all $x, y, z \in X$.

Thus ρ is a metric on X, and we see that a normed linear space is also a metric space; in particular, it is a topological space. Functional analysis (in normed linear spaces) is essentially the study of the interaction between the algebraic (linear) structure and the topological (metric) structure of such spaces.

In a normed linear space X we will denote by $B(x,r)$ the **open ball** with center at x and radius r, i.e.,

$$B(x,r) = \{y \in X : \|y - x\| < r\}.$$

8.8. DEFINITION. A normed linear space is a **Banach space** if it is a complete metric space with respect to the metric induced by its norm.

8.9. EXAMPLES. (i) $\mathbb{R}^n$ is a Banach space with respect to the norm

$$\|(x_1, x_2, \ldots, x_n)\| = (x_1^2 + x_2^2 + \cdots + x_n^2)^{\frac{1}{2}}.$$

(ii) For $1 \le p \le \infty$ the linear spaces $L^p(X,\mu)$ are Banach spaces with respect to the norms

$$\|f\|_{L^p(X,\mu)} = \Big(\int_X |f|^p \, d\mu\Big)^{\frac{1}{p}} \quad \text{if} \quad 1 \le p < \infty,$$
$$\|f\|_{L^p(X,\mu)} = \inf\{M : \mu(\{x \in X : |f(x)| > M\}) = 0\} \quad \text{if} \quad p = \infty.$$

This is a rephrasing of Theorem 6.29.

(iii) For $1 \leq p \leq \infty$ the linear spaces l^p of Examples 8.6 (v), (vi) are Banach spaces with respect to the norms

$$\|\{a_k\}\|_{l^p} = (\sum_{k=1}^{\infty} |a_k|^p)^{\frac{1}{p}} \quad \text{if} \quad 1 \leq p < \infty,$$
$$\|\{a_k\}\|_{l^p} = \sup_{k \geq 1} |a_k| \quad \text{if} \quad p = \infty.$$

This is a consequence of Theorem 6.29 with appropriate choices of X and μ.

(iv) If X is a compact metric space, then the linear space $C(X)$ of all continuous real-valued functions on X is a Banach space with respect to the norm

$$\|f\| = \sup_{x \in X} |f(x)|.$$

If $\{x_k\}$ is a sequence in a Banach space X, the series $\sum_{k=1}^{\infty} x_k$ converges to $x \in X$ if the sequence of partial sums $s_m = \sum_{k=1}^{m} x_k$ converges to x, i.e.,

$$\|x - s_m\| = \left\| x - \sum_{k=1}^{m} x_k \right\| \to 0$$

as $m \to \infty$.

8.10. Proposition. *Suppose X is a Banach space. Then every absolutely convergence series is convergent, i.e., if the series $\sum_{k=1}^{\infty} \|x_k\|$ converges in $\mathbb{R}$, then the series $\sum_{k=1}^{\infty} x_k$ converges in X.*

Proof. If $m > l$, then

$$\left\| \sum_{k=1}^{m} x_k - \sum_{k=1}^{l} x_k \right\| = \left\| \sum_{k=l+1}^{m} x_k \right\| \leq \sum_{k=l+1}^{m} \|x_k\|.$$

Thus if $\sum_{k=1}^{\infty} \|x_k\|$ converges in $\mathbb{R}$, the sequence of partial sums $\{\sum_{k=1}^{m} x_k\}_{m=1}^{\infty}$ is Cauchy in X and therefore converges to some element. □

8.11. Definitions. Suppose X and Y are linear spaces. A mapping $T : X \to Y$ is **linear** if for every $x, y \in X$ and $\alpha, \beta \in \mathbb{R}$, one has

$$T(\alpha x + \beta y) = \alpha T(x) + \beta T(y).$$

If X and Y are normed linear spaces and $T : X \to Y$ is a linear mapping, then T is **bounded** if there exists a constant M such that

$$\|T(x)\| \leq M \ \|x\|$$

for each $x \in X$.

8.12. Theorem. *Suppose X and Y are normed linear spaces and $T : X \to Y$ is linear. Then*

(i) *The linear mapping T is bounded if and only if*

$$\sup\{\|T(x)\| : x \in X,\ \|x\| = 1\} < \infty.$$

(ii) *The linear mapping T is continuous if and only if it is bounded.*

PROOF. (i) If M is a constant such that

$$\|T(x)\| \le M\ \|x\|$$

for each $x \in X$, then for all $x \in X$ with $\|x\| = 1$ we have $\|T(x)\| \le M$.

On the other hand, if

$$K = \sup\{\|T(x)\| : x \in X,\ \|x\| = 1\} < \infty,$$

then for all $0 \neq x \in X$ we have

$$\|T(x)\| = \|x\| \left\| T\left(\frac{x}{\|x\|}\right)\right\| \le K\ \|x\|.$$

(ii) If T is continuous at 0, then there exists $\delta > 0$ such that $\|T(x)\| \le 1$ whenever $x \in B(0, 2\delta)$. If $\|x\| = 1$, then

$$\|T(x)\| = \frac{1}{\delta}\|T(\delta x)\| \le \frac{1}{\delta}.$$

Thus

$$\sup\{\|T(x)\| : x \in X,\ \|x\| = 1\} \le \frac{1}{\delta}.$$

On the other hand, if there is a constant M such that

$$\|T(x)\| \le M\ \|x\|$$

for each $x \in X$, then for every $\varepsilon > 0$, one has

$$\|T(x)\| < \varepsilon$$

whenever $x \in B\left(0, \frac{\varepsilon}{M}\right)$. Let $x_0 \in X$. If $\|y - x_0\| < \frac{\varepsilon}{M}$, then

$$\|T(y) - T(x_0)\| = \|T(y - x_0)\| < \varepsilon.$$

Thus T is continuous on X. □

8.13. DEFINITION. If $T : X \to Y$ is a linear mapping from a normed linear space X into a normed linear space Y, we set

$$\|T\| = \sup\{\|T(x)\| : x \in X,\ \|x\| = 1\}. \tag{8.1}$$

This choice of notation will be justified by Theorem 8.16, where we will show that $\|T\|$ is a norm on an appropriate linear space.

8.14. PROPOSITION. *Suppose X and Y are normed linear spaces, $\{T_k\}$ is a sequence of bounded linear mappings of X into Y, and $T : X \to Y$ is a mapping such that*

$$\lim_{k\to\infty} \|T_k(x) - T(x)\| = 0$$

for each $x \in X$. Then T is a linear mapping and

$$\|T\| \le \liminf_{k\to\infty} \|T_k\|.$$

PROOF. For $x, y \in X$ and $\alpha, \beta \in \mathbb{R}$,

$$\begin{aligned} \|T(\alpha x + \beta y) - \alpha T(x) - \beta T(y)\| &\le \|T(\alpha x + \beta y) - T_k(\alpha x + \beta y)\| \\ &\quad + \|\alpha(T_k(x) - T(x)) + \beta(T_k(y) - T(y))\| \\ &\le \|T(\alpha x + \beta y) - T_k(\alpha x + \beta y)\| \\ &\quad + |\alpha|\, \|T_k(x) - T(x)\| + |\beta|\, \|T_k(y) - T(y)\| \end{aligned}$$

for all $k \ge 1$. Thus T is a linear mapping.

If $x \in X$ with $\|x\| = 1$, then

$$\|T(x)\| \le \|T_k(x)\| + \|T(x) - T_k(x)\| \le \|T_k\| + \|T(x) - T_k(x)\|$$

for all $k \ge 1$. Thus

$$\|T(x)\| \le \liminf_{k\to\infty} \|T_k\|$$

for all $x \in X$ with $\|x\| = 1$ and hence

$$\|T\| = \sup\{\|T(x)\| : x \in X,\ \|x\| = 1\} \le \liminf_{k\to\infty} \|T_k\| .$$

□

Suppose X and Y are linear spaces, and let $\mathcal{L}(X, Y)$ denote the set of all linear mappings of X into Y. For $T, S \in \mathcal{L}(X, Y)$ and $\alpha, \beta \in \mathbb{R}$ define

$$(T + S)(x) = T(x) + S(x),$$
$$(\alpha T)(x) = \alpha T(x),$$

for $x \in X$. Note that these are the "usual" definitions of the sum and scalar multiple of functions. It is left as an exercise to show that with these operations $\mathcal{L}(X, Y)$ is a linear space.

8.15. NOTATION. If X and Y are normed linear spaces, denote by $\mathcal{B}(X, Y)$ the set of all **bounded** linear mappings of X into Y. Clearly $\mathcal{B}(X, Y)$ is a subspace of $\mathcal{L}(X, Y)$. If $Y = \mathbb{R}$, we will refer to the elements of $\mathcal{L}(X, \mathbb{R})$ as **linear functionals** on X.

8.16. THEOREM. *Suppose X and Y are normed linear spaces. Then* (8.1) *defines a norm on $\mathcal{B}(X, Y)$. If Y is a Banach space, then $\mathcal{B}(X, Y)$ is a Banach space with respect to this norm.*

PROOF. Clearly $\|T\| \ge 0$. If $\|T\| = 0$, then $T(x) = 0$ for all $x \in X$ with $\|x\| = 1$. Thus if $0 \ne x \in X$, then

$$T(x) = \|x\|\, T\left(\frac{x}{\|x\|}\right) = 0,$$

i.e., $T = 0$.

If $T, S \in \mathcal{B}(X, Y)$ and $\alpha \in \mathbb{R}$, then

$$\|\alpha T\| = \sup\{|\alpha|\, \|T(x)\| : x \in X,\ \|x\| = 1\} = |\alpha|\, \|T\|$$

and

$$\begin{aligned}\|T+S\| &= \sup\{\|T(x)+S(x)\| : x \in X,\ \|x\|=1\}\\ &\le \sup\{\|T(x)\|+\|S(x)\| : x \in X,\ \|x\|=1\}\\ &\le \|T\|+\|S\|\,.\end{aligned}$$

Thus (8.1) defines a norm on $\mathcal{B}(X,Y)$.

Suppose Y is a Banach space and $\{T_k\}$ is a Cauchy sequence in $\mathcal{B}(X,Y)$. Then $\{\|T_k\|\}$ is bounded, i.e., there is a constant M such that $\|T_k\| \le M$ for all k.

For all $x \in X$ and $k, m \ge 1$, one has

$$\|T_k(x) - T_m(x)\| \le \|T_k - T_m\|\ \|x\|\,.$$

Thus $\{T_k(x)\}$ is a Cauchy sequence in Y. Since Y is a Banach space, there is an element $T(x) \in Y$ such that

$$\|T_k(x) - T(x)\| \to 0$$

as $k \to \infty$. In view of Proposition 8.14 we know that T is a linear mapping of X into Y. Moreover, again by Proposition 8.14,

$$\|T\| = \liminf_{k\to\infty} \|T_k\| \le M,$$

whence $T \in \mathcal{B}(X,Y)$. □

Exercises for Section 8.1

1. Use the Hausdorff maximal principle to show that every linear space has a basis. Hint: Observe that a linearly independent subset of a linear space X spans X if and only if it is maximal with respect to set inclusion, i.e., if and only if it is not contained in any other linearly independent subset of X.
2. For $i = 1, 2, \ldots, m$, let X_i be a Banach space with norm $\|\cdot\|_i$. The Cartesian product

$$X = \prod_{i=1}^{m} X_i$$

consisting of points $x = (x_1, x_2, \ldots, x_m)$ with $x_i \in X_i$ is a vector space under the definitions

$$x + y = (x_1 + y_1, \ldots, x_m + y_m), \quad cx = (cx_1, \ldots, cx_m).$$

Prove that X is a Banach space with respect to each of the equivalent norms

$$\|x\| = \left(\sum_{i=1}^{m} \|x_i\|_i^p\right)^{1/p}, \quad 1 \le p < \infty.$$

8.2. Hahn–Banach Theorem

In this section we prove the existence of extensions of linear functionals from a subspace Y of X to all of X satisfying various conditions.

8.17. THEOREM (Hahn–Banach theorem: seminorm version). *Suppose X is a linear space and p is a seminorm on X. Let Y be a subspace of X and $f : Y \to \mathbb{R}$ a linear functional such that $f(x) \le p(x)$ for all $x \in Y$. Then there exists a linear functional $g : X \to \mathbb{R}$ such that $g(x) = f(x)$ for all $x \in Y$ and $g(x) \le p(x)$ for all $x \in X$.*

PROOF. Let $\mathcal{F}$ denote the family of all pairs (W, h), where W is a subspace with $Y \subset W \subset X$ and h is a linear functional on W such that $h = f$ on Y and $h \le p$ on W. For each $(W, h) \in \mathcal{F}$ let

$$G(W, h) = \{(x, r) : x \in W, r = h(x)\} \subset X \times \mathbb{R}.$$

Observe that if $(W_1, h_1), (W_2, h_2) \in \mathcal{F}$, then $G(W_1, h_1) \subset G(W_2, h_2)$ if and only if

$$W_1 \subset W_2 \tag{8.2}$$

$$h_2 = h_1 \text{ on } W_1. \tag{8.3}$$

Set $\mathcal{E} = \{G(W, h) : (W, h) \in \mathcal{F}\}$. If $\mathcal{T}$ is a subfamily of $\mathcal{E}$ that is linearly ordered by inclusion, set

$$W_\infty = \bigcup\{W : G(W, h) \in \mathcal{T} \text{ for some } (W, h) \in \mathcal{T}\}.$$

Clearly W_∞ is a subspace with $Y \subset W_\infty \subset X$. In view of (8.2) we can define a linear functional h_∞ on W_∞ by setting

$$h_\infty(x) = h(x) \quad \text{if} \quad G(W, h) \in \mathcal{T} \text{ and } x \in W.$$

Thus $(W_\infty, h_\infty) \in \mathcal{F}$ and $G(W, h) \subset G(W_\infty, h_\infty)$ for each $G(W, h) \in \mathcal{T}$. Hence, we may apply Zorn's lemma (see p. 8), to conclude that $\mathcal{E}$ contains a maximal element. This means that there is a pair $(W_0, h_0) \in \mathcal{F}$ such that $G(W_0, h_0)$ is not contained in any other set $G(W, h) \in \mathcal{E}$.

Since from the definition of $\mathcal{F}$ we have $h_0 = f$ on Y and $h_0 \le p$ on W_0, the proof will be complete when we show that $W_0 = X$.

To the contrary, suppose $W_0 \ne X$, and let $x_0 \in X - W_0$. Set

$$W' = \{x + \alpha x_0 : x \in W_0, \alpha \in \mathbb{R}\}.$$

Then W' is a subspace of X containing W_0. If $x, x' \in W_0$, $\alpha, \alpha' \in \mathbb{R}$, and

$$x + \alpha x_0 = x' + \alpha' x_0,$$

then

$$x - x' = (\alpha' - \alpha) x_0.$$

If $\alpha' - \alpha \neq 0$, this would imply that $x_0 \in W_0$. Thus $x = x'$ and $\alpha' = \alpha$. Thus each element of W' has a unique representation in the form $x + \alpha x_0$, and hence we can define a linear functional h' on W' by fixing $c \in \mathbb{R}$ and setting

$$h'(x + \alpha x_0) = h_0(x) + \alpha c$$

for each $x \in W_0$, $\alpha \in \mathbb{R}$. Clearly $h' = h_0$ on W_0. We now choose c such that $h' \leq p$ on W'.

Observe that for $x, x' \in W_0$, we have

$$h_0(x) + h_0(x') = h_0(x + x_0 + x' - x_0) \leq p(x + x_0) + p(x' - x_0),$$

and thus

$$h_0(x') - p(x' - x_0) \leq p(x + x_0) - h_0(x)$$

for all $x, x' \in W_0$.

In view of the last inequality there is a $c \in \mathbb{R}$ such that

$$\sup\{h_0(x) - p(x - x_0) : x \in W_0\} \leq c \leq \inf\{p(x + x_0) - h_0(x) : x \in W_0\}.$$

With this choice of c we see that for all $x \in W_0$ and $\alpha \neq 0$,

$$h'(x + \alpha x_0) = \alpha h'(\frac{x}{\alpha} + x_0) = \alpha(h_0(\frac{x}{\alpha}) + c);$$

if $\alpha > 0$, then $\alpha c \leq p(x + x_0) - h_0(x)$, and therefore

$$\begin{aligned} h'(x + \alpha x_0) &\leq \alpha(h_0(\frac{x}{\alpha}) + p(\frac{x}{\alpha} + x_0) - h_0(\frac{x}{\alpha})) \\ &= \alpha p(\frac{x}{\alpha} + x_0) = p(x + \alpha x_0). \end{aligned}$$

If $\alpha < 0$, then

$$\begin{aligned} h'(x + \alpha x_0) &= \alpha(h_0(\frac{x}{\alpha}) + c) \\ &= |\alpha| (h_0(\frac{x}{|\alpha|}) - c) \\ &\leq |\alpha| (h_0(\frac{x}{|\alpha|}) + p(\frac{x}{|\alpha|} - x_0) - h_0(\frac{x}{|\alpha|})) \\ &= |\alpha| p(\frac{x}{|\alpha|} - x_0) = p(x + \alpha x_0). \end{aligned}$$

Thus $(W', h') \in \mathcal{F}$, and $G(W_0, h_0)$ is a proper subset of $G(W', h')$, contradicting the maximality of $G(W_0, h_0)$. This implies that our assumption that $W_0 \neq X$ must be false, and so it follows that $g := h_0$ is a linear functional on X such that $g = f$ on Y and $g \leq p$ on X. □

8.18. Remark. The proof of Theorem 8.17 actually gives more than is asserted in the statement of the theorem. A careful reading of the proof shows that the function p need not be a seminorm; it suffices that p be **subadditive**, i.e.,

$$p(x + y) \leq p(x) + p(y),$$

and **positively homogeneous**, i.e.,

$$p(\alpha x) = \alpha p(x),$$

whenever $\alpha \geq 0$. In particular, p need not be nonnegative.

As an immediate consequence of Theorem 8.17 we obtain the following result.

8.19. THEOREM (Hahn–Banach theorem: norm version). *Suppose X is a normed linear space and Y is a subspace of X. If f is a linear functional on Y and M is a positive constant such that*

$$|f(x)| \leq M \|x\|$$

for each $x \in Y$, then there is a linear functional g on X such that $g = f$ on Y and

$$|g(x)| \leq M \|x\|$$

for each $x \in X$.

PROOF. Observe that $p(x) = M \|x\|$ is a seminorm on X. Thus by Theorem 8.17 there is a linear functional g on X that extends f to X and such that

$$g(x) \leq M \|x\|$$

for each $x \in X$. Since $g(-x) = -g(x)$ and $\|-x\| = \|x\|$, it follows immediately that

$$|g(x)| \leq M \|x\|$$

for each $x \in X$. □

The following is a useful consequence of Theorem 8.19,

8.20. THEOREM. *Suppose X is a normed linear space, Y is a subspace of X, and $x_0 \in X$ is such that*

$$\rho = \inf_{y \in Y} \|x_0 - y\| > 0,$$

i.e., the distance from x_0 to Y is positive. Then there is a bounded linear functional f on X such that

$$f(y) = 0 \text{ for all } y \in Y, \quad f(x_0) = 1 \quad \text{and} \quad \|f\| = \frac{1}{\rho}.$$

PROOF. We will use the following observation throughout the proof, namely, that since Y is a vector space, it follows that

$$\rho = \inf_{y \in Y} \|x_0 - y\| = \inf_{y \in Y} \|x_0 - (-y)\| = \inf_{y \in Y} \|x_0 + y\| .$$

Set

$$W = \{y + \alpha x_0 : y \in Y, \alpha \in \mathbb{R}\}.$$

Then, as noted in the proof of Theorem 8.17, W is a subspace of X containing Y, and each element of W has a unique representation of the form $y + \alpha x_0$ with $y \in Y$ and $\alpha \in \mathbb{R}$. Thus we can define a linear functional g on W by

$$g(y + \alpha x_0) = \alpha.$$

If $\alpha \neq 0$, then

$$\|y + \alpha x_0\| = |\alpha| \left\| \frac{y}{\alpha} + x_0 \right\| \geq |\alpha| \, \rho,$$

whence

$$|g(y + \alpha x_0)| \leq \frac{1}{\rho} \|y + \alpha x_0\| .$$

Thus g is a bounded linear functional on W such that $g(y) = 0$ for $y \in Y$, $|g(w)| \leq \frac{1}{\rho} \|w\|$ for $w \in W$, and $g(x_0) = 1$. In view of Theorem 8.19 there is a bounded linear functional f on X such that $f = g$ on W and $\|f\| \leq \frac{1}{\rho}$. There is a sequence $\{y_k\}$ in Y such that

$$\lim_{k \to \infty} \|y_k + x_0\| = \rho.$$

Let $x_k = \dfrac{y_k + x_0}{\|y_k + x_0\|}$. Then $\|x_k\| = 1$ and

$$\|f(x_k)\| = \|g(x_k)\| = \frac{1}{\|y_k + x_0\|}$$

for all k, whence $\|f\| = \dfrac{1}{\rho}$. □

Exercises for Section 8.2

1. Suppose Y is a closed subspace of a normed linear space X, $Y \neq X$, and $\varepsilon > 0$. Show that there is an element $x \in X$ such that $\|x\| = 1$ and
$$\inf_{y \in Y} \|x - y\| > 1 - \varepsilon.$$
2. Let $f \colon X \to Y$ be a linear mapping of a normed linear space X into a normed linear space Y. Show that f is bounded if and only if f is continuous at one point.
3. The kernel of a linear mapping $f \colon X \to \mathbb{R}^1$ is the set $\{x : f(x) = 0\}$. Prove that f is bounded if and only if the kernel of f is closed in X.

8.3. Continuous Linear Mappings

In this section we deduce from the Baire category theorem three important results concerning continuous linear mappings between Banach spaces.

We first prove a "linear" version of the uniform boundedness principle, Theorem 3.35.

8.21. Theorem (Uniform boundedness principle). *Let $\mathcal{F}$ be a family of continuous linear mappings from a Banach space X into a normed linear space Y such that*

$$\sup_{T \in \mathcal{F}} \|T(x)\| < \infty$$

for each $x \in X$. Then

$$\sup_{T\in\mathcal{F}} \|T\| < \infty.$$

PROOF. Observe that for each $T \in \mathcal{F}$ the real-valued function $x \mapsto \|T(x)\|$ is continuous. In view of Theorem 3.35, there exist a nonempty open subset U of X and a number $M > 0$ such that

$$\|T(x)\| \leq M$$

for each $x \in U$ and $T \in \mathcal{F}$.

Fix $x_0 \in U$ and let $r > 0$ be such that $B(x_0, r) \subset U$. If $z = x_0 + y \in B(x_0, r)$ and $T \in \mathcal{F}$, then

$$\|T(z)\| = \|T(x_0) + T(y)\| \geq \|T(y)\| - \|T(x_0)\|,$$

i.e.,

$$\|T(y)\| \leq \|T(z)\| + \|T(x_0)\| \leq 2M$$

for each $y \in B(0, r)$.

If $x \in X$ with $\|x\| = 1$, then $\rho x \in B(0, r)$ for all $0 < \rho < r$, in particular for $\rho = r/2$. Therefore, for each $T \in \mathcal{F}$, we have

$$\|T(x)\| = \frac{2}{r}\left\|T(\frac{r}{2}x)\right\| \leq \frac{4M}{r}.$$

Thus

$$\sup_{T\in\mathcal{F}} \|T\| \leq \frac{4M}{r}.$$

□

8.22. COROLLARY. *Suppose X is a Banach space, Y is a normed linear space, $\{T_k\}$ is a sequence of bounded linear mappings of X into Y, and $T : X \to Y$ is a mapping such that*

$$\lim_{k\to\infty} \|T_k(x) - T(x)\| = 0$$

for each $x \in X$. Then T is a bounded linear mapping and

$$\|T\| \leq \liminf_{k\to\infty} \|T_k\| < \infty.$$

PROOF. In view of Proposition 8.14 we have only to show that $\{\|T_k\|\}$ is bounded. Since for each $x \in X$ the sequence $\{T_k(x)\}$ converges, it is bounded. From Theorem 8.21 we see that $\{\|T_k\|\}$ is bounded. □

8.23. DEFINITION. Let X and Y be normed linear spaces. A mapping $T : X \to Y$ is said to be open if $T(U)$ is an open subset of Y whenever U is an open subset of X.

8.24. THEOREM (Open mapping theorem). *If T is a bounded linear mapping of a Banach space X onto a Banach space Y, then T is an open mapping.*

PROOF. Fix $\varepsilon > 0$. Since T maps X onto Y, we have

$$\begin{aligned} Y &= \bigcup_{k=1}^{\infty} T(B(0, k\varepsilon)). \\ &= \bigcup_{k=1}^{\infty} \overline{T(B(0, k\varepsilon))}. \end{aligned}$$

Since Y is a complete metric space, the Baire category theorem asserts that one of these closed sets has a nonempty interior; that is, there exist $k_0 \geq 1$, $y \in Y$, and $\delta > 0$ such that

$$\overline{T(B(0, k_0\varepsilon)} \supset B(y, \delta).$$

First, we will show that the origin is in the interior of $\overline{T(B(0, 2\varepsilon))}$. For this purpose, note that if $z \in B(\frac{y}{k_0}, \frac{\delta}{k_0})$, then

$$\left\| z - \frac{y}{k_0} \right\| < \frac{\delta}{k_0},$$

i.e.,

$$\|k_0 z - y\| < \delta.$$

Thus $k_0 z \in B(y, \delta) \subset \overline{T(B(0, k_0\varepsilon))}$, which implies that $z \in \overline{T(B(0, \varepsilon))}$. Setting $y_0 = \frac{y}{k_0}$ and $\delta_0 = \frac{\delta}{k_0}$, we have

$$B(y_0, \delta_0) \subset \overline{T(B(0, \varepsilon))}.$$

If $w \in B(0, \delta_0)$, then $z = y_0 + w \in B(y_0, \delta_0)$ and there exist sequences $\{x_k\}$ and $\{x'_k\}$ in $B(0, \varepsilon)$ such that

$$\begin{aligned} &\|T(x_k) - y_0\| \to 0 \\ &\|T(x'_k) - z\| \to 0 \end{aligned}$$

and hence

$$\|T(x'_k - x_k) - w\| \to 0$$

as $k \to \infty$. Thus, since $\|x'_k - x_k\| < 2\varepsilon$, it follows that

$$B(0, \delta_0) \subset \overline{T(B(0, 2\varepsilon))}. \tag{8.4}$$

Now we will show that the origin is interior to $T(B(0, 2\varepsilon))$. So fix $0 < \varepsilon_0 < \varepsilon$ and let $\{\varepsilon_k\}$ be a decreasing sequence of positive numbers such that $\sum_{k=1}^{\infty} \varepsilon_k < \varepsilon_0$. In view of (8.4), for each $k \geq 0$ there is a $\delta_k > 0$ such that

$$B(0, \delta_k) \subset \overline{T(B(0, \varepsilon_k))}.$$

We may assume that $\delta_k \to 0$ as $k \to \infty$. Fix $y \in B(0, \delta_0)$. Then y is arbitrarily close to elements of $T(B(0, \varepsilon_0))$, and therefore there is an $x_0 \in B(0, \varepsilon_0)$ such that

$$\|y - T(x_0)\| < \delta_1,$$

i.e.,

$$y - T(x_0) \in B(0, \delta_1) \subset \overline{T(B(0, \varepsilon_1))}.$$

Thus there is an $x_1 \in B(0, \varepsilon_1)$ such that

$$\|y - T(x_0) - T(x_1)\| < \delta_2,$$

whence $y - T(x_0) - T(x_1) \in \overline{T(B(0, \varepsilon_2))}$. By induction there is a sequence $\{x_k\}$ such that $x_k \in B(0, \varepsilon_k)$ and

$$\left\| y - T(\sum_{k=0}^{m} x_k) \right\| < \delta_{m+1}.$$

Thus $T(\sum_{k=0}^{m} x_k)$ converges to y as $m \to \infty$. For all $m > 0$, we have

$$\sum_{k=0}^{m} \|x_k\| < \sum_{k=0}^{\infty} \varepsilon_k < \varepsilon_0,$$

which implies that the series $\sum_{k=0}^{\infty} x_k$ converges absolutely; since X is a Banach space and since $x_k \in B(0, \varepsilon_0)$ for all k, the series converges to an element x in the closure of $B(0, \varepsilon_0)$, which is contained in $B(0, 2\varepsilon_0)$; see Proposition 8.10. The continuity of T implies $T(x) = y$. Since y is an arbitrary point in $B(0, \delta_0)$, we conclude that

$$B(0, \delta_0) \subset T(B(0, 2\varepsilon_0)) \subset T(B(0, 2\varepsilon)), \tag{8.5}$$

which shows that the origin is interior to $T(B(0, 2\varepsilon)$.

Finally, suppose U is an open subset of X and $y = T(x)$ for some $x \in U$. Let $\varepsilon > 0$ be such that $B(x, \varepsilon) \subset U$. Then (8.5) states that there exists $\delta > 0$ such that

$$B(0, \delta) \subset T(B(0, \varepsilon)).$$

From the linearity of T, we have

$$\begin{aligned} T(B(x, \varepsilon)) &= \{T(x) + T(w); w \in B(0, \varepsilon)\} \\ &= \{y + T(w) : w \in B(0, \varepsilon)\} \\ &\supset \{y + z : z \in B(0, \delta)\} = B(y, \delta). \end{aligned}$$

□

As an immediate consequence of the open mapping theorem we have the following corollary.

8.25. COROLLARY. *If T is a one-to-one bounded linear mapping of a Banach space X onto a Banach space Y, then $T^{-1} : Y \to X$ is a bounded linear mapping.*

PROOF. The existence and linearity of T^{-1} are evident. If U is an open subset of X, then the inverse image of U under T^{-1} is simply $T(U)$, which is open by Theorem 8.24. Thus, T^{-1} is bounded, by Theorem 8.12. □

8.26. DEFINITION. The **graph** of a mapping $T: X \to Y$ is the set

$$\{(x, T(x)) : x \in X\} \subset X \times Y.$$

It is left as an exercise to show that the graph of a continuous linear mapping T of a normed linear space X into a normed linear space Y is a closed subset of $X \times Y$. The following theorem shows that when X and Y are Banach spaces, the converse is true (cf. Exercise 5, Section 8.3).

8.27. THEOREM (Closed graph theorem.). *If T is a linear mapping of a Banach space X into a Banach space Y and the graph of T is closed in $X \times Y$, then T is continuous.*

PROOF. For each $x \in X$ set

$$\|x\|_1 := \|x\| + \|T(x)\|.$$

It is readily verified that $\|\cdot\|_1$ is a norm on X. Let us show that it is complete. Suppose $\{x_k\}$ is a Cauchy sequence in X with respect to $\|\cdot\|_1$. Then $\{x_k\}$ is a Cauchy sequence in X and $\{T(x_k)\}$ is a Cauchy sequence in Y. Since X and Y are Banach spaces, there exist $x \in X$ and $y \in Y$ such that $\|x_k - x\| \to 0$ and $\|T(x_k) - y\| \to 0$ as $k \to \infty$. Since the graph of T is closed, we must have

$$(x, y) = (x, T(x)).$$

This implies $\|T(x_k) - T(x)\| \to 0$ as $k \to \infty$, and hence that $\|x - x_k\|_1 \to 0$ as $k \to \infty$. Thus X is a Banach space with respect to the norm $\|\cdot\|_1$.

Consider two copies of X, the first one with X equipped with the norm $\|\cdot\|_1$ and the second with X equipped with $\|\cdot\|$. Then the identity mapping $I\colon (X, \|\cdot\|_1) \to (X, \|\cdot\|)$ is continuous, since $\|x\| \le \|x\|_1$ for all $x \in X$. According to Corollary 8.25, I is an open map, which means that the inverse mapping of I is also continuous. Hence, there is a constant C such that

$$\|x\| + \|T(x)\| = \|x\|_1 \le C \|x\|$$

for all $x \in X$. Evidently $C \ge 1$. Thus

$$\|T(x)\| \le (C - 1) \|x\|,$$

from which we conclude that T is continuous. □

Exercises for Section 8.3

1. Suppose $T : X \to Y$ is a continuous linear mapping of a normed linear space X into a normed linear space Y. Show that the graph of T is closed in $X \times Y$.
2. Suppose $T : X \to Y$ is a univalent, continuous linear mapping of a Banach space X into a Banach space Y. Prove that $T(X)$ is closed in Y if and only if

$$\|x\| \le C \|T(x)\|$$

for each $x \in X$.

3. Let T_k be a sequence of bounded linear operators $T_k\colon X \to Y$, where X is a Banach space and Y is a normed linear space. If $\lim_{k\to\infty} T_k(x)$ exists for each $x \in X$, Corollary 8.22 yields the existence of a bounded linear operator $T\colon X \to Y$ such that $\lim_{k\to\infty} T_k(x) = T(x)$ for each $x \in X$. Give an example that shows that T fails to be bounded if X is not assumed to be a Banach space.
4. Let M be an arbitrary closed subspace of a normed linear space X. Let us say that $x, y \in X$ are *equivalent*, written $x \sim y$, if $x - y \in M$. We will denote by $[x]$ the coset comprising all elements $y \in X$ such that $x \sim y$.
   (a) With $[x] + [y] := [x + y]$ and $[cx] := c[x]$, where $c \in \mathbb{R}$, prove that these operations are well defined and that these cosets form a vector space.
   (b) Let us define
   $$\|[x]\| := \inf_{y\in M} \|x - y\|.$$
   Prove that $\|[\cdot]\|$ is a norm on the space $\mathcal{M}$ of all cosets $[x]$.
   (c) The space $\mathcal{M}$ is called the quotient space and is denoted by X/M. Prove that if M is a closed subspace of a Banach space X, then X/M is also a Banach space.
5. Let X denote the set of all sequences $\{a_k\}_{k=1}^{\infty}$ such that all but finitely many of the a_k are equal to zero.
   (a) Show that X is a linear space under the usual definitions of addition and scalar multiplication.
   (b) Show that $\|\{a_k\}\| = \max_{k\geq 1} |a_k|$ is a norm on X.
   (c) Define a mapping $T : X \to X$ by
   $$T(\{a_k\}) = \{ka_k\}.$$
   Show that T is a linear mapping.
   (d) Show that the graph of T is closed in $X \times X$.
   (e) Show that T is not continuous.

8.4. Dual Spaces

Here we introduce the important concept of the dual space of a normed linear space and the associated notion of weak topology.

8.28. Definition. Let X be a normed linear space. The **dual space** X^* of X is the linear space of all bounded linear functionals on X equipped with the norm
$$\|f\| = \sup\{|f(x)| : x \in X, \|x\| = 1\}.$$
In view of Theorem 8.16, we know that $X^* = \mathcal{B}(X, \mathbb{R})$ is a Banach space.

We will begin with a result concerning the relationship between the topologies of X and X^*. Recall that a topological space is **separable** if it contains a countable dense subset.

8.29. THEOREM. *X is separable if X^* is separable.*

PROOF. Let $\{f_k\}_{k=1}^{\infty}$ be a countable dense subset of X^*. For each k there is an element $x_k \in X$ with $\|x_k\| = 1$ such that

$$|f_k(x_k)| \geq \frac{1}{2} \|f_k\| .$$

Let W denote the set of all finite linear combinations of elements of $\{x_k\}$ with rational coefficients. Then it is easily verified that $\overline{W}$ is a subspace of X.

If $\overline{W} \neq X$, then there exists an element $x_0 \in X - \overline{W}$, and

$$\inf_{w \in \overline{W}} \|x_0 - w\| > 0.$$

By Theorem 8.20 there exists an $f \in X^*$ such that

$$f(w) = 0 \text{ for all } w \in \overline{W} \quad \text{and} \quad f(x_0) = 1.$$

Since $\{f_k\}$ is dense in X^*, there is a subsequence $\{f_{k_j}\}$ for which

$$\lim_{j \to \infty} \left\| f_{k_j} - f \right\| = 0.$$

However, since $\left\| x_{k_j} \right\| = 1$, it follows that

$$\left\| f_{k_j} - f \right\| \geq \left\| f_{k_j}(x_{k_j}) - f(x_{k_j}) \right\| = \left\| f_{k_j}(x_{k_j}) \right\| \geq \frac{1}{2} \left\| f_{k_j} \right\|$$

for each j. Thus $\left\| f_{k_j} \right\| \to 0$ as $j \to \infty$, which implies that $f = 0$, contradicting the fact that $f(x_0) = 1$. Thus $\overline{W} = X$. □

8.30. DEFINITION. If X and Y are linear spaces and $T : X \to Y$ is a one-to-one linear mapping of X onto Y, we will call T a **linear isomorphism** and say that X and Y are **linearly isomorphic**. If, in addition, X and Y are normed linear spaces and $\|T(x)\| = \|x\|$ for each $x \in X$, then T is an **isometric isomorphism** and X and Y are **isometrically isomorphic**.

Denote by X^{**} the dual space of X^*. Suppose X is a normed linear space. For each $x \in X$ let $\Phi(x)$ be the linear functional on X^* defined by

$$\Phi(x)(f) = f(x) \tag{8.6}$$

for each $f \in X^*$. Since

$$|\Phi(x)(f)| \leq \|f\| \, \|x\| ,$$

the linear functional $\Phi(x)$ is bounded; in fact, $\|\Phi(x)\| \leq \|x\|$. Thus $\Phi(x) \in X^{**}$. It is readily verified that Φ is a bounded linear mapping of X into X^{**} with $\|\Phi\| \leq 1$.

The following result is the key to understanding the relationship between X and X^*.

8.31. PROPOSITION. *Suppose X is a normed linear space. Then*

$$\|x\| = \sup\{|f(x)| : f \in X^*, \ \|f\| = 1\},$$

for each $x \in X$.

PROOF. Fix $x \in X$. If $f \in X^*$ with $\|f\| = 1$, then

$$|f(x)| \leq \|f\| \; \|x\| \leq \|x\| .$$

If $x \neq 0$, then the distance from x to the subspace $\{0\}$ is $\|x\|$, and according to Theorem 8.20, there is an element $g \in X^*$ such that $g(x) = 1$ and $\|g\| = \frac{1}{\|x\|}$. Set $f = \|x\| \, g$. Then $\|f\| = 1$ and $f(x) = \|x\|$. Thus

$$\sup\{|f(x)| : f \in X^*, \; \|f\| = 1\} = \|x\|$$

for each $x \in X$. □

8.32. THEOREM. *The mapping Φ is an isometric isomorphism of X onto $\Phi(X)$.*

PROOF. In view of Proposition 8.31, we have

$$\|\Phi(x)\| = \sup\{|f(x)| : f \in X^*, \; \|f\| = 1\} = \|x\|$$

for each $x \in X$. □

The mapping Φ is called the **natural embedding** of X in X^{**}.

8.33. DEFINITION. A normed linear space X is said to be **reflexive** if

$$\Phi(X) = X^{**},$$

in which case X is isometrically isomorphic to X^{**}.

Since X^{**} is a Banach space (see Theorem 8.16), it follows that every reflexive normed linear space is in fact a Banach space.

8.34. EXAMPLES. (i) The Banach space $\mathbb{R}^n$ is reflexive.

(ii) If $1 \leq p < \infty$ and $\dfrac{1}{p} + \dfrac{1}{p'} = 1$, then the linear mapping

$$\Psi \colon L^{p'}(X, \mu) \to (L^p(X, \mu))^*$$

defined by

$$\Psi(g)(f) = \int_X gf \, d\mu$$

for $g \in L^{p'}(X, \mu)$ and $f \in L^p(X, \mu)$ is an isometric isomorphism of $L^{p'}(X, \mu)$ onto $(L^p(X, \mu))^*$. This is a rephrasing of Theorem 6.48 (note that for $p = 1$, μ needs to be σ-finite).

(iii) For $1 < p < \infty$, $L^p(X, \mu)$ is reflexive. In order to show this, we fix $1 < p < \infty$. We need to show that the natural embedding $\Phi : L^p(X, \mu) \to (L^p(X, \mu)^{**}$ is onto. From (ii) we have the isometric isomorphisms

$$\Psi_1 \colon L^{p'}(X, \mu) \to (L^p(X, \mu))^*, \tag{8.7}$$

$$\Psi_1(g)(f) = \int_X gf d\mu \;\; , \; f \in L^p,$$

and

$$\Psi_2 \colon L^p(X, \mu) \to (L^{p'}(X, \mu))^*, \tag{8.8}$$

$$\Psi_2(f)(g) = \int_X fgd\mu \quad , g \in L^{p'}.$$

Let $w \in L^p(X,\mu)^{**}$. From (8.7) we have $w \circ \Psi_1 \in (L^{p'}(X,\mu))^*$. Thus, (8.8) implies that there exists $f \in L^p(X,\mu)$ such that $\Psi_2(f) = w \circ \Psi_1$. Therefore,

$$\Psi_2(f)(g) = w \circ \Psi_1(g) = \int_X fgd\mu = \Psi_1(g)(f) \text{ for all } g \in L^{p'}. \tag{8.9}$$

We now proceed to check that

$$\Phi(f) = w. \tag{8.10}$$

Let $\alpha \in (L^p(X,\mu))^*$. Then from (8.7) we obtain $g \in L^{p'}(X,\mu)$ such that $\Psi_1(g) = \alpha$, and therefore from (8.9) we conclude that

$$\Phi(f)(\alpha) = \alpha(f) = \Psi_1(g)(f) = w(\Psi_1(g)) = w(\alpha),$$

which proves (8.10).

(iv) Let $\Omega \subset \mathbb{R}^n$ be an open set. We recall that λ denotes the Lebesgue measure in $\mathbb{R}^n$. We will prove that $L^1(\Omega,\lambda)$ is not reflexive. Proceeding by contradiction, if $L^1(\Omega,\lambda)$ were reflexive, and since $L^1(\Omega,\lambda)$ is separable (see Exercise 8, Section 8.4), it would follow that $L^1(\Omega,\lambda)^{**}$ is separable, and hence $L^1(\Omega,\lambda)^*$ would also be separable. From (iii) we know that $L^1(\Omega,\lambda)^*$ is isometrically isomorphic to $L^\infty(\Omega,\lambda)$. Therefore, we would conclude that $L^\infty(\Omega,\lambda)$ would be separable, which contradicts Exercise 9, Section 8.4.

In addition to the topology induced by the norm on a normed linear space it is useful to consider a smaller, i.e., "weaker," topology. The **weak topology** on a normed linear space X is the smallest topology on X with respect to which each $f \in X^*$ is continuous. That such a weak topology exists may be seen by observing that the intersection of any family of topologies for X is a topology for X. In particular, the intersection of all topologies for X that contains all sets of the form $f^{-1}(U)$, where $f \in X^*$ and U is an open subset of $\mathbb{R}$, is precisely the weak topology. Every topology for X with respect to which each $f \in X^*$ is continuous must contain the weak topology for X. We temporarily denote the weak topology by $\mathcal{T}_w$. Then $f^{-1}(U) \in \mathcal{T}_w$ whenever $f \in X^*$ and U is open in $\mathbb{R}$. Consequently the family of all subsets of the form

$$\{x : |f_i(x) - f_i(x_0)| < \varepsilon_i,\ 1 \le i \le m\},$$

where $x_0 \in X$, m is a positive integer, and $\varepsilon_i > 0$, $f_i \in X^*$ for $1 \le i \le m$, forms a basis for $\mathcal{T}_w$. From this observation it is evident that a sequence $\{x_k\}$ in X **converges weakly** (i.e., with respect to the topology $\mathcal{T}_w$) to $x \in X$ if and only if

$$\lim_{k\to\infty} f(x_k) = f(x)$$

for each $f \in X^*$.

In order to distinguish the weak topology from the topology induced by the norm, we will refer to the latter as the **strong topology**.

8.35. THEOREM. *Suppose X is a normed linear space and the sequence $\{x_k\}$ converges weakly to $x \in X$. Then the following assertions hold:*

(i) *The sequence $\{\|x_k\|\}$ is bounded.*

(ii) *Let W denote the subspace of X spanned by $\{x_k : k = 1, 2, \ldots\}$. Then x belongs to the closure of W in the strong topology.*

(iii)

$$\|x\| \leq \liminf_{k\to\infty} \|x_k\|.$$

PROOF. (i) Let $f \in X^*$. Since $\{f(x_k)\}$ is a convergent sequence in $\mathbb{R}$, we have

$$\sup\{|f(x_k)| : k = 1, 2, \ldots\} < \infty,$$

which may be written as

$$\sup_{1\leq k<\infty} |\Phi(x_k)(f)| < \infty.$$

Since this is true for each $f \in X^*$ and since X^* is a Banach space (see Theorem 8.16), it follows from the uniform boundedness principle, Theorem 8.21, that

$$\sup_{1\leq k<\infty} \|\Phi(x_k)\| < \infty.$$

In view of Theorem 8.32, this means that

$$\sup_{1\leq k<\infty} \|x_k\| < \infty.$$

(ii) Let $\overline{W}$ denote the closure of W in the strong topology. If $x \notin \overline{W}$, then by Theorem 8.20 there is an element $f \in X^*$ such that $f(x) = 1$ and $f(w) = 0$ for all $w \in \overline{W}$. But since $f(x_k) = 0$ for all k, we must have

$$f(x) = \lim_{k\to\infty} f(x_k) = 0,$$

which contradicts the fact that $f(x) = 1$. Thus $x \in \overline{W}$.

(iii) If $f \in X^*$ and $\|f\| = 1$, then

$$|f(x)| = \lim_{k\to\infty} |f(x_k)| \leq \liminf_{k\to\infty} \|x_k\|.$$

Since this is true for every such f, we have

$$\|x\| \leq \liminf_{k\to\infty} \|x_k\|.$$

□

8.36. THEOREM. *If X is a reflexive Banach space and Y is a closed subspace of X, then Y is a reflexive Banach space.*

PROOF. For $f \in X^*$, let f_Y denote the restriction of f to Y. Then evidently $f_Y \in Y^*$ and $\|f_Y\| \leq \|f\|$. For $\omega \in Y^{**}$ let $\omega_Y : X^* \to \mathbb{R}$ be given by

$$\omega_Y(f) = \omega(f_Y)$$

for each $f \in X^*$. Then ω_Y is a linear functional on X^* and $\omega_Y \in X^{**}$, since

$$|\omega_Y(f)| = |\omega(f_Y)| \leq \|\omega\| \ \|f_Y\| \leq \|\omega\| \ \|f\|$$

for each $f \in X^*$. Since X is reflexive, there is an element $x_0 \in X$ such that $\Phi(x_0) = \omega_Y$, where Φ is as in Definition 8.33, and therefore

$$\omega_Y(f) = \Phi(x_0)(f) = f(x_0)$$

for each $f \in X^*$. If $x_0 \notin Y$, then by Theorem 8.20 there exists $f \in X^*$ such that $f(x_0) = 1$ and $f(y) = 0$ for each $y \in Y$. This implies that $f_Y = 0$, and hence we arrive at the contradiction

$$1 = f(x_0) = \omega_Y(f) = \omega(f_Y) = 0.$$

Thus $x_0 \in Y$.

For every $g \in Y^*$ there is, by Theorem 8.19, an $f \in X^*$ such that $f_Y = g$. Thus

$$g(x_0) = f(x_0) = \omega_Y(f) = \omega(f_Y) = \omega(g).$$

Thus the image of x_0 under the natural embedding of Y into Y^{**} is ω. Since $\omega \in Y^{**}$ is arbitrary, Y is reflexive. □

8.37. THEOREM. *If X is a reflexive Banach space, then for every $R > 0$, the closed ball $B := \{x \in X : \|x\| \leq R\}$ is sequentially compact in the weak topology.*

PROOF. Assume first that X is separable. Then since X is reflexive, X^{**} is separable, and by Theorem 8.29, X^* is separable. Let $\{f_m\}_{m=1}^\infty$ be dense in X^* and $\{x_k\} \in B$. Since $\{f_1(x_k)\}$ is bounded in $\mathbb{R}$, there must be a subsequence $\{x_k^1\}$ of $\{x_k\}$ such that $\{f_1(x_k^1)\}$ converges in $\mathbb{R}$. Since the sequence $\{f_2(x_k^1)\}$ is bounded in $\mathbb{R}$, there is a subsequence $\{x_k^2\}$ of $\{x_k^1\}$ such that $\{f_2(x_k^2)\}$ converges in $\mathbb{R}$. Continuing in this way we obtain a sequence of subsequences of $\{x_k\}$ such that $\{x_k^m\}$ is a subsequence of $\{x_k^{m-1}\}$ for $m > 1$ and $\{f_m(x_k^m)\}$ converges as $k \to \infty$ for each $m \geq 1$. Set $y_k = x_k^k$. Then $\{y_k\}$ is a subsequence of $\{x_k\}$ such that $\{f_m(y_k)\}$ converges as $k \to \infty$ for each $m \geq 1$. For arbitrary $f \in X^*$,

$$\begin{aligned} |f(y_k) - f(y_l)| &\leq |f(y_k) - f_m(y_k)| + |f_m(y_k) - f_m(y_l)| + |f_m(y_l) - f(y_l)| \\ &\leq \|f_m - f\| \left(\|y_k\| + \|y_l\|\right) + |f_m(y_k) - f_m(y_l)| \end{aligned}$$

for all k, l, m. Given $\varepsilon > 0$, there exists an m such that

$$\|f_m - f\| < \frac{\varepsilon}{4M},$$

where $M = \sup_{k\geq 1} \|x_k\| < \infty$. Since $\{f_m(y_k)\}$ is a Cauchy sequence, there is a positive integer K such that

$$|f_m(y_k) - f_m(y_l)| < \frac{\varepsilon}{2}$$

whenever $k, l > K$. Thus, from the previous three inequalities, we have that

$$|f(y_k) - f(y_l)| \leq \frac{\varepsilon}{4M}\, 2M + \frac{\varepsilon}{2} < \varepsilon$$

whenever $k, l > K$, and we see that $\{f(y_k)\}$ is a Cauchy sequence in $\mathbb{R}$. Set

$$\alpha(f) = \lim_{k\to\infty} f(y_k)$$

for each $f \in X^*$. Evidently α is a linear functional on X^*, and since

$$|\alpha(f)| \leq \|f\|\; M$$

for each $f \in X^*$, we have $\alpha \in X^{**}$. Since X is reflexive, we know that the isometry Φ (see (8.6)) is onto X^{**}. Thus, there exists $x \in X$ such that $\Phi(x) = \alpha$ and therefore

$$\Phi(x)(f) := f(x) = \alpha(f) = \lim_{k\to\infty} f(y_k)$$

for each $f \in X^*$. Thus $\{y_k\}$ converges to x in the weak topology, and Theorem 8.35 gives $x \in B$.

Now suppose that X is a reflexive Banach space, not necessarily separable, and suppose $\{x_k\} \in B$. It suffices to show that there exist $x \in B$ and a subsequence such that $\{x_k\} \to x$ weakly. Let Y denote the closure in the strong topology of the subspace of X spanned by $\{x_k\}$. Then Y is obviously separable, and by Theorem 8.36, Y is a reflexive Banach space. Thus there is a subsequence $\{x_{k_j}\}$ of $\{x_k\}$ that converges weakly in Y to an element $x \in Y$, i.e.,

$$g(x) = \lim_{j\to\infty} g(x_{k_j})$$

for each $g \in Y^*$. For $f \in X^*$ let f_Y denote the restriction of f to Y. As in the proof of Theorem 8.36 we see that $f_Y \in Y^*$. Thus

$$f(x) = f_Y(x) = \lim_{j\to\infty} f_Y(x_{k_j}) = \lim_{j\to\infty} f(x_{k_j}).$$

Thus $\{x_{k_j}\}$ converges weakly to x in X. Furthermore, since $\|x_k\| \leq 1$, the same is true for x by Theorem 8.35, and so we have $x \in B$. □

8.38. EXAMPLE. Let $1 < p < \infty$. Since $L^p(\mathbb{R}^n, \lambda)$ is a reflexive Banach space, Theorem 8.37 implies that the ball

$$B = \{f \in L^p : \|f\|_p \leq R\} \subset L^p(\mathbb{R}^n, \lambda)$$

is sequentially compact in the weak topology. Thus if $\{f_k\}$ is a sequence in $L^p(\mathbb{R}^n, \lambda)$ such that

$$\|f_k\|_p \leq R \;\; k = 1, 2, 3, \ldots,$$

then there exist a subsequence $\{f_{k_j}\}$ of $\{f_k\}$ and a function $f \in B$ such that

$$f_{k_j} \to f \text{ weakly.} \tag{8.11}$$

But (8.11) is equivalent to

$$F(f_{k_j}) \to F(f) \quad \text{for all } F \in L^p(\mathbb{R}^n, \lambda)^*.$$

Therefore, Example 8.34 (ii) yields

$$\int_{\mathbb{R}^n} f_{k_j} g d\lambda \to \int_{\mathbb{R}^n} f g d\lambda \quad \text{for all } g \in L^{p'}(\mathbb{R}^n, \lambda).$$

8.39. DEFINITION. If X is a normed linear space, we may also consider the weak topology on X^*, i.e., the smallest topology on X^* with respect to which each linear functional $\omega \in X^{**}$ is continuous. It turns out to be convenient to consider an even weaker topology on X^*. The **weak* topology** on X^* is defined as the smallest topology on X^* with respect to which each linear functional $\omega \in \Phi(X) \subset X^{**}$ is continuous. Here Φ is the natural embedding of X into X^{**}. As in the case of the weak topology on X, we can, utilizing the natural embedding, describe a basis for the weak* topology on X^* as the family of all sets of the form

$$\{f \in X^* \ : \ |f(x_i) - f_0(x_i)| < \varepsilon_i \text{ for } 1 \le i \le m\},$$

where m is any positive integer, $f_0 \in X^*$, and $x_i \in X$, $\varepsilon_i > 0$ for $1 \le i \le m$. Thus a sequence $\{f_k\}$ in X^* converges in the weak* topology to an element $f \in X^*$ if and only if

$$\lim_{k\to\infty} f_k(x) = f(x)$$

for each $x \in X$. Of course, if X is a reflexive Banach space, the weak and weak* topologies on X^* coincide.

We remark that the basis for the weak* topology on X^* is similar in form to the basis for the weak topology on X except that the roles of X and X^* are interchanged.

The importance of the weak* topology is indicated by the following theorem.

8.40. THEOREM (Alaoglu's theorem). *Suppose X is a normed linear space. The unit ball $B := \{f \in X^* : \|f\| \le 1\}$ of X^* is compact in the weak* topology.*

PROOF. If $f \in B$, then $f(x) \in [-\|x\|, \|x\|]$ for each $x \in X$. Set $I_x = [-\|x\|, \|x\|]$ for $x \in X$. Then according to Tychonoff's theorem, Theorem 3.43, the product

$$P = \prod_{x\in X} I_x,$$

with the product topology, is compact. Recall that B is by definition the set of all functions f defined on X with the property that $f(x) \in I_x$ for each $x \in X$. Thus the set B can be viewed as a subset B' of P. Moreover, the relative topology induced on B' by the product topology is easily seen to coincide with the relative topology induced on B by the weak* topology. Thus the proof will be complete if we show that B' is a closed subset of P in the product topology.

Let f be an element in the closure of B'. Then given $\varepsilon > 0$ and $x \in X$, there is a $g \in B'$ such that $|f(x) - g(x)| < \varepsilon$. Thus

$$|f(x)| \leq \varepsilon + |g(x)| \leq \varepsilon + \|x\|,$$

since $g \in B'$. Since ε and x are arbitrary, we must have

$$|f(x)| \leq \|x\| \tag{8.12}$$

for each $x \in X$.

Now suppose that $x, y \in X$, $\alpha, \beta \in \mathbb{R}$, and set $z = \alpha x + \beta y$. Then given $\varepsilon > 0$, there is a $g \in B'$ such that

$$|f(x) - g(x)| < \varepsilon, \quad |f(y) - g(y)| < \varepsilon, \quad |f(z) - g(z)| < \varepsilon.$$

Thus since g is linear, we have

$$\begin{aligned} &|f(z) - \alpha f(x) - \beta f(y)| \\ &\quad \leq |f(z) - g(z)| + |\alpha|\,|f(x) - g(x)| + |\beta|\,|f(y) - g(y)| \\ &\quad \leq \varepsilon(1 + |\alpha| + |\beta|), \end{aligned}$$

from which it follows that

$$f(\alpha x + \beta y) = \alpha f(x) + \beta f(y),$$

i.e., f is linear. In view of (8.12), we have $f \in B'$. Thus B' is closed and hence compact in the product topology, from which it follows immediately that B is compact in the weak* topology. □

The proof of the following corollary of Theorem 8.40 is left as an exercise. Also, see Exercise 5, Section 8.4.

8.41. COROLLARY. *The unit ball in a reflexive Banach space is both compact and sequentially compact in the weak topology.*

8.42. EXAMPLE. We apply Alaoglu's theorem with $X = L^1(\mathbb{R}^n, \lambda)$, which is a normed linear space. Therefore, for all $R > 0$, the ball $B = \{f \in L^1(\mathbb{R}^n, \lambda)^* : \|f\| \leq R\} \subset L^1(\mathbb{R}^n, \lambda)^*$ is compact in the weak* topology. We recall the isometric isomorphism

$$\Psi : L^\infty(\mathbb{R}^n, \lambda) \to L^1(\mathbb{R}^n, \lambda)^*,$$

given by

$$\Psi(g)(f) = \int_{\mathbb{R}^n} gf, \;\; f \in L^1(\mathbb{R}^n, \lambda).$$

Using Ψ, we can rewrite the conclusion of Alaoglu's theorem as saying that

$$B = \{\Psi(f) \in L^1(\mathbb{R}^n, \lambda)^* : \|f\|_\infty \leq R\} \subset L^1(\mathbb{R}^n, \lambda)^*$$

is compact in the weak* topology of $L^1(\mathbb{R}^n, \lambda)^*$. From this it follows that if $\{f_k\} \subset L^\infty(\mathbb{R}^n, \lambda)$ satisfies

$$\|f_k\|_\infty = \|\Psi(f_k)\| \leq R, \;\; k = 1, 2, 3, \ldots,$$

then there exist a subsequence f_{k_j} of $\{f_k\}$ and $f \in L^\infty(\mathbb{R}^n, \lambda)$ such that $\Psi(f_{k_j}) \to \Psi(f)$ in the weak* topology of $L^1(\mathbb{R}^n, \lambda)^*$. This is equivalent to

$$\Psi(f_{k_j})(g) \to \Psi(f)(g) \quad \text{for all } g \in L^1(\mathbb{R}^n, \lambda),$$

that is,

$$\int_{\mathbb{R}^n} f_{k_j} g d\lambda \to \int_{\mathbb{R}^n} f g d\lambda \quad \text{for all } g \in L^1(\mathbb{R}^n, \lambda).$$

Exercises for Section 8.4

1. Suppose X is a normed linear space and $\{f_k\}$ is a sequence in X^* that converges in the weak* topology to $f \in X^*$. Show that
$$\sup_{k \geq 1} \|f_k\| < \infty$$
and
$$\|f\| \leq \liminf_{k \to \infty} \|f_k\| .$$
2. Show that a subspace of a normed linear space is closed in the strong topology if and only if it is closed in the weak topology.
3. Prove that no subspace of a normed linear space can be open.
4. Show that a finite-dimensional subspace of a normed linear space is closed in the strong topology.
5. Show that if X is an infinite-dimensional normed linear space, then there is a bounded sequence $\{x_k\}$ in X no subsequence of which is convergent in the strong topology. (Hint: Use Exercise 4, Section 8.4, and Exercise 1, Section 8.2.) Thus conclude that the unit ball in an infinite-dimensional normed linear space is not compact.
6. Show that every Banach space X is isometrically isomorphic to a closed linear subspace of $C(\Gamma)$ (cf. Example 8.9(iv)), where Γ is a compact Hausdorff space. (Hint: Set
$$\Gamma = \{f \in X^* : \|f\| \leq 1\}$$
with the weak* topology. Use the natural embedding of X into X^{**}.)
7. As usual, let $C[0,1]$ denote the space of continuous functions on $[0,1]$ endowed with the sup norm. Prove that if f_k is a sequence of functions in $C[0,1]$ that converge weakly to f, then the sequence is bounded and $f_k(t) \to f(t)$ for each $t \in [0,1]$.
8. Suppose Ω is an open subset of $\mathbb{R}^n$, and let λ denote Lebesgue measure on Ω. Set
$$\mathcal{P} = \{x = (x_1, x_2, \ldots, x_n) \in \mathbb{R}^n : x_j \quad \text{is rational for each} \quad 1 \leq j \leq n\}$$
and let $\mathcal{Q}$ denote the set of all open cubes in $\mathbb{R}^n$ with edges parallel to the coordinate axes and vertices in $\mathcal{P}$. (i) Show that if E is a Lebesgue

measurable subset of $\mathbb{R}^n$ with $\lambda(E) < \infty$ and $\varepsilon > 0$, then there exists a disjoint finite sequence $\{Q_k\}_{k=1}^m$ with each $Q_k \in \mathcal{Q}$ such that

$$\left\| \chi_E - \sum_{k=1}^{m} \chi_{Q_k} \right\|_{L^p(\Omega,\lambda)} < \varepsilon.$$

(ii) Show that the set of all finite linear combinations of elements of $\{\chi_Q : Q \in \mathcal{Q}\}$ with rational coefficients is dense in $L^p(\Omega, \lambda)$.
(iii) Conclude that $L^p(\Omega, \lambda)$ is separable.

9. Suppose Ω is an open subset of $\mathbb{R}^n$ and let λ denote Lebesgue measure on Ω. Show that $L^\infty(\Omega, \lambda)$ is not separable. (Hint: If $B(x, r) \in \Omega$ and $0 < r_1 < r_2 \leq r$, then

$$\left\| \chi_{B(x,r_1)} - \chi_{B(x,r_2)} \right\|_{L^\infty(\Omega,\lambda)} = 1.)$$

10. Referring to Exercise 2, Section 8.1, prove that there is a natural isomorphism between X^* and $\prod_{i=1}^m X_i^*$. Thus conclude that X is reflexive if each X_i is reflexive.
11. Let X be a normed linear space and suppose that the sequence $\{x_k\}$ converges to $x \in X$ in the strong topology. Show that $\{x_k\}$ converges weakly to x.

8.5. Hilbert Spaces

We consider in this section Hilbert spaces, i.e., Banach spaces in which the norm is induced by an inner product. This additional structure allows us to study the representation of elements of the space in terms of orthonormal systems.

8.43. DEFINITION. An **inner product** on a linear space X is a real-valued function $(x, y) \mapsto \langle x, y \rangle$ on $X \times X$ such that for $x, y, z \in X$ and $\alpha, \beta \in \mathbb{R}$, one has

$$\begin{aligned} \langle x, y \rangle &= \langle y, x \rangle, \\ \langle \alpha x + \beta y, z \rangle &= \alpha \langle x, z \rangle + \beta \langle y, z \rangle, \\ \langle x, x \rangle &\geq 0, \\ \langle x, x \rangle &= 0 \text{ if, and only if, } x = 0. \end{aligned}$$

8.44. THEOREM. *Suppose that X is a linear space on which an inner product $\langle \cdot, \cdot \rangle$ is defined. We can define a norm on X as follows:*

$$\|x\| = \sqrt{\langle x, x \rangle}.$$

PROOF. It follows immediately from the definition of an inner product that

$$\begin{aligned} \|x\| &\geq 0 \text{ for all } x \in X, \\ \|x\| &= 0 \text{ if, and only if } x = 0, \\ \|\alpha x\| &= |\alpha| \, \|x\| \text{ for all } \alpha \in \mathbb{R}, \ x \in X. \end{aligned}$$

Only the triangle inequality remains to be proved. To do this, we first prove the Schwarz inequality

$$|\langle x, y\rangle| \leq \|x\| \ \|y\| .$$

Suppose $x, y \in X$ and $\lambda \in \mathbb{R}$. From the properties of an inner product, we have

$$\begin{aligned} 0 \leq \|x - \lambda y\|^2 &= \langle x - \lambda y, x - \lambda y\rangle \\ &= \|x\|^2 - 2\lambda\langle x, y\rangle + \lambda^2 \|y\|^2 , \end{aligned}$$

and thus

$$2\langle x, y\rangle \leq \frac{1}{\lambda} \|x\|^2 + \lambda \|y\|^2$$

for all $\lambda > 0$. Assuming $y \neq 0$ and setting $\lambda = \frac{\|x\|}{\|y\|}$, we see that

$$\langle x, y\rangle \leq \|x\| \ \|y\| . \tag{8.13}$$

Note that (8.13) also holds if $y = 0$. Since

$$-\langle x, y\rangle = \langle x, -y\rangle \leq \|x\| \ \|y\| ,$$

we see that

$$|\langle x, y\rangle| \leq \|x\| \ \|y\| \tag{8.14}$$

for all $x, y \in X$.

For the triangle inequality, observe that

$$\begin{aligned} \|x + y\|^2 &= \|x\|^2 + 2\langle x, y\rangle + \|y\|^2 \\ &\leq \|x\|^2 + 2 \|x\| \ \|y\| + \|y\|^2 \\ &= (\|x\| + \|y\|)^2. \end{aligned}$$

Thus

$$\|x + y\| \leq \|x\| + \|y\|$$

for all $x, y \in X$, from which we see that $\|\cdot\|$ is a norm on X. □

Thus we see that a linear space equipped with an inner product is a normed linear space. The inequality (8.14) is called the **Schwarz inequality**.

8.45. DEFINITION. A **Hilbert space** is a linear space with an inner product that is a Banach space with respect to the norm induced by the inner product (as in Theorem 8.44).

8.46. DEFINITION. We will say that two elements x, y in a Hilbert space H are **orthogonal** if $\langle x, y\rangle = 0$. If M is a subspace of H, we set

$$M^{\perp} = \{x \in H : \langle x, y\rangle = 0 \text{ for all } y \in M\}.$$

It is easily seen that $M^{\perp}$ is a subspace of H.

We next investigate the "geometry" of a Hilbert space.

8.47. THEOREM. *Suppose M is a closed subspace of a Hilbert space H. Then for each $x_0 \in H$ there exists a unique $y_0 \in M$ such that*

$$\|x_0 - y_0\| = \inf_{y \in M} \|x_0 - y\|.$$

Moreover, y_0 is the unique element of M such that $x_0 - y_0 \in M^{\perp}$.

PROOF. If $x_0 \in M$, the assertion is obvious, so assume $x_0 \in H - M$. Since M is closed and $x_0 \notin M$, we have

$$d = \inf_{y \in M} \|x_0 - y\| > 0.$$

There is a sequence $\{y_k\}$ in M such that

$$\lim_{k \to \infty} \|x_0 - y_k\| = d.$$

For all k, l, we have

$$\begin{aligned} &\|(x_0 - y_k) - (x_0 - y_l)\|^2 + \|(x_0 - y_k) + (x_0 - y_l)\|^2 \\ &= 2\|x_0 - y_k\|^2 + 2\|x_0 - y_l\|^2, \end{aligned}$$

$$\begin{aligned} \|y_k - y_l\|^2 &= 2(\|x_0 - y_k\|^2 + \|x_0 - y_l\|^2) - 4\left\|x_0 - \frac{1}{2}(y_k + y_l)\right\|^2 \\ &\le 2(\|x_0 - y_k\|^2 + \|x_0 - y_l\|^2) - 4d^2. \end{aligned}$$

Since the right-hand side of the last inequality above tends to 0 as $k, l \to \infty$, we see that $\{y_k\}$ is a Cauchy sequence in H, which consequently converges to an element y_0. Since M is closed, $y_0 \in M$.

Now let $y \in M$, $\lambda \in \mathbb{R}$, and compute

$$\begin{aligned} d^2 &\le \|x_0 - (y_0 + \lambda y)\|^2 \\ &= \|x_0 - y_0\|^2 - 2\lambda\langle x_0 - y_0, y\rangle + \lambda^2 \|y\|^2 \\ &= d^2 - 2\lambda\langle x_0 - y_0, y\rangle + \lambda^2 \|y\|^2, \end{aligned}$$

whence

$$\langle x_0 - y_0, y\rangle \le \frac{\lambda}{2} \|y\|^2$$

for all $\lambda > 0$. Since λ is otherwise arbitrary, we conclude that

$$\langle x_0 - y_0, y\rangle \le 0$$

for each $y \in M$. But then

$$-\langle x_0 - y_0, y\rangle = \langle x_0 - y_0, -y\rangle \leq 0$$

for each $y \in M$, and thus $\langle x_0 - y_0, y\rangle = 0$ for each $y \in M$, i.e., $x_0 - y_0 \in M^\perp$.

If $y_1 \in M$ is such that

$$\|x_0 - y_1\| = \inf_{y \in M} \|x_0 - y\|,$$

then the above argument shows that $x_0 - y_1 \in M^\perp$. Hence, $y_1 - y_0 = (x_0 - y_0) - (x_0 - y_1) \in M^\perp$. Since we also have $y_1 - y_0 \in M$, this implies

$$\|y_1 - y_0\|^2 = \langle y_1 - y_0, y_1 - y_0\rangle = 0,$$

i.e., $y_1 = y_0$. □

8.48. THEOREM. *Suppose M is a closed subspace of a Hilbert space H. Then for each $x \in H$ there exists a unique pair of elements $y \in M$ and $z \in M^\perp$ such that $x = y + z$.*

PROOF. We may assume that $M \neq H$. Let $x \in H$. According to Theorem 8.47 there is a $y \in M$ such that $z = x - y \in M^\perp$. This establishes the existence of $y \in M$, $z \in M^\perp$ such that $x = y + z$.

To show uniqueness, suppose that $y_1, y_2 \in M$ and $z_1, z_2 \in M^\perp$ are such that $y_1 + z_1 = y_2 + z_2$. Then

$$y_1 - y_2 = z_2 - z_1,$$

which means that $y_1 - y_2 \in M \cap M^\perp$. Thus

$$\|y_1 - y_2\|^2 = \langle y_1 - y_2, y_1 - y_2\rangle = 0,$$

whence $y_1 = y_2$. This, in turn, implies that $z_1 = z_2$. □

If $y \in H$ is fixed and we define the function

$$f(x) = \langle y, x\rangle$$

for $x \in H$, then f is a linear functional on H. Furthermore, from Schwarz's inequality,

$$|f(x)| = |\langle y, x\rangle| \leq \|y\| \ \|x\|,$$

which implies that $\|f\| \leq \|y\|$. If $y = 0$, then $\|f\| = 0$. If $y \neq 0$, then

$$f(\frac{y}{\|y\|}) = \|y\|.$$

Thus $\|f\| = \|y\|$. Using Theorem 8.48, we will show that every continuous linear functional on H is of this form.

8.49. THEOREM (Riesz representation theorem). *Suppose H is a Hilbert space. Then for each $f \in H^*$ there exists a unique $y \in H$ such that*

$$f(x) = \langle y, x\rangle \tag{8.15}$$

for each $x \in H$. Moreover, under this correspondence H and H^* are isometrically isomorphic.

PROOF. Suppose $f \in H^*$. If $f = 0$, then (8.15) holds with $y = 0$. So assume that $f \neq 0$. Then

$$M = \{x \in H : f(x) = 0\}$$

is a closed subspace of H and $M \neq H$. We infer from Theorem 8.48 that there is an element $x_0 \in M^\perp$ with $x_0 \neq 0$. Since $x_0 \notin M$, we have $f(x_0) \neq 0$. Since for each $x \in H$ we have

$$f\left(x - \frac{f(x)}{f(x_0)}x_0\right) = 0,$$

we see that

$$x - \frac{f(x)}{f(x_0)}x_0 \in M$$

for each $x \in H$. Thus

$$\langle x - \frac{f(x)}{f(x_0)}x_0, x_0\rangle = 0$$

for each $x \in H$. This last equation may be rewritten as

$$f(x) = \frac{f(x_0)}{\|x_0\|^2}\langle x_0, x\rangle.$$

Thus if we set $y = \dfrac{f(x_0)}{\|x_0\|^2}x_0$, we see that (8.15) holds. We have already observed that the norm of a linear functional f satisfying (8.15) is $\|y\|$.

If $y_1, y_2 \in H$ are such that $\langle y_1, x\rangle = \langle y_2, x\rangle$ for all $x \in H$, then

$$\langle y_1 - y_2, x\rangle = 0$$

for all $x \in X$. Thus in particular,

$$\|y_1 - y_2\|^2 = \langle y_1 - y_2, y_1 - y_2\rangle = 0,$$

whence $y_1 = y_2$. This shows that the y that represents f in (8.15) is unique.

We may rephrase the above results as follows. Let $\Psi : H \to H^*$ be defined for each $x \in H$ by

$$\Psi(x)(y) = \langle x, y\rangle$$

for all $y \in H$. Then Ψ is a one-to-one linear mapping of H onto H^*. Furthermore, $\|\Psi(x)\| = \|x\|$ for each $x \in H$. Thus Ψ is an isometric isomorphism of H onto H^*. □

8.50. THEOREM. *Every Hilbert space H is a reflexive Banach space, and consequently the set $\{x \in H : \|x\| \leq 1\}$ is compact in the weak topology.*

PROOF. We first show that H^* is a Hilbert space. Let Ψ be as in the proof of Theorem 8.49 above and define

$$\langle f, g\rangle = \langle \Psi^{-1}(f), \Psi^{-1}(g)\rangle \tag{8.16}$$

for each pair $f, g \in H^*$. Note that the right-hand member of the equation above is the inner product in H. That (8.16) defines an inner product on H^* follows immediately from the properties of Ψ. Furthermore,

$$\langle f, f\rangle = \langle \Psi^{-1}(f), \Psi^{-1}(f)\rangle = \|f\|^2$$

for each $f \in H^*$. Thus the norm on H^* is induced by this inner product. If $\omega \in H^{**}$, then by Theorem 8.49, there is an element $g \in H^*$ such that

$$\omega(f) = \langle g, f\rangle$$

for each $f \in H^*$. Again by Theorem 8.49 there is an element $x \in H$ such that $\Psi(x) = g$. Thus

$$\omega(f) = \langle g, f\rangle = \langle \Psi(x), f\rangle = \langle x, \Psi^{-1}(f)\rangle = f(x)$$

for each $f \in H^*$, from which we conclude that H is reflexive. Thus $\{x \in H : \|x\| \leq 1\}$ is compact in the weak topology by Corollary 8.41. □

We next consider the representation of elements of a Hilbert space by "Fourier series."

8.51. DEFINITION. A subset $\mathcal{F}$ of a Hilbert space is an **orthonormal family** if for each pair of elements $x, y \in \mathcal{F}$, we have

$$\langle x, y\rangle = 0 \text{ if } x \neq y,$$
$$\langle x, y\rangle = 1 \text{ if } x = y.$$

An orthonormal family $\mathcal{F}$ in H is **complete** if the only element $x \in H$ for which $\langle x, y\rangle = 0$ for all $y \in \mathcal{F}$ is $x = 0$.

8.52. THEOREM. *Every Hilbert space contains a complete orthonormal family.*

PROOF. This assertion follows from the Hausdorff maximal principle (see Exercise 1, Section 8.5). □

8.53. THEOREM. *If H is a separable Hilbert space and $\mathcal{F}$ is an orthonormal family in H, then $\mathcal{F}$ is at most countable.*

PROOF. If $x, y \in \mathcal{F}$ and $x \neq y$, then

$$\|x - y\|^2 = \|x\|^2 - 2\langle x, y\rangle + \|y\|^2 = 2.$$

Thus

$$B(x, \frac{1}{2}) \bigcap B(y, \frac{1}{2}) = \emptyset$$

whenever x, y are distinct elements of $\mathcal{F}$. If $\mathcal{E}$ is a countable dense subset of H, then for each $x \in \mathcal{F}$ the set $B(x, \frac{1}{2})$ must contain an element of $\mathcal{E}$. □

We next study the properties of orthonormal families, beginning with countable orthonormal families.

8.54. THEOREM. *Suppose $\{x_k\}_{k=1}^{\infty}$ is an orthonormal sequence in a Hilbert space H. Then the following assertions hold:*

(i) *For each $x \in H$,*

$$\sum_{k=1}^{\infty} \langle x, x_k \rangle^2 \leq \|x\|^2 .$$

(ii) *If $\{\alpha_k\}$ is a sequence of real numbers, then*

$$\left\| x - \sum_{k=1}^{m} \langle x, x_k \rangle x_k \right\| \leq \left\| x - \sum_{k=1}^{m} \alpha_k x_k \right\|$$

for each $m \geq 1$.

(iii) *If $\{\alpha_k\}$ is a sequence of real numbers, then $\sum_{k=1}^{\infty} \alpha_k x_k$ converges in H if and only if $\sum_{k=1}^{\infty} \alpha_k^2$ converges in $\mathbb{R}$, in which case the sum is independent of the order in which the terms are arranged, i.e., the series converges* **unconditionally**, *and*

$$\left\| \sum_{k=1}^{\infty} \alpha_k x_k \right\|^2 = \sum_{k=1}^{\infty} \alpha_k^2 .$$

PROOF. (i) For every positive integer m,

$$\begin{aligned} 0 &\leq \left\| x - \sum_{k=1}^{m} \langle x, x_k \rangle x_k \right\|^2 \\ &= \|x\|^2 - 2\langle x, \sum_{k=1}^{m} \langle x, x_k \rangle x_k \rangle + \left\| \sum_{k=1}^{m} \langle x, x_k \rangle x_k \right\|^2 \\ &= \|x\|^2 - 2\sum_{k=1}^{m} \langle x, x_k \rangle^2 + \sum_{k=1}^{m} \langle x, x_k \rangle^2 \\ &= \|x\|^2 - \sum_{k=1}^{m} \langle x, x_k \rangle^2 . \end{aligned}$$

Thus

$$\sum_{k=1}^{m} \langle x, x_k \rangle^2 \leq \|x\|^2$$

for all $m \geq 1$, from which assertion (i) follows.

(ii) Fix a positive integer m and let M denote the subspace of H spanned by $\{x_1, x_2, \dots, x_m\}$. Then M is finite-dimensional and hence closed; see Exercise 4, Section 8.4. Since for all $1 \le k \le m$, we have

$$\langle x_k, x - \sum_{k=1}^{m} \langle x, x_k \rangle x_k \rangle = 0,$$

we see that

$$x - \sum_{k=1}^{m} \langle x, x_k \rangle x_k \in M^{\perp}.$$

In view of Theorem 8.47, this implies assertion (ii), since

$$\sum_{k=1}^{m} \alpha_k x_k \in M.$$

(iii) For all positive integers m and l with $m > l$, we have

$$\left\| \sum_{k=1}^{m} \alpha_k x_k - \sum_{k=1}^{l} \alpha_k x_k \right\|^2 = \left\| \sum_{k=l+1}^{m} \alpha_k x_k \right\|^2 = \sum_{k=l+1}^{m} \alpha_k^2.$$

Thus the sequence $\{\sum_{k=1}^{m} \alpha_k x_k\}_{m=1}^{\infty}$ is a Cauchy sequence in H if and only if $\sum_{k=1}^{\infty} \alpha_k^2$ converges in $\mathbb{R}$.

Suppose $\sum_{k=1}^{\infty} \alpha_k^2 < \infty$. Since for for all m we have

$$\left\| \sum_{k=1}^{m} \alpha_k x_k \right\|^2 = \sum_{k=1}^{m} \alpha_k^2,$$

we see that

$$\left\| \sum_{k=1}^{\infty} \alpha_k x_k \right\|^2 = \sum_{k=1}^{\infty} \alpha_k^2.$$

Let $\{\alpha_{k_j}\}$ be any rearrangement of the sequence $\{\alpha_k\}$. Then

$$\left\| \sum_{k=1}^{m} \alpha_k x_k - \sum_{j=1}^{m} \alpha_{k_j} x_{k_j} \right\|^2 = \sum_{k=1}^{m} \alpha_k^2 - 2 \sum_{k_j \le m} \alpha_{k_j}^2 + \sum_{j=1}^{m} \alpha_{k_j}^2, \tag{8.17}$$

for every m. Since the sum of the series $\sum_{k=1}^{\infty} \alpha_k^2$ is independent of the order of the terms, the last member of (8.17) converges to 0 as $m \to \infty$. □

8.55. THEOREM. *Suppose $\mathcal{F}$ is an orthonormal system in a Hilbert space H and $x \in H$. Then:*

(i) *The set $\{y \in \mathcal{F} : \langle x, y \rangle \neq 0\}$ is at most countable.*

(ii) *The series $\sum_{y \in \mathcal{F}} \langle x, y \rangle y$ converges unconditionally in H.*

PROOF. (i) Let $\varepsilon > 0$ and set

$$\mathcal{F}_\varepsilon = \{y \in \mathcal{F} : |\langle x, y \rangle| > \varepsilon\}.$$

In view of Theorem 8.54(a), the number of elements in $\mathcal{F}_\varepsilon$ cannot exceed $\frac{|x|^2}{\varepsilon^2}$, and thus $\mathcal{F}_\varepsilon$ is a finite set. Since

$$\{y \in \mathcal{F} : \langle x, y\rangle \neq 0\} = \bigcup_{k=1}^{\infty} \{y \in \mathcal{F} : |\langle x, y\rangle| > \frac{1}{k}\},$$

we see that $\{y \in \mathcal{F} : \langle x, y\rangle \neq 0\}$ is at most countable. (ii) Set $\{y_k\}_{k=1}^{\infty} = \{y \in \mathcal{F} : \langle x, y\rangle \neq 0\}$. Then, according to Theorem 8.54(i), (iii), the series $\sum_{k=1}^{\infty}\langle x, y_k\rangle y_k$ converges unconditionally in H. □

8.56. THEOREM. *Suppose $\mathcal{F}$ is a complete orthonormal system in a Hilbert space H. Then*

$$x = \sum_{y\in\mathcal{F}} \langle x, y\rangle y,$$

for each $x \in H$, where all but countably many terms in the series are equal to 0 and the series converges unconditionally in H.

PROOF. Let Y denote the subspace of H spanned by $\mathcal{F}$ and let M denote the closure of Y. If $z \in M^\perp$, then $\langle z, y\rangle = 0$ for each $y \in \mathcal{F}$. Since $\mathcal{F}$ is complete, this implies that $z = 0$. Thus $M^\perp = \{0\}$.

Fix $x \in H$. By Theorem 8.55 (ii), the series $\sum_{y\in\mathcal{F}}\langle x, y\rangle y$ converges unconditionally to an element $\overline{x} \in H$.

Set $\{y_k\}_{k=1}^{\infty} = \{y \in \mathcal{F} : \langle x, y\rangle \neq 0\}$. If $y \in \mathcal{F}$ and $\langle x, y\rangle = 0$, then

$$\langle x - \sum_{k=1}^{m}\langle x, y_k\rangle y_k, y\rangle = 0$$

for each $m \geq 1$. Then

$$\begin{aligned} |\langle x - \overline{x}, y\rangle| &= \left|\langle x - \sum_{k=1}^{m}\langle x, y_k\rangle y_k, y\rangle - \langle x - \overline{x}, y\rangle\right| \\ &= \left|\langle \overline{x} - \sum_{k=1}^{m}\langle x, y_k\rangle y_k, y\rangle\right| \\ &\leq \left\|\overline{x} - \sum_{k=1}^{m}\langle x, y_k\rangle y_k\right\| \|y\|, \end{aligned}$$

and we see that $\langle x - \overline{x}, y\rangle = 0$.

If $1 \leq l \leq m$, then

$$\langle x - \sum_{k=1}^{m}\langle x, y_k\rangle y_k, y_l\rangle = 0,$$

and hence $\langle x - \overline{x}, y_l\rangle = 0$ for every $l \geq 1$.

We have thus shown that

$$\langle x - \overline{x}, y\rangle = 0 \tag{8.18}$$

for each $y \in \mathcal{F}$, from which it follows immediately that (8.18) holds for each $y \in Y$.

If $w \in M$, then there is a sequence $\{w_k\}$ in Y such that $\|w - w_k\| \to 0$ as $k \to \infty$. Thus

$$\langle x - \overline{x}, w \rangle = \lim_{k\to\infty} \langle x - \overline{x}, w_k \rangle = 0,$$

which means that $x - \overline{x} \in M^{\perp} = \{0\}$. □

8.57. EXAMPLES. (i) The Banach space $\mathbb{R}^n$ is a Hilbert space with inner product

$$\langle (x_1, x_2, \ldots, x_n), (y_1, y_2, \ldots, y_n) \rangle = x_1 y_1 + x_2 y_2 + \cdots + x_n y_n.$$

(ii) The Banach space $L^2(X, \mu)$ is a Hilbert space with inner product

$$\langle f, g \rangle = \int_X fg \, d\mu.$$

(iii) The Banach space l^2 is a Hilbert space with inner product

$$\langle \{a_k\}, \{b_k\} \rangle = \sum_{k=1}^{\infty} a_k b_k.$$

Suppose H is a separable Hilbert space containing a countable orthonormal system $\{x_k\}_{k=1}^{\infty}$. For all $x \in H$ the sequence $\{\langle x, x_k \rangle\}$ is in l^2 and

$$\sum_{k=1}^{\infty} \langle x, x_k \rangle^2 = \|x\|^2.$$

On the other hand, if $\{a_k\} \in l^2$, then according to Theorem 8.54, the series $\sum_{k=1}^{\infty} a_k x_k$ converges in H. Set $x = \sum_{k=1}^{\infty} a_k x_k$. Then

$$\langle x, x_l \rangle = \lim_{m\to\infty} \langle \sum_{k=1}^{m} a_k x_k, x_l \rangle = a_l$$

for each l, and hence

$$\sum_{k=1}^{\infty} a_k^2 = \|x\|^2.$$

Thus the linear mapping $T : H \to l^2$ given by

$$T(x) = \{\langle x, x_k \rangle\}$$

is an isometric isomorphism of H onto l^2.

If Ω is a open subset of $\mathbb{R}^n$ and λ denotes Lebesgue measure on Ω, then $L^2(\Omega, \lambda)$ is separable and hence isometrically isomorphic to l^2.

Exercises for Section 8.5

1. Show that every Hilbert space contains a complete orthonormal system. (Hint: Observe that an orthonormal system in a Hilbert space H is complete if and only if it is maximal with respect to set inclusion, i.e., it is not contained in any other orthonormal system.)

2. Suppose T is a linear mapping of a Hilbert space H into a Hilbert space E such that $\|T(x)\| = \|x\|$ for each $x \in H$. Show that

$$\langle T(x), T(y)\rangle = \langle x, y\rangle$$

for all $x, y \in H$.
3. Show that a Hilbert space that contains a countable complete orthonormal system is separable.
4. Fix $1 < p < \infty$. Set $x^m = \{x_k^m\}_{k=1}^\infty$, where

$$x_k^m = \begin{cases} 1 & \textit{if } k = m, \\ 0 & \text{otherwise.} \end{cases}$$

Show that the sequence $\{x^m\}_{m=1}^\infty$ in l^p (cf. Example 8.9(iii)) converges to 0 in the weak topology but does not converge in the strong topology.

8.6. Weak and Strong Convergence in L^p

Although it is easily seen that strong convergence implies weak convergence in L^p, it is shown below that under certain conditions, weak convergence implies strong convergence.

We now apply some of the results of this chapter in the setting of L^p spaces. To begin, we note that if $1 \le p < \infty$, then in view of Example 8.34 (ii), a sequence $\{f_k\}_{k=1}^\infty$ in $L^p(X, \mathcal{M}, \mu)$ converges weakly to $f \in L^p(X, \mathcal{M}, \mu)$ if and only if

$$\lim_{k\to\infty} \int_X f_k g \, d\mu = \int_X fg \, d\mu$$

for each $g \in L^{p'}(X, \mathcal{M}, \mu)$.

8.58. THEOREM. *Let $(X, \mathcal{M}, \mu)$ be a measure space and suppose f and $\{f_k\}_{k=1}^\infty$ are functions in $L^p(X, \mathcal{M}, \mu)$. If $1 \le p < \infty$ and $\|f_k - f\|_p \to 0$, then $f_k \to f$ weakly in L^p.*

PROOF. This is a consequence of the fact that in every normed linear space, strong convergence implies week convergence (see Exercise 11, Section 8.4). In this theorem, in which the normed linear space is L^p, the result also follows from Hölder's inequality. □

If $\{f_k\}_{k=1}^\infty$ is a sequence of functions with $\|f_k\|_p \le M$ for some M and all k, then since $L^p(X)$ is reflexive for $1 < p < \infty$, Theorem 8.37 asserts that there is a subsequence that converges weakly to some $f \in L^p(X)$. The next result shows that if it is also known that $f_k \to f$ μ-a.e., then the full sequence converges weakly to f.

8.59. THEOREM. *Let $1 < p < \infty$. If $f_k \to f$ μ-a.e., then $f_k \to f$ weakly in L^p if and only if $\{\|f_k\|_p\}$ is a bounded sequence.*

PROOF. Necessity follows immediately from Theorem 8.35.

To prove sufficiency, let $M \geq 0$ be such that $\|f_k\|_p \leq M$ for all positive integers k. Then Fatou's lemma implies

$$\begin{aligned} \|f\|_p^p &= \int_X |f|^p \, d\mu = \int_X \lim_{k\to\infty} |f_k|^p \, d\mu \\ &\leq \liminf_{k\to\infty} \int_X |f_k|^p \, d\mu \leq M^p. \end{aligned} \tag{8.19}$$

Let $\varepsilon > 0$ and $g \in L^{p'}(X)$. Refer to Theorem 6.40 to obtain $\delta > 0$ such that

$$\left(\int_E |g|^{p'} \, d\mu\right)^{1/p'} < \frac{\varepsilon}{6M} \tag{8.20}$$

whenever $E \in \mathcal{M}$ and $\mu(E) < \delta$. We claim that there exists a set $F \in \mathcal{M}$ such that $\mu(F) < \infty$ and

$$\left(\int_{\widetilde{F}} |g|^{p'} \, d\mu\right)^{1/p'} < \frac{\varepsilon}{6M}. \tag{8.21}$$

To verify the claim, set $A_t\!: \; = \{x : |g(x)|^{p'} \geq t\}$ and observe that by the monotone convergence theorem,

$$\lim_{t\to 0^+} \int_X \chi_{A_t} |g|^{p'} \, d\mu = \int_X |g|^{p'} \, d\mu,$$

and thus (8.21) holds with $F = A_t$ for sufficiently small positive t.

We can now apply Egorov's theorem (Theorem 5.18) on F to obtain $A \in \mathcal{M}$ such that $A \subset F$, $\mu(F - A) < \delta$, and $f_k \to f$ uniformly on A. Let k_0 be such that $k \geq k_0$ implies

$$\left(\int_A |f - f_k|^p \, d\mu\right)^{1/p} \|g\|_{p'} < \frac{\varepsilon}{3}. \tag{8.22}$$

Setting $E = F - A$ in (8.20), we obtain from (8.19), (8.21), (8.22), and Hölder's inequality that

$$\begin{aligned} \left|\int_X fg \, d\mu - \int_X f_k g \, d\mu\right| &\leq \int_X |f - f_k| \, |g| \, d\mu \\ &= \int_A |f - f_k| \, |g| \, d\mu + \int_{F-A} |f - f_k| \, |g| \, d\mu \\ &\quad + \int_{\widetilde{F}} |f - f_k| \, |g| \, d\mu \\ &\leq \|f - f_k\|_{p,A} \, \|g\|_{p'} \\ &\quad + \|f - f_k\|_p \left(\|g\|_{p',F-A} + \|g\|_{p',\widetilde{F}}\right) \\ &\leq \frac{\varepsilon}{3} + 2M\left(\frac{\varepsilon}{6M} + \frac{\varepsilon}{6M}\right) \\ &= \varepsilon \end{aligned}$$

for all $k \geq k_0$. □

If the hypotheses of the last result are changed to include that $\|f_k\|_p \to \|f\|_p$, then we can prove that $\|f_k - f\|_p \to 0$. This is an immediate consequence of the following theorem.

8.60. THEOREM. *Let $1 \le p < \infty$. Suppose f and $\{f_k\}_{k=1}^\infty$ are functions in $L^p(X, \mathcal{M}, \mu)$ such that $f_k \to f$ μ-a.e. and the sequence $\{\|f_k\|_p\}_{k=1}^\infty$ is bounded. Then*

$$\lim_{k\to\infty} (\|f_k\|_p^p - \|f_k - f\|_p^p) = \|f\|_p^p .$$

PROOF. Set

$$M: = \sup_{k\ge 1} \|f_k\|_p < \infty$$

and note that by Fatou's lemma, $\|f\|_p \le M$.

Fix $\varepsilon > 0$ and observe that the function

$$h_\varepsilon(t): = ||t+1|^p - |t|^p| - \varepsilon\,|t|^p$$

is continuous on $\mathbb{R}$ and

$$\lim_{|t|\to\infty} h_\varepsilon(t) = -\infty.$$

Thus there is a constant $C_\varepsilon > 0$ such that $h_\varepsilon(t) < C_\varepsilon$ for all $t \in \mathbb{R}$. It follows that

$$||a+b|^p - |a|^p| \le \varepsilon\,|a|^p + C_\varepsilon\,|b|^p \tag{8.23}$$

for all real numbers a and b.

Set

$$G_k^\varepsilon: = [|\,|f_k|^p - |f_k - f|^p - |f|^p\,| - \varepsilon\,|f_k - f|^p]^+$$

and note that $G_k^\varepsilon \to 0$ μ-a.e. as $k \to \infty$.

Setting $a = f_k - f$ and $b = f$ in (8.23), we see that

$$\begin{aligned} |\,|f_k|^p - |f_k - f|^p - |f|^p\,| &\le |\,|f_k|^p - |f_k - f|^p\,| + |f|^p \\ &\le \varepsilon\,|f_k - f|^p + (C_\varepsilon + 1)\,|f|^p , \end{aligned}$$

from which it follows that

$$G_k^\varepsilon \le (C_\varepsilon + 1)\,|f|^p ,$$

and by the dominated convergence theorem,

$$\lim_{k\to\infty} \int_X G_k^\varepsilon \, d\mu = 0.$$

Since

$$||f_k|^p - |f_k - f|^p - |f|^p| \le G_k^\varepsilon + \varepsilon\,|f_k - f|^p ,$$

we see that

$$\limsup_{k\to\infty} \int_X ||f_k|^p - |f_k - f|^p - |f|^p| \; d\mu \le \varepsilon(2M)^p.$$

Since ε is arbitrary, the proof is complete. □

We have the following corollary.

8.61. COROLLARY. *Let* $1 < p < \infty$. *If* $f_k \to f$ μ*-a.e. and* $\|f_k\|_p \to \|f\|_p$, *then* $f_k \to f$ *weakly in* L^p *and* $\|f_k - f\|_p \to 0$.

The following provides a summary of our results.

8.62. THEOREM. *The following statements hold in a general measure space* $(X, \mathcal{M}, \mu)$.

(i) *If* $f_k \to f$ μ*-a.e., then*

$$\|f\|_p \leq \liminf_{k\to\infty} \|f_k\|_p, \quad 1 \leq p < \infty.$$

(ii) *If* $f_k \to f$ μ*-a.e. and* $\{\|f_k\|_p\}$ *is a bounded sequence, then* $f_k \to f$ *weakly in* L^p, $1 < p < \infty$.

(iii) *If* $f_k \to f$ μ*-a.e. and there exists a function* $g \in L^p$ *such that for each* k, $|f_k| \leq g$ μ*-a.e. on* X, *then* $\|f_k - f\|_p \to 0$, $1 \leq p < \infty$.

(iv) *If* $f_k \to f$ μ*-a.e. and* $\|f_k\|_p \to \|f\|_p$, *then* $\|f_k - f\|_p \to 0$, $1 \leq p < \infty$.

PROOF. Fatou's lemma implies (i), (ii) follows from Theorem 8.59, (iii) follows from Lebesgue's dominated convergence theorem, and (iv) is a restatement of Theorem 8.60. □

We conclude this chapter with another proof of the Radon–Nikodym theorem, which is based on the Riesz representation theorem for Hilbert spaces (Theorem 8.49). Since $L^2(X, \mu)$ is a Hilbert space, Theorem 8.49 applied to $H = L^2(X, \mu)$ is a particular case of Theorem 6.48, which is the Riesz representation theorem for L^p spaces. The proof of Theorem 6.48 is based on Theorem 6.43, which proves the Radon–Nikodym theorem. We note that the shorter proof of the Radon–Nikodym theorem that we now present does not rely on Theorem 6.48.

8.63. THEOREM. *Let* μ *and* ν *be* σ*-finite measures on* $(X, \mathcal{M})$ *with* $\nu \ll \mu$. *Then there exists a function* $h \in L^1_{\text{loc}}(X, \mathcal{M}, \mu)$ *such that*

$$\nu(E) = \int_X h \, d\mu$$

for all $E \in \mathcal{M}$.

PROOF. First, we will assume $\nu \geq 0$ and that both measures are finite. We will prove that there is a measurable function g on X such that $0 \leq g < 1$ and

$$\int_X f(1-g) \, d\nu = \int_X fg \, d\mu$$

for all $f \in L^2(X, \mathcal{M}, \mu + \nu)$. For this purpose, define

$$T(f) = \int_X f \, d\nu \tag{8.24}$$

for $f \in L^2(X, \mathcal{M}, \mu + \nu)$. By Hölder's inequality, it follows that

$$f \in L^1(X, \mathcal{M}, \mu + \nu)$$

and therefore $f \in L^1(X, \mathcal{M}, \nu)$ also. Hence, we see that T is finite for $f \in L^2(X, \mathcal{M}, \mu+\nu)$ and thus is a well-defined linear functional. Furthermore, it is a bounded linear functional, because

$$\begin{aligned} |T(f)| &\leq \left(\int_X |f|^2 \, d\nu\right)^{1/2} (\nu(X))^{1/2} \\ &= \|f\|_{2;\nu} (\nu(X))^{1/2} \\ &\leq \|f\|_{2;\nu+\mu} (\nu(X))^{1/2}. \end{aligned}$$

Referring to Theorem 8.49, there is a function $\varphi \in L^2(\mu + \nu)$ such that

$$T(f) = \int_X f\varphi \, d(\nu + \mu) \tag{8.25}$$

for all $f \in L^2(\mu + \nu)$. Observe that $\varphi \geq 0$ $(\mu + \nu)$-a.e., for otherwise, we would obtain

$$T(\chi_A) < 0,$$

where $A\colon = \{\varphi < 0\}$, which is impossible. Now, (8.24) and (8.25) imply

$$\int_X f(1 - \varphi) \, d\nu = \int_X f\varphi \, d\mu \tag{8.26}$$

for $f \in L^2(\mu + \nu)$. If f is taken as χ_E, where $E\colon = \{\varphi \geq 1\}$, we obtain

$$0 \leq \mu(E) = \int_X \chi_E \, d\mu \leq \int_X \chi_E\varphi \, d\mu = \int_X \chi_E(1 - \varphi) \, d\nu \leq 0.$$

Hence, we have $\mu(E) = 0$, and consequently, $\nu(E) = 0$. Setting $g = \varphi\chi_{\widetilde{E}}$, we have $0 \leq g < 1$ and $g = \varphi$ almost everywhere with respect to both μ and ν. Reference to (8.26) yields

$$\int_X f(1 - g) \, d\nu = \int_X fg \, d\mu.$$

Since g is bounded, we can replace f by $(1 + g + g^2 + \cdots + g^k)\chi_E$ in this equation for every positive integer k and E measurable. Then we obtain

$$\int_E (1 - g^{k+1}) \, d\nu = \int_E g(1 + g + g^2 + \cdots + g^k) d\mu.$$

Since $0 \leq g < 1$ almost everywhere with respect to both μ and ν, the left-hand side tends to $\nu(E)$, while the integrands on the right-hand side increase monotonically to some measurable function h. Thus, by the monotone convergence theorem, we obtain

$$\nu(E) = \int_E h \, d\mu$$

for every measurable set E. This gives us the desired result in the case that both μ and ν are finite measures.

The proof of the general case proceeds as in Theorem 6.43. □

Exercises for Section 8.6

1. Suppose $f_i \to f$ weakly in $L^p(X, \mathcal{M}, \mu)$, $1 < p < \infty$, and that $f_i \to f$ pointwise μ-a.e. Prove that $f_i^+ \to f^+$ and $f_i^- \to f^-$ weakly in L^p.
2. Show that the previous exercise is false if the hypothesis of pointwise convergence is dropped.
3. Let $C := C[0,1]$ denote the space of continuous functions on $[0,1]$ endowed with the usual sup norm and let X be a linear subspace of C that is closed relative to the L^2 norm.
   (i) Prove that X is also closed in C.
   (ii) Show that $\|f\|_2 \leq \|f\|_\infty$ for each $f \in X$.
   (iii) Prove that there exists $M > 0$ such that $\|f\|_\infty \leq M \|f\|_2$ for all $f \in X$.
   (iv) For each $t \in [0,1]$, show that there is a function $g_t \in L^2$ such that
   $$f(t) = \int g_t(x) f(x)\, dx$$
   for all $f \in X$.
   (v) Show that if $f_k \to f$ weakly in L^2, where $f_k \in X$, then $f_k(x) \to f(x)$ for each $x \in [0,1]$.
   (vi) As a consequence of the above, show that if $f_k \to f$ weakly in L^2, where $f_k \in X$, then $f_k \to f$ strongly in L^2; that is, $\|f_k - f\|_2 \to 0$.
4. A subset $K \subset X$ in a linear space is called **convex** if for every two points $x, y \in K$, the line segment $tx + (1-t)y$, $t \in [0,1]$, also belongs to K.
   (i) Prove that the unit ball in a normed linear space is convex.
   (ii) If K is a convex set, a point $x_0 \in K$ is called an **extreme point** of K if x_0 is not in the interior of any line segment that lies in K; that is, if $x_0 = ty + (1-t)z$, where $0 < t < 1$, then either y or z is not in K. Show that every $f \in L^p[0,1]$, $1 < p < \infty$, with $\|f\|_p = 1$ is an extreme point of the unit ball.
   (iii) Show that the extreme points of the unit ball in $L^\infty[0,1]$ are the functions f with $|f(x)| = 1$ for a.e. x.
   (iv) Show that the unit ball in $L^1[0,1]$ has no extreme points.

CHAPTER 9

Measures and Linear Functionals

9.1. The Daniell Integral

Theorem 6.48 states that a function in $L^{p'}$ can be regarded as a bounded linear functional on L^p. Here we show that a large class of measures can be represented as bounded linear functionals on the space of continuous functions. This is a very important result that has many useful applications and provides a fundamental connection between measure theory and functional analysis.

Suppose $(X, \mathcal{M}, \mu)$ is a measure space. Integration defines an operation that is linear, order-preserving, and continuous relative to increasing convergence. Specifically, we have

(i) $\int (kf)\, d\mu = k \int f\, d\mu$ whenever $k \in \mathbb{R}$ and f is an integrable function;
(ii) $\int (f+g)\, d\mu = \int f\, d\mu + \int g\, d\mu$ whenever f, g are integrable;
(iii) $\int f\, d\mu \leq \int g\, d\mu$ whenever f, g are integrable functions with $f \leq g$ μ-a.e.;
(iv) if $\{f_i\}$ is a nondecreasing sequence of integrable functions, then

$$\lim_{i\to\infty} \int f_i\, d\mu = \int \lim_{i\to\infty} f_i\, d\mu.$$

The main objective of this section is to show that if a linear functional defined on an appropriate space of functions possesses the four properties above, then it can be expressed as the operation of integration with respect to some measure. Thus, we will have shown that these properties completely characterize the operation of integration.

It turns out that the proof of our main result is no more difficult when cast in a very general framework, so we proceed by introducing the concept of a lattice.

9.1. Definition. If f, g are real-valued functions defined on a space X, we define

$$(f \wedge g)(x) := \min[f(x), g(x)],$$
$$(f \vee g)(x) := \max[f(x), g(x)].$$

A collection L of real-valued functions defined on an abstract space X is called a **lattice** if the following conditions are satisfied: If $0 \leq c < \infty$ and f

W.P. Ziemer, *Modern Real Analysis*, Graduate Texts in Mathematics 278,
https://doi.org/10.1007/978-3-319-64629-9_9

and g are elements of L, then so are the functions $f+g$, cf, $f \wedge g$, and $f \wedge c$. Furthermore, if $f \leq g$, then $g-f$ is required to belong to L. Note that if f and g belong to L with $g \geq 0$, then so does $f \vee g = f+g-f \wedge g$. Therefore, f^+ belongs to L. We define $f^- = f^+ - f$, and so $f^- \in L$. We let L^+ denote the functions f in L for which $f \geq 0$. Clearly, if L is a lattice, then so is L^+.

For example, the space of continuous functions on a metric space is a lattice, as is the space of integrable functions, but the collection of lower semicontinuous functions is not a lattice, because it is not closed under the $\wedge$-operation (see Exercise 9.1).

9.2. THEOREM. *Suppose L is a lattice of functions on X and let $T\colon L \to \mathbb{R}$ be a functional satisfying the following conditions for all functions in L:*
(i) $T(f+g) = T(f) + T(g)$;
(ii) $T(cf) = cT(f)$ *whenever* $0 \leq c < \infty$;
(iii) $T(f) \geq T(g)$ *whenever* $f \geq g$;
(iv) $T(f) = \lim_{i\to\infty} T(f_i)$ *whenever* $f\colon = \lim_{i\to\infty} f_i$ *is a member of* L;
(v) $\{f_i\}$ *is nondecreasing.*
Then there exists an outer measure μ on X such that for each $f \in L$, f is μ-measurable and
$$T(f) = \int_X f \, d\mu.$$
In particular, $\{f > t\}$ is μ-measurable whenever $f \in L$ and $t \in \mathbb{R}$.

PROOF. First, observe that since $T(0) = T(0 \cdot f) = 0 \cdot T(f) = 0$, it follows that (iii) above implies $T(f) \geq 0$ whenever $f \in L^+$. Next, with the convention that the infimum of the empty set is ∞, for an arbitrary set $A \subset X$, we define
$$\mu(A) = \inf\left\{\lim_{i\to\infty} T(f_i)\right\},$$
where the infimum is taken over all sequences of functions $\{f_i\}$ with the property

$$f_i \in L^+,\ \{f_i\}_{i=1}^{\infty} \text{ is nondecreasing, and } \lim_{i\to\infty} f_i \geq \chi_A. \tag{9.1}$$

Such sequences of functions are called **admissible** for A. In accordance with our convention, if there is no admissible sequence of functions for A, we define $\mu(A) = \infty$. It follows from the definition that if $f \in L^+$ and $f \geq \chi_A$, then

$$T(f) \geq \mu(A). \tag{9.2}$$

On the other hand, if $f \in L^+$ and $f \leq \chi_A$, then

$$T(f) \leq \mu(A). \tag{9.3}$$

This is true because if $\{f_i\}$ is admissible for A, then $g_i\colon = f_i \wedge f$ is a nondecreasing sequence with $\lim_{i\to\infty} g_i = f$. Hence,
$$T(f) = \lim_{i\to\infty} T(g_i) \leq \lim_{i\to\infty} T(f_i)$$

and therefore

$$T(f) \leq \mu(A).$$

The first step is to show that μ is an outer measure on X. For this, the only nontrivial property to be established is countable subadditivity. For this purpose, let

$$A \subset \bigcup_{i=1}^{\infty} A_i.$$

For each fixed i and arbitrary $\varepsilon > 0$, let $\{f_{i,j}\}_{j=1}^{\infty}$ be an admissible sequence of functions for A_i with the property that

$$\lim_{j\to\infty} T(f_{i,j}) < \mu(A_i) + \frac{\varepsilon}{2^i}.$$

Now define

$$g_k = \sum_{i=1}^{k} f_{i,k}$$

and obtain

$$\begin{aligned} T(g_k) &= \sum_{i=1}^{k} T(f_{i,k}) \\ &\leq \sum_{i=1}^{k} \lim_{m\to\infty} T(f_{i,m}) \\ &\leq \sum_{i=1}^{k} \left(\mu(A_i) + \frac{\varepsilon}{2^i}\right). \end{aligned}$$

After we show that $\{g_k\}$ is admissible for A, we can take limits of both sides as $k \to \infty$ and conclude that

$$\mu(A) \leq \sum_{i=1}^{\infty} \mu(A_i) + \varepsilon.$$

Since ε is arbitrary, this will prove the countable subadditivity of μ and thus establish that μ is an outer measure on X.

To see that $\{g_k\}$ is admissible for A, select $x \in A$. Then $x \in A_i$ for some i, and with $k\colon = \max(i,j)$, we have $g_k(x) \geq f_{i,j}(x)$. Hence,

$$\lim_{k\to\infty} g_k(x) \geq \lim_{j\to\infty} f_{i,j}(x) \geq 1.$$

The next step is to prove that each element $f \in L$ is a μ-measurable function. Since $f = f^+ - f^-$, it suffices to show that f^+ is μ-measurable, the proof involving f^- being similar. For this we appeal to Theorem 5.13, which asserts that it is sufficient to prove

$$\mu(A) \geq \mu(A \cap \{f^+ \leq a\}) + \mu(A \cap \{f^+ \geq b\}) \tag{9.4}$$

whenever $A \subset X$ and $a < b$ are real numbers. Since $f^+ \geq 0$, we may as well take $a \geq 0$. Let $\{g_i\}$ be an admissible sequence of functions for A and define

$$h = \frac{[f^+ \wedge b - f^+ \wedge a]}{b - a}, \quad k_i = g_i \wedge h.$$

Observe that $h = 1$ on $\{f^+ \geq b\}$. Since $\{g_i\}$ is admissible for A, it follows that $\{k_i\}$ is admissible for $A \cap \{f^+ \geq b\}$. Furthermore, since $h = 0$ on $\{f^+ \leq a\}$, we have that $\{g_i - k_i\}$ is admissible for $A \cap \{f^+ \leq a\}$. Consequently,

$$\lim_{i\to\infty} T(g_i) = \lim_{i\to\infty} [T(k_i) + T(g_i - k_i)] \geq \mu(A \cap \{f^+ \geq b\}) + \mu(A \cap \{f^+ \leq a\}).$$

Since $\{g_i\}$ is an arbitrary admissible sequence for A, we obtain

$$\mu(A) \geq \mu(A \cap \{f^+ \geq b\}) + \mu(A \cap \{f^+ \leq a\}),$$

which proves (9.4).

The last step is to prove that

$$T(f) = \int f \, d\mu \quad \text{for} \quad f \in L.$$

We begin by considering $f \in L^+$. With $f_t \colon = f \wedge t$, $t \geq 0$, note that if $\varepsilon > 0$ and k is a positive integer, then

$$0 \leq f_{k\varepsilon}(x) - f_{(k-1)\varepsilon}(x) \quad \text{for} \quad x \in X,$$

and

$$f_{k\varepsilon}(x) - f_{(k-1)\varepsilon}(x) = \begin{cases} \varepsilon & \text{for } f(x) \geq k\varepsilon, \\ 0 & \text{for } f(x) \leq (k-1)\varepsilon. \end{cases}$$

Therefore,

$$\begin{aligned} T(f_{k\varepsilon} - f_{(k-1)\varepsilon}) &\geq \varepsilon\mu(\{f \geq k\varepsilon\}) && \text{by (9.2)} \\ &\geq \textstyle\int (f_{(k+1)\varepsilon} - f_{k\varepsilon}) \, d\mu, && \text{since } f_{(k+1)\varepsilon} - f_{k\varepsilon} \leq \varepsilon\chi_{\{f\geq k\varepsilon\}} \\ &\geq \varepsilon\mu(\{f \geq (k+1)\varepsilon\}) && \text{since } f_{(k+1)\varepsilon} - f_{k\varepsilon} = \varepsilon\chi_{\{f\geq k\varepsilon\}} \\ & && \text{on } \{f \geq (k+1)\varepsilon\} \\ &\geq T(f_{(k+2)\varepsilon} - f_{(k+1)\varepsilon}). && \text{by (9.3)} \end{aligned}$$

Now taking the sum as k ranges from 1 to n, we obtain

$$T(f_{n\varepsilon}) \geq \int (f_{(n+1)\varepsilon} - f_\varepsilon) \, d\mu \geq T(f_{(n+2)\varepsilon} - f_{2\varepsilon}).$$

Since $\{f_{n\varepsilon}\}$ is a nondecreasing sequence with $\lim_{n\to\infty} f_{n\varepsilon} = f$, we have

$$T(f) \geq \int (f - f_\varepsilon) \, d\mu \geq T(f - f_{2\varepsilon}).$$

Also, $f_\varepsilon \to 0$ as $\varepsilon \to 0^+$, so that

$$T(f) = \int f \, d\mu.$$

Finally, if $f \in L$, then $f^+ \in L^+$, $f^- \in L^+$, thus yielding

$$T(f) = T(f^+) - T(f^-) = \int f^+ \, d\mu - \int f^- \, d\mu = \int f \, d\mu. \qquad \square$$

Now that we have established the existence of an outer measure μ corresponding to the functional T, we address the question of its uniqueness. For this purpose, we need the following lemma, which asserts that μ possesses a type of outer regularity.

9.3. LEMMA. *Under the assumptions of the preceding theorem, let $\mathcal{M}$ denote the σ-algebra generated by all sets of the form $\{f > t\}$, where $f \in L$ and $t \in \mathbb{R}$. Then for every $A \subset X$ there is a set $W \in \mathcal{M}$ such that*

$$A \subset W \quad \textit{and} \quad \mu(A) = \mu(W).$$

PROOF. If $\mu(A) = \infty$, take $W = X$. If $\mu(A) < \infty$, we proceed as follows. For each positive integer i let $\{f_{i,j}\}_{j=1}^{\infty}$ be an admissible sequence for A with the property

$$\lim_{j\to\infty} T(f_{i,j}) < \mu(A) + \frac{1}{i},$$

and define, for each positive integer j,

$$\begin{aligned} g_{i,j} &:= \inf\{f_{1,j}, f_{2,j}, \dots, f_{i,j}\}, \\ B_{i,j} &:= \{x : g_{i,j}(x) > 1 - \frac{1}{i}\}. \end{aligned}$$

Then,

$$g_{i+1,j} \leq g_{i,j} \leq g_{i,j+1} \quad \text{and} \quad B_{i+1,j} \subset B_{i,j} \subset B_{i,j+1}.$$

Furthermore, with the help of (9.2), we have

$$(1 - 1/i)\mu(B_{i,j}) \leq T(g_{i,j}) \leq T(f_{i,j}).$$

Now let

$$V_i = \bigcup_{j=1}^{\infty} B_{i,j}.$$

Observe that $V_i \supset V_{i+1} \supset A$ and $V_i \in \mathcal{M}$ for all i. Indeed, to verify that $V_i \supset A$, it is sufficient to show that $g_{i,j}(x_0) > 1 - 1/i$ for $x_0 \in A$ whenever j is sufficiently large. This is accomplished by observing that there exist positive integers $j_1, j_2, \dots, j_i$ such that $f_{1,j}(x_0) > 1 - 1/i$ for $j \geq j_1$, $f_{2,j}(x_0) > 1 - 1/i$ for $j \geq j_2$, $\dots$, $f_{i,j}(x_0) > 1 - 1/i$ for $j \geq j_i$. Thus, for j larger than $\max\{j_1, j_2, \dots, j_i\}$, we have $g_{i,j}(x_0) > 1 - 1/i$. For each positive integer i,

$$\begin{aligned} \mu(A) \leq \mu(V_i) &= \lim_{j\to\infty} \mu(B_{i,j}) \\ &\leq (1 - 1/i)^{-1} \lim_{j\to\infty} T(f_{i,j}) \qquad (9.5) \\ &\leq (1 - 1/i)^{-1}[\mu(A) + 1/i], \end{aligned}$$

and therefore,

$$\mu(A) = \lim_{i\to\infty} \mu(V_i). \qquad (9.6)$$

Note that (9.5) implies $\mu(V_i) < \infty$. Therefore, if we take

$$W = \bigcap_{i=1}^{\infty} V_i,$$

we obtain $A \subset W$, $W \in \mathcal{M}$, and

$$\begin{aligned}\mu(W) &= \mu\left(\bigcap_{i=1}^{\infty} V_i\right) \\ &= \lim_{i\to\infty} \mu(V_i) \\ &= \mu(A).\end{aligned}$$ □

The preceding proof reveals the manner in which T uniquely determines μ. Indeed, for $f \in L^+$ and for $t, h > 0$, define

$$f_h = \frac{f \wedge (t+h) - f \wedge t}{h}. \tag{9.7}$$

Let $\{h_i\}$ be a nonincreasing sequence of positive numbers such that $h_i \to 0$. Observe that $\{f_{h_i}\}$ is admissible for the set $\{f > t\}$, while $f_{h_i} \leq \chi_{\{f>t\}}$ for all small h_i. From Theorem 9.2 we know that

$$T(f_{h_i}) = \int f_{h_i}\, d\mu.$$

Thus, the monotone convergence theorem implies that

$$\lim_{i\to\infty} T(f_{h_i}) = \mu(\{f > t\}),$$

and this shows that the value of μ on the set $\{f > t\}$ is uniquely determined by T. From this it follows that $\mu(B_{i,j})$ is uniquely determined by T, and therefore by (9.5), the same is true for $\mu(V_i)$. Finally, referring to (9.6), where it is assumed that $\mu(A) < \infty$, we have that $\mu(A)$ is uniquely determined by T. The requirement that $\mu(A) < \infty$ will be ensured if $\chi_A \leq f$ for some $f \in L$; see (9.2). Thus, we have the following corollary.

9.4. Corollary. *If T is as in Theorem 9.2, the corresponding outer measure μ is uniquely determined by T on all sets A with the property that $\chi_A \leq f$ for some $f \in L$.*

Functionals that satisfy the conditions of Theorem 9.2 are called **monotone** (or alternatively, **positive**). In the spirit of the Jordan decomposition theorem, we now investigate the question of determining what functionals can be written as the difference of monotone functionals.

9.5. Theorem. *Suppose L is a lattice of functions on X and let $T\colon L \to \mathbb{R}$ be a functional satisfying the following four conditions for all functions in L:*

$$T(f+g) = T(f) + T(g),$$

$$T(cf) = cT(f) \text{ whenever } 0 \leq c < \infty,$$

$$\sup\{T(k) : 0 \leq k \leq f\} < \infty \text{ for all } f \in L^+,$$

$$T(f) = \lim_{i\to\infty} T(f_i) \text{ whenever } f = \lim_{i\to\infty} f_i \text{ and } \{f_i\} \text{ is nondecreasing.}$$

Then there exist positive functionals T^+ and T^- defined on the lattice L^+ satisfying the conditions of Theorem 9.2 and the property

$$T(f) = T^+(f) - T^-(f) \tag{9.8}$$

for all $f \in L^+$.

PROOF. We define T^+ and T^- on L^+ as follows:

$$T^+(f) = \sup\{T(k) : 0 \leq k \leq f\},$$
$$T^-(f) = -\inf\{T(k) : 0 \leq k \leq f\}.$$

To prove (9.8), let $f, g \in L^+$ with $f \geq g$. Then $f \geq f - g$ and therefore $-T^-(f) \leq T(f-g)$. Hence, $T(g) - T^-(f) \leq T(g) + T(f-g) = T(f)$. Taking the supremum over all g with $g \leq f$ yields

$$T^+(f) - T^-(f) \leq T(f).$$

Similarly, since $f - g \leq f$, we have $T(f-g) \leq T^+(f)$, so that $T(f) = T(g) + T(f-g) \leq T(g) + T^+(f)$. Taking the infimum over all $0 \leq g \leq f$ implies

$$T(f) \leq -T^-(f) + T^+(f).$$

Hence we obtain

$$T(f) = T^+(f) - T^-(f).$$

Now we will prove that T^+ satisfies the conditions of Theorem 9.2 on L^+. For this, let $f, g, h \in L^+$ with $f + g \geq h$ and set $k = \inf\{f, h\}$. Then $f \geq k$ and $g \geq h - k$; consequently,

$$T^+(f) + T^+(g) \geq T(k) + T(h-k) = T(h).$$

Since this holds for all $h \leq f + g$ with $h \in L^+$, we have

$$T^+(f) + T^+(g) \geq T^+(f+g).$$

It is easy to verify the opposite inequality and also that T^+ is both positively homogeneous and monotone.

To show that T^+ satisfies the last condition of Theorem 9.2, let $\{f_i\}$ be a nondecreasing sequence in L^+ such that $f_i \to f$. If $k \in L^+$ and $k \leq f$, then the sequence $g_i = \inf\{f_i, k\}$ is nondecreasing and converges to k as $i \to \infty$. Hence

$$T(k) = \lim_{i\to\infty} T(g_i) \leq \lim_{i\to\infty} T^+(f_i),$$

and therefore

$$T^+(f) \leq \lim_{i\to\infty} T^+(f_i).$$

The opposite inequality is obvious, and thus equality holds.

That T^- also satisfies the conditions of Theorem 9.2 is almost immediate. Indeed, let $R = -T$ and observe that R satisfies the first, second, and fourth conditions of our current theorem. It also satisfies the third, because $R^+(f) = (-T)^+(f) = T^-(f) = T(f) - T^+(f) < \infty$ for each $f \in L^+$. Thus what we have just proved for T^+ applies to R^+ as well. In particular, R^+ satisfies the conditions of Theorem 9.2. $\square$

9.6. COROLLARY. *Let T be as in the previous theorem. Then there exist outer measures μ^+ and μ^- on X such that for each $f \in L$, f is both μ^+ and μ^- measurable and*

$$T(f) = \int f \, d\mu^+ - \int f \, d\mu^-.$$

PROOF. Theorem 9.2 supplies outer measures μ^+, μ^- such that

$$T^+(f) = \int f \, d\mu^+,$$
$$T^-(f) = \int f \, d\mu^-,$$

for all $f \in L^+$. Since each $f \in L$ can be written as $f = f^+ - f^-$, the result follows. □

A functional T satisfying the conditions of Theorem 9.2 is called a **monotone Daniell integral**. By a **Daniell integral** we mean a functional T of the type in Theorem 9.5.

Exercises for Section 9.1

1. Show that the space of lower semicontinuous functions on a metric space is not a lattice because it is not closed under the $\wedge$ operation.
2. Let L denote the family of all functions of the form $u \circ p$, where $u: \mathbb{R} \to \mathbb{R}$ is continuous and $p: \mathbb{R}^2 \to \mathbb{R}$ is the orthogonal projection defined by $p(x_1, x_2) = x_1$. Define a functional T on L by

$$T(u \circ p) = \int_0^1 u \, d\lambda.$$

Prove that T meets the conditions of Theorem 9.2 and that the measure μ representing T satisfies

$$\mu(A) = \lambda[p(A) \cap [0,1]]$$

for all $A \subset \mathbb{R}^2$. Also, show that not all Borel sets are μ-measurable.
3. Prove that the sequence f_{h_i} is admissible for the set $\{f > t\}$, where f_h is defined by (9.7).
4. Let $L = C_c(\mathbb{R})$.
   (i) Prove **Dini's theorem:** If $\{f_i\}$ is a nonincreasing sequence in L that converges pointwise to 0, then $f_i \to 0$ uniformly.
   (ii) Define $T: L \to \mathbb{R}$ by

$$T(f) = \int f,$$

where the integral sign denotes the Riemann integral. Prove that T is a Daniell integral.

(iii) Roughly speaking, it can be shown that the measures μ^+ and μ^- that occur in Corollary 9.6 are carried by disjoint sets. This can be made precise in the following way. Prove that for an arbitrary $f \in L^+$, there exists a μ-measurable function g on X such that

$$g(x) = \begin{cases} f(x) & \text{for } \mu^+ - \text{a.e. } x, \\ 0 & \text{for } \mu^- - \text{a.e. } x. \end{cases}$$

9.2. The Riesz Representation Theorem

As a major application of the theorem in the previous section, we will prove the Riesz representation theorem, which asserts that a large class of measures on a locally compact Hausdorff space can be identified with positive linear functionals on the space of continuous functions with compact support.

We recall that a Hausdorff space X is said to be **locally compact** if for each $x \in X$, there is an open set U containing x such that $\overline{U}$ is compact. We let

$$C_c(X)$$

denote the set of all continuous maps $f: X \to \mathbb{R}$ whose support

$$\operatorname{spt} f := \text{closure}\{x : f(x) \neq 0\}$$

is compact. The class of **Baire sets** is defined as the smallest σ-algebra containing the sets $\{f > t\}$ for all $f \in C_c(X)$ and all real numbers t. Since each $\{f > t\}$ is an open set, it follows that the Borel sets contain the Baire sets. If X is a locally compact Hausdorff space satisfying the second axiom of countability, the converse is true (Exercise 3, Section 9.2). We will call an outer measure μ on X a **Baire outer measure** if all Baire sets are μ-measurable.

We first establish two results that are needed for our development.

9.7. LEMMA. (Urysohn's lemma) *Let $K \subset U$ be compact and open sets, respectively, in a locally compact Hausdorff space X. Then there exists a function $f \in C_c(X)$ such that*

$$\chi_K \leq f \leq \chi_U.$$

PROOF. Let $r_1, r_2, \ldots$ be an enumeration of the rationals in $(0, 1)$, where we take $r_1 = 0$ and $r_2 = 1$. By Theorem 3.18, there exist open sets V_0 and V_1 with compact closures such that

$$K \subset V_1 \subset \overline{V}_1 \subset V_0 \subset \overline{V}_0 \subset V.$$

Proceeding by induction, we will associate an open set V_{r_k} with each rational r_k in the following way. Assume that $V_{r_1}, \ldots, V_{r_k}$ have been chosen in such a manner that $\overline{V}_{r_j} \subset V_{r_i}$ whenever $r_i < r_j$. Among the numbers $r_1, r_2, \ldots, r_k$ let r_m denote the smallest that is larger than r_{k+1} and let r_n denote the largest that is smaller than r_{k+1}. Referring again to Theorem 3.18, there exists $V_{r_{k+1}}$ such that

$$\overline{V}_{r_n} \subset V_{r_{k+1}} \subset \overline{V}_{r_{k+1}} \subset V_{r_m}.$$

Continuing this process, for each rational number $r \in [0,1]$ there is a corresponding open set V_r. The countable collection $\{V_r\}$ satisfies the following properties: $K \subset V_1$, $\overline{V}_0 \subset U$, $\overline{V}_r$ is compact, and

$$\overline{V}_s \subset V_r \quad \text{for} \quad r < s. \tag{9.9}$$

For rational numbers r and s, define

$$f_r := r\,\chi_{V_r} \quad \text{and} \quad g_s := \chi_{\overline{V}_s} + s\,\chi_{X-\overline{V}_s}.$$

Referring to Definition 3.68, it is easy to see that each f_r is lower semicontinuous and each g_s is upper semicontinuous. Consequently, with

$$f(x) := \sup\{f_r(x) : r \in \mathbb{Q} \cap (0,1)\} \quad \text{and} \quad g := \inf\{g_s(x) : r \in \mathbb{Q} \cap (0,1)\}$$

it follows that f is lower semicontinuous and g is upper semicontinuous. Note that $0 \le f \le 1$, $f = 1$ on K, and that spt $f \subset \overline{V}_0$.

The proof will be completed when we show that f is continuous. This is established by showing that $f = g$. To this end, note that $f \le g$, for otherwise, there exists $x \in X$ such that $f_r(x) > g_s(x)$ for some rational numbers r and s. But this is possible only if $r > s$, $x \in V_r$, and $x \notin \overline{V}_s$. However, $r > s$ implies $V_r \subset V_s$.

Finally, we observe that if it were true that $f(x) < g(x)$ for some x, then there would exist rational numbers r and s such that

$$f(x) < r < s < g(x).$$

This would imply $x \notin V_r$ and $x \in \overline{V}_s$, contradicting (9.9). Hence, $f = g$. □

9.8. THEOREM. *Suppose K is compact and $V_1, \ldots, V_n$ are open sets in a locally compact Hausdorff space X such that*

$$K \subset V_1 \cup \cdots \cup V_n.$$

Then there are continuous functions g_i with $0 \le g_i \le 1$ and spt $g_i \subset V_i$, $i = 1, 2, \ldots, n$, such that

$$g_1(x) + g_2(x) + \cdots + g_n(x) = 1 \quad \text{for all} \quad x \in K.$$

PROOF. By Theorem 3.18, each $x \in K$ is contained in an open set U_x with compact closure such that $\overline{U}_x \subset V_i$ for some i. Since K is compact, there exist finitely many points $x_1, x_2, \ldots, x_k$ in K such that

$$K \subset \bigcup_{i=1}^{k} U_{x_i}.$$

For $1 \le j \le n$, let G_j denote the union of the $\overline{U}_{x_i}$ that lie in V_j. Urysohn's lemma, Lemma 9.7, provides continuous functions f_j with $0 \le f_j \le 1$ such that

$$\chi_{G_j} \le f_j \le \chi_{V_j}.$$

Now $\sum_{j=1}^{n} f_j \geq 1$ on K, so applying Urysohn's lemma again, there exists $f \in C_c(X)$ with $f = 1$ on K and $\operatorname{spt} f \subset \{\sum_{j=1}^{n} f_j > 0\}$. Let $f_{n+1} = 1 - f$, so that $\sum_{j=1}^{n+1} f_j > 0$ everywhere. Now define

$$g_i = \frac{f_i}{\sum_{j=1}^{n+1} f_j}$$

to obtain the desired conclusion. □

The theorem on the existence of the Daniell integral provides the following as an immediate application.

9.9. THEOREM. (Riesz representation theorem) *Let X be a locally compact Hausdorff space and let L denote the lattice $C_c(X)$ of continuous functions with compact support. If $T\colon L \to \mathbb{R}$ is a linear functional such that*

$$\sup\{T(g) : 0 \leq g \leq f\} < \infty \tag{9.10}$$

whenever $f \in L^+$, then there exist Baire outer measures μ^+ and μ^- such that for each $f \in C_c(X)$,

$$T(f) = \int_X f\, d\mu^+ - \int_X f\, d\mu^-.$$

The outer measures μ^+ and μ^- are uniquely determined on the family of compact sets.

PROOF. If we can show that T satisfies the hypotheses of Theorem 9.5, then Corollary 9.6 provides outer measures μ^+ and μ^- satisfying our conclusion.

All the conditions of Theorem 9.5 are easily verified except perhaps the last one. Thus, let $\{f_i\}$ be a nondecreasing sequence in L^+ whose limit is $f \in L^+$, and refer to Urysohn's lemma to find a function $g \in C_c(X)$ such that $g \geq 1$ on $\operatorname{spt} f$.

Choose $\varepsilon > 0$ and define compact sets

$$K_i = \{x : f(x) \geq f_i(x) + \varepsilon\},$$

such that $K_1 \supset K_2 \supset \ldots$. Now

$$\bigcap_{i=1}^{\infty} K_i = \emptyset,$$

and therefore the $\widetilde{K_i}$ form an open covering of X, and in particular, of $\operatorname{spt} f$. Since $\operatorname{spt} f$ is compact, there is an index i_0 such that

$$\operatorname{spt} f \subset \bigcup_{i=1}^{i_0} \widetilde{K_i} = \widetilde{K_{i_0}}.$$

Since $\{f_i\}$ is a nondecreasing sequence and $g \geq 1$ on $\operatorname{spt} f$, this implies that

$$f(x) < f_i(x) + \varepsilon g(x)$$

for all $i \geq i_0$ and all $x \in X$. Note that (9.10) implies that

$$M: = \sup\{|T(k)| : k \in L, 0 \leq k \leq g\} < \infty.$$

Thus, we obtain

$$0 \leq f - f_i \leq \varepsilon g, \qquad |T(f - f_i)| \leq \varepsilon M < \infty.$$

Since ε is arbitrary, we conclude that $T(f_i) \to T(f)$, as required.

Concerning the assertion of uniqueness, select a compact set K and use Theorem 3.18 to find an open set $U \supset K$ with compact closure. Consider the lattice

$$L_U: = C_c(X) \cap \{f : \operatorname{spt} f \subset U\}$$

and let $T_U(f): = T(f)$ for $f \in L_U$. Then we obtain outer measures μ_U^+ and μ_U^- such that

$$T_U(f) = \int f \, d\mu_U^+ - \int f \, d\mu_U^-$$

for all $f \in L_U$. With the help of Urysohn's lemma and Corollary 9.4, we find that μ_U^+ and μ_U^- are uniquely determined, and so $\mu_U^+(K) = \mu^+(K)$ and $\mu_U^-(K) = \mu^-(K)$. □

9.10. Remark. In the previous result, it might be tempting to define a signed measure μ by $\mu: = \mu^+ - \mu^-$ and thereby reach the conclusion that

$$T(f) = \int_X f \, d\mu$$

for each $f \in C_c(X)$. However, this is not possible, because $\mu^+ - \mu^-$ may be undefined. That is, both $\mu^+(E)$ and $\mu^-(E)$ may possibly assume the value $+\infty$ for some set E. See Exercise 6, Section 9.2, to obtain a resolution of this problem in some situations.

If X is a compact Hausdorff space, then all continuous functions are bounded, and $C(X)$ becomes a normed linear space with the norm $\|f\| : = \sup\{|f(x)| : x \in X\}$. We will show that there is a very useful characterization of the dual of $C(X)$ that results as a direct consequence of the Riesz representation theorem. First, recall that the norm of a linear functional T on $C(X)$ is defined in the usual way by

$$\|T\| = \sup\{T(f) : \|f\| \leq 1\}.$$

If T is written as the difference of two positive functionals, $T = T^+ - T^-$, then

$$\|T\| \leq \|T^+\| + \|T^-\| = T^+(1) + T^-(1).$$

The opposite inequality is also valid, for if $f \in C(X)$ is any function with $0 \leq f \leq 1$, then $|2f - 1| \leq 1$ and

$$\|T\| \geq T(2f - 1) = 2T(f) - T(1).$$

Taking the supremum over all such f yields

$$\begin{aligned}\|T\| &\geq 2T^{+}(1) - T(1) \\ &= T^{+}(1) + T^{-}(1).\end{aligned}$$

Hence,

$$\|T\| = T^{+}(1) + T^{-}(1). \tag{9.11}$$

9.11. COROLLARY. *Let X be a compact Hausdorff space. Then for every bounded linear functional $T\colon C(X) \to \mathbb{R}$, there exists a unique signed Baire outer measure μ on X such that*

$$T(f) = \int_X f \, d\mu$$

for each $f \in C(X)$. Moreover, $\|T\| = \|\mu\|(X)$. Thus, the dual of $C(X)$ is isometrically isomorphic to the space of signed Baire outer measures on X.

PROOF. A bounded linear functional T on $C(X)$ is easily seen to imply property (9.10), and therefore Theorem 9.9 implies there exist Baire outer measures μ^{+} and μ^{-} such that with $\mu\colon = \mu^{+} - \mu^{-}$, we have

$$T(f) = \int_X f \, d\mu$$

for all $f \in C(X)$. Thus,

$$\begin{aligned}|T(f)| &\leq \int |f| \, d\|\mu\| \\ &\leq \|f\| \, \|\mu\|(X),\end{aligned}$$

and therefore, $\|T\| \leq \|\mu\|(X)$. On the other hand, with

$$T^{+}(f)\colon = \int_X f \, d\mu^{+} \quad \text{and} \quad T^{-}(f)\colon = \int_X f \, d\mu^{-},$$

we have, with the help of (9.11),

$$\begin{aligned}\|\mu\|(X) &\leq \mu^{+}(X) + \mu^{-}(X) \\ &= T^{+}(1) + T^{-}(1) = \|T\|.\end{aligned}$$

Hence, $\|T\| = \|\mu\|(X)$. □

Baire sets arise naturally in our development because they constitute the smallest σ-algebra that contains sets of the form $\{f > t\}$, where $f \in C_c(X)$ and $t \in \mathbb{R}$. These are precisely the sets that occur in Theorem 9.2 when L is taken as $C_c(X)$. In view of Exercise 2, Section 9.2, note that the outer measure obtained in the preceding corollary is Borel. In general, it is possible to have access to Borel outer measures rather than merely Baire measures, as seen in the following result.

9.12. THEOREM. *Assume the hypotheses and notation of the Riesz representation theorem. Then there is a signed Borel outer measure* $\bar{\mu}$ *on* X *with the property that*

$$\int f\, d\bar{\mu} = T(f) = \int f\, d\mu \tag{9.12}$$

for all $f \in C_c(X)$.

PROOF. Assuming first that T satisfies the conditions of Theorem 9.2, we will prove there is a Borel outer measure $\bar{\mu}$ such that (9.12) holds for every $f \in L^+$. Then referring to the proof of the Riesz representation theorem, it follows that there is a signed outer measure $\bar{\mu}$ that satisfies (9.12).

We define $\bar{\mu}$ in the following way. For every open set U let

$$\alpha(U) := \sup\{T(f) : f \in L^+, f \le 1 \text{ and spt } f \subset U\}, \tag{9.13}$$

and for every $A \subset X$, define

$$\bar{\mu}(A) := \inf\{\alpha(U) : A \subset U, U \text{ open}\}.$$

Observe that $\bar{\mu}$ and α agree on all open sets.

First, we verify that $\bar{\mu}$ is countably subadditive. Let $\{U_i\}$ be a sequence of open sets and let

$$V = \bigcup_{i=1}^{\infty} U_i.$$

If $f \in L^+, f \le 1$ and spt $f \subset V$, the compactness of spt f implies that

$$\text{spt } f \subset \bigcup_{i=1}^{N} U_i$$

for some positive integer N. Now appeal to Lemma 11.20 to obtain functions $g_1, g_2, \ldots, g_N \in L^+$ with $g_i \le 1$, spt $g_i \subset U_i$, and

$$\sum_{i=1}^{N} g_i(x) = 1 \quad \text{whenever} \quad x \in \text{spt } f.$$

Thus,

$$f(x) = \sum_{i=1}^{N} g_i(x) f(x),$$

and therefore

$$T(f) = \sum_{i=1}^{N} T(g_i f) \le \sum_{i=1}^{N} \alpha(U_i) < \sum_{i=1}^{\infty} \alpha(U_i).$$

This implies

$$\alpha\left(\bigcup_{i=1}^{\infty} U_i\right) \le \sum_{i=1}^{\infty} \alpha(U_i).$$

From this and the definition of $\bar{\mu}$, it follows easily that $\bar{\mu}$ is countably subadditive on all sets.

Next, to show that $\bar{\mu}$ is a Borel outer measure, it suffices to show that each open set U in $\bar{\mu}$ is measurable. For this, let $\varepsilon > 0$ and let $A \subset X$ be an arbitrary set with $\bar{\mu}(A) < \infty$. Choose an open set $V \supset A$ such that $\alpha(V) < \bar{\mu}(A) + \varepsilon$ and a function $f \in L^+$ with $f \leq 1$, $\operatorname{spt} f \subset V \cap U$, and $T(f) > \alpha(V \cap U) - \varepsilon$. Also, choose $g \in L^+$ with $g \leq 1$ and $\operatorname{spt} g \subset V - \operatorname{spt} f$ such that $T(g) > \alpha(V - \operatorname{spt} f) - \varepsilon$. Then, since $f + g \leq 1$, we obtain

$$\begin{aligned}\bar{\mu}(A) + \varepsilon \geq \alpha(V) &\geq T(f+g) = T(f) + T(g)\\ &\geq \alpha(V \cap U) + \alpha(V - \operatorname{spt} f)\\ &\geq \bar{\mu}(A \cap U) + \bar{\mu}(A - U) - 2\varepsilon.\end{aligned}$$

This shows that U is $\bar{\mu}$-measurable, since ε is arbitrary.

Finally, to establish (9.12), by Theorem 6.60 it suffices to show that

$$\bar{\mu}(\{f > s\}) = \mu(\{f > s\})$$

for all $f \in L^+$ and all $s \in \mathbb{R}$. It follows from the definitions that

$$\bar{\mu}(U) = \alpha(U) \leq \mu(U)$$

whenever U is an open set. In particular, we have $\bar{\mu}(U_s) \leq \mu(U_s)$, where $U_s :=\{f > s\}$. To prove the opposite inequality, note that if $f \in L^+$ and $0 < s < t$, then

$$\alpha(U_s) \geq \frac{T[f \wedge (t+h) - f \wedge t]}{h} \quad \text{for} \quad h > 0,$$

because each

$$f_h := \frac{f \wedge (t+h) - f \wedge t}{h} = \begin{cases} 1 & \text{if } f \geq t+h,\\ \frac{f-t}{h} & \text{if } t < f < t+h,\\ 0 & \text{if } f \leq t,\end{cases}$$

is a competitor in (9.13). Furthermore, since $f_{h_1} \geq f_{h_2}$ when $h_1 \leq h_2$ and $\lim_{h \to 0^+} f_h = \chi_{\{f>t\}}$, we may apply the monotone convergence theorem to obtain

$$\begin{aligned}\mu(\{f > t\}) &= \lim_{h \to 0^+} \int_X f_h \, d\mu\\ &= \lim_{h \to 0^+} \frac{T[f \wedge (t+h) - f \wedge t]}{h}\\ &\leq \alpha(U_s) = \bar{\mu}(U_s).\end{aligned}$$

Since

$$\mu(U_s) = \lim_{t \to s^+} \mu(\{f > t\}),$$

we have $\mu(U_s) \leq \bar{\mu}(U_s)$. □

From Corollary 9.4 we see that $\bar{\mu}$ and μ agree on compact sets, and therefore $\bar{\mu}$ is finite on compact sets. Furthermore, it follows from Lemma 9.3 that for an arbitrary set $A \subset X$, there exists a Borel set $B \supset A$ such that $\bar{\mu}(B) = \bar{\mu}(A)$. Recall that an outer measure with these properties is called a **Radon outer measure** (see Definition 4.14). Note that this definition is compatible with that of **Radon measure** given in Definition 4.47.

Exercises for Section 9.2

1. Provide an alternative (and simpler) proof of Urysohn's lemma (Lemma 9.7) in the case that X is a locally compact metric space.
2. Suppose X is a locally compact Hausdorff space.
   (i) If $f \in C_c(X)$ is nonnegative, then $\{a \leq f \leq \infty\}$ is a compact G_δ set for all $a > 0$.
   (ii) If $K \subset X$ is a compact G_δ set, then there exists $f \in C_c(X)$ with $0 \leq f \leq 1$ and $K = f^{-1}(1)$.
   (iii) The Baire sets are generated by compact G_δ sets.
3. Prove that the Baire sets and Borel sets are the same in a locally compact Hausdorff space satisfying the second axiom of countability.
4. Consider $(X, \mathcal{M}, \mu)$, where X is a compact Hausdorff space and $\mathcal{M}$ is the family of Borel sets. If $\{\nu_i\}$ is a sequence of Radon measures with $\nu_i(X) \leq M$ for some positive M, then Alaoglu's theorem (Theorem 8.40) implies that there is a subsequence ν_{i_j} that converges weak* to some Radon measure ν. Suppose $\{f_i\}$ is a sequence of functions in $L^1(X, \mathcal{M}, \mu)$. What can be concluded if $\|f_i\|_{1;\mu} \leq M$?
5. With the same notation as in the previous problem, suppose that ν_i converges weak* to ν. Prove the following:
   (i) $\limsup_{i\to\infty} \nu_i(F) \leq \nu(F)$ whenever F is a closed set.
   (ii) $\liminf_{i\to\infty} \nu_i(U) \geq \nu(U)$ whenever U is an open set.
   (iii) $\lim_{i\to\infty} \nu_i(E) = \nu(E)$ whenever E is a measurable set with $\nu(\partial E) = 0$.
6. Assume that X is a locally compact Hausdorff space that can be written as a countable union of compact sets. Then, under the hypotheses and notation of the Riesz representation theorem, prove that there exist a (nonnegative) Baire outer measure μ and a μ-measurable function g such that
   (i) $|g(x)| = 1$ for μ-a.e. x, and
   (ii) $T(f) = \int_X fg\, d\mu$ for all $f \in C_c(X)$.

CHAPTER 10

Distributions

10.1. The Space $\mathcal{D}$

In the previous chapter, we saw how a bounded linear functional on the space $C_c(\mathbb{R}^n)$ can be identified with a measure. In this chapter, we will pursue this idea further by considering linear functionals on a smaller space, thus producing objects called distributions that are more general than measures. Distributions are of fundamental importance in many areas such as partial differential equations, the calculus of variations, and harmonic analysis. Their importance was formally acknowledged by the mathematics community when Laurent Schwartz, who initiated and developed the theory of distributions, was awarded the Fields Medal at the 1950 International Congress of Mathematicians. In the next chapter, some applications of distributions will be given. In particular, it will be shown how distributions are used to obtain a solution to a fundamental problem in partial differential equations, namely, the Dirichlet problem. We begin by introducing the space of functions on which distributions are defined.

Let $\Omega \subset \mathbb{R}^n$ be an open set. We begin by investigating a space that is much smaller than $C_c(\Omega)$, namely, the space $\mathcal{D}(\Omega)$ of all infinitely differentiable functions whose supports are contained in Ω. We let $C^k(\Omega)$, $1 \leq k \leq \infty$, denote the class of functions defined on Ω whose partial derivatives of all orders up to and including k are continuous. Also, we denote by $C_c^k(\Omega)$ the set of functions in $C^k(\Omega)$ whose supports are contained in Ω.

It is not immediately obvious that such functions exist, so we begin by analyzing the following function defined on $\mathbb{R}$:

$$f(x) = \begin{cases} e^{-1/x}, & x > 0, \\ 0, & x \leq 0. \end{cases}$$

Observe that f is C^∞ on $\mathbb{R} - \{0\}$. It remains to show that all derivatives exist and are continuous at $x = 0$. Now,

$$f'(x) = \begin{cases} \frac{1}{x^2} e^{-1/x}, & x > 0, \\ 0, & x < 0, \end{cases}$$

W.P. Ziemer, *Modern Real Analysis*, Graduate Texts in Mathematics 278,
https://doi.org/10.1007/978-3-319-64629-9_10

and therefore

$$\lim_{x\to 0} f'(x) = 0. \tag{10.1}$$

Also,

$$\lim_{h\to 0+} \frac{f(h)-f(0)}{h} = 0. \tag{10.2}$$

Note that (10.2) implies that $f'(0)$ exists and $f'(0) = 0$. Moreover, (10.1) gives that f' is continuous at $x = 0$.

A similar argument establishes the same conclusion for all higher derivatives of f. Indeed, a direct calculation shows that the kth derivatives of f for $x \neq 0$ are of the form $P_{2k}(\frac{1}{x})f(x)$, where P is a polynomial of degree $2k$. Thus

$$\begin{aligned} \lim_{x\to 0+} f^{(k)}(x) &= \lim_{x\to 0+} P\left(\frac{1}{x}\right) e^{\frac{-1}{x}} \\ &= \lim_{x\to 0+} \frac{P(\frac{1}{x})}{e^{\frac{1}{x}}} \\ &= \lim_{t\to\infty} \frac{P(t)}{e^t} = 0, \end{aligned} \tag{10.3}$$

where this limit is computed by using L'Hôpital's rule repeatedly. Finally,

$$\begin{aligned} \lim_{h\to 0+} \frac{f^{(k-1)}(h) - f^{(k-1)}(0)}{h} &= \lim_{h\to 0+} \frac{f^{(k-1)}(h)}{h} \\ &= \lim_{h\to 0+} \frac{P_{2(k-1)}\left(\frac{1}{h}\right) e^{\frac{-1}{h}}}{h} \\ &= \lim_{h\to 0+} Q\left(\frac{1}{h}\right) e^{\frac{-1}{h}} \\ &= \lim_{h\to 0+} \frac{Q(t)}{e^t} = 0, \end{aligned} \tag{10.4}$$

where Q is a polynomial of degree $2k-1$. From (10.4) we conclude that $f^{(k)}$ exists at $x = 0$ and $f^{(k)}(0) = 0$. Moreover, (10.3) shows that $f^{(k)}$ is continuous at $x = 0$.

We can now construct a C^∞ function with compact support in $\mathbb{R}^n$. For this, let

$$F(x) = f(1 - |x|^2),\ x \in \mathbb{R}^n.$$

With $x = (x_1, x_2, \ldots, x_n)$, observe that $1 - |x|^2$ is the polynomial $1 - (x_1^2 + x_2^2 + \cdots + x_n^2)$ and therefore, that F is an infinitely differentiable function of x. Moreover, F is nonnegative and is zero for $|x| \geq 1$.

It is traditional in many parts of analysis to denote C^∞ functions with compact support by φ. We will adopt this convention. Now for some notation. The **partial derivative operators** are denoted by $D_i = \partial/\partial x_i$ for

$1 \leq i \leq n$. Thus,

$$D_i\varphi = \frac{\partial \varphi}{\partial x_i}. \tag{10.5}$$

If $\alpha = (\alpha_1, \alpha_2, \ldots, \alpha_n)$ is an n-tuple of nonnegative integers, then α is called a **multi-index**, and the **length** of α is defined as

$$|\alpha| = \sum_{i=1}^{n} \alpha_i.$$

Using this notation, higher-order derivatives are denoted by

$$D^\alpha \varphi = \frac{\partial^{|\alpha|}\varphi}{\partial x_1^{\alpha_1} \ldots \partial x_n^{\alpha_n}}.$$

For example,

$$\frac{\partial^2 \varphi}{\partial x \partial y} = D^{(1,1)}\varphi \qquad \text{and} \qquad \frac{\partial^3 \varphi}{\partial^2 x \partial y} = D^{(2,1)}\varphi.$$

Also, we let $\nabla\varphi(x)$ denote the **gradient** of φ at x, that is,

$$\begin{aligned}\nabla\varphi(x) &= \left(\frac{\partial \varphi}{\partial x_1}(x), \frac{\partial \varphi}{\partial x_2}(x), \ldots, \frac{\partial \varphi}{\partial x_n}(x)\right) \\ &= (D_1\varphi(x), D_2\varphi(x), \ldots, D_n\varphi(x)).\end{aligned} \tag{10.6}$$

We have shown that there exist C^∞ functions φ with the property that $\varphi(x) > 0$ for $x \in B(0,1)$ and that $\varphi(x) = 0$ whenever $|x| \geq 1$. By multiplying φ by a suitable constant, we can assume that

$$\int_{B(0,1)} \varphi(x)\, d\lambda(x) = 1. \tag{10.7}$$

By employing an appropriate scaling of φ, we can duplicate these properties on the ball $B(0,\varepsilon)$ for every $\varepsilon > 0$. For this purpose, let

$$\varphi_\varepsilon(x) = \varepsilon^{-n}\varphi\left(\frac{x}{\varepsilon}\right).$$

Then φ_ε has the same properties on $B(0,\varepsilon)$ as does φ on $B(0,1)$. Furthermore,

$$\int_{B(0,\varepsilon)} \varphi_\varepsilon(x)\, d\lambda(x) = 1. \tag{10.8}$$

Thus, we have shown that $\mathcal{D}(\Omega)$ is nonempty; we will often call functions in $\mathcal{D}(\Omega)$ **test functions**.

We now show how the functions φ_ε can be used to generate more C^∞ functions with compact support. Given a function $f \in L^1_{\text{loc}}(\mathbb{R}^n)$, recall the definition of convolution first introduced in Section 6.11:

$$f * \varphi_\varepsilon(x) = \int_{\mathbb{R}^n} f(x-y)\varphi_\varepsilon(y)\, d\lambda(y) \tag{10.9}$$

for all $x \in \mathbb{R}^n$. We will use the notation $f_\varepsilon \colon = f * \varphi_\varepsilon$. The function f_ε is called a **mollifier** of f.

10.1. THEOREM.

(i) *If* $f \in L^1_{\text{loc}}(\mathbb{R}^n)$, *then for every* $\varepsilon > 0, f_\varepsilon \in C^\infty(\mathbb{R}^n)$.

(ii)

$$\lim_{\varepsilon \to 0} f_\varepsilon(x) = f(x)$$

whenever x *is a Lebesgue point for* f. *If* f *is continuous, then* f_ε *converges to* f *uniformly on compact subsets of* $\mathbb{R}^n$.

(iii) *If* $f \in L^p(\mathbb{R}^n)$, $1 \le p < \infty$, *then* $f_\varepsilon \in L^p(\mathbb{R}^n)$, $\|f_\varepsilon\|_p \le \|f\|_p$, *and* $\lim_{\varepsilon \to 0} \|f_\varepsilon - f\|_p = 0$.

PROOF. Recall that $D_i f$ denotes the partial derivative of f with respect to the ith variable. The proof that f_ε is C^∞ will be established if we can show that

$$D_i(\varphi_\varepsilon * f) = (D_i \varphi_\varepsilon) * f \tag{10.10}$$

for each $i = 1, 2, \ldots, n$. To see this, assume (10.10) for the moment. The right-hand side of the equation is a continuous function (Exercise 2, Section 10.1), thus showing that f_ε is C^1. Then if α denotes the n-tuple with $|\alpha| = 2$ and with 1 in the ith and j^{th} positions, it follows that

$$\begin{aligned} D^\alpha(\varphi_\varepsilon * f) &= D_i[D_j(\varphi_\varepsilon * f)] \\ &= D_i[(D_j\varphi_\varepsilon) * f] \\ &= (D^\alpha \varphi_\varepsilon) * f, \end{aligned}$$

which proves that $f_\varepsilon \in C^2$, since again the right-hand side of the equation is continuous. Proceeding in this way by induction, it follows that $f \in C^\infty$.

We now turn to the proof of (10.10). Let $e_1, \ldots, e_n$ be the standard basis of $\mathbb{R}^n$ and consider the partial derivative with respect to the ith variable. For every real number h, we have

$$f_\varepsilon(x + he_i) - f_\varepsilon(x) = \int_{\mathbb{R}^n} [\varphi_\varepsilon(x - z + he_i) - \varphi_\varepsilon(x - z)] f(z)\, d\lambda(z).$$

Let $\alpha(t)$ denote the integrand:

$$\alpha(t) = \varphi_\varepsilon(x - z + te_i) f(z).$$

Then α is a C^∞ function of t. The chain rule implies

$$\begin{aligned} \alpha'(t) &= \nabla \varphi_\varepsilon(x - z + te_i) \cdot e_i f(z) \\ &= D_i \varphi_\varepsilon(x - z + te_i) f(z). \end{aligned}$$

Since

$$\alpha(h) - \alpha(0) = \int_0^h \alpha'(t)\, dt,$$

we have

$$
\begin{aligned}
&f_\varepsilon(x+he_i)-f_\varepsilon(x)\\
&=\int_{\mathbb{R}^n}[\varphi_\varepsilon(x-z+he_i)-\varphi_\varepsilon(x-z)]f(z)\,d\lambda(z)\\
&=\int_{\mathbb{R}^n}\alpha(h)-\alpha(0)\,d\lambda(z)\\
&=\int_{\mathbb{R}^n}\int_0^h D_i\varphi_\varepsilon(x-z+te_i)f(z)\,dt\,d\lambda(z)\\
&=\int_0^h\int_{\mathbb{R}^n} D_i\varphi_\varepsilon(x-z+te_i)f(z)\,d\lambda(z)\,dt.
\end{aligned}
$$

It follows from Lebesgue's dominated convergence theorem that the inner integral is a continuous function of t. Now divide both sides of the equation by h and take the limit as $h\to 0$ to obtain

$$D_if_\varepsilon(x)=\int_{\mathbb{R}^n}D_i\varphi_\varepsilon(x-z)f(z)\,d\lambda(z)=(D_i\varphi_\varepsilon)*f(x),$$

which establishes (10.10) and therefore (i).

In case (ii), observe that

$$
\begin{aligned}
&|f_\varepsilon(x)-f(x)|\\
&\le\int_{\mathbb{R}^n}\varphi_\varepsilon(x-y)\,|f(y)-f(x)|\;d\lambda(y) \qquad (10.11)\\
&\le\max_{\mathbb{R}^n}\varphi\,\varepsilon^{-n}\int_{B(x,\varepsilon)}|f(x)-f(y)|\;d\lambda(y)\to 0
\end{aligned}
$$

as $\varepsilon\to 0$ whenever x is a Lebesgue point for f. Now consider the case that f is continuous and $K\subset\mathbb{R}^n$ is compact. Then f is uniformly continuous on K and also on the closure of the open set $U=\{x:\operatorname{dist}(x,K)<1\}$. For each $\eta>0$, there exists $0<\varepsilon<1$ such that $|f(x)-f(y)|<\eta$ whenever $|x-y|<\varepsilon$ and whenever $x,y\in U$, and in particular when $x\in K$. Consequently, it follows from (10.11) that whenever $x\in K$,

$$|f_\varepsilon(x)-f(x)|\le M\eta,$$

where M is the product of $\max_{\mathbb{R}^n}\varphi$ and the Lebesgue measure of the unit ball. Since η is arbitrary, this shows that f_ε converges uniformly to f on K.

The first part of (iii) follows from Theorem 6.57, since $\|\varphi_\varepsilon\|_1=1$.

Finally, addressing the second part of (iii), for each $\eta>0$, select a continuous function g with compact support on $\mathbb{R}^n$ such that

$$\|f-g\|_p<\eta \qquad (10.12)$$

(see Exercise 12, Section 6.5). Because g has compact support, it follows from (ii) that $\|g-g_\varepsilon\|_p<\eta$ for all ε sufficiently small. Now apply Theorem 10.1 (iii) and (10.12) to the difference $g-f$ to obtain

$$\|f-f_\varepsilon\|_p\le\|f-g\|_p+\|g-g_\varepsilon\|_p+\|g_\varepsilon-f_\varepsilon\|_p\le 3\eta.$$

This shows that $\|f - f_\varepsilon\|_p \to 0$ as $\varepsilon \to 0$. □

Exercises for 10.1

1. Let K be a compact subset of an open set Ω. Prove that there exists $f \in C_c^\infty(\Omega)$ such that $f \equiv 1$ on K.
2. Suppose $f \in L^1_{\text{loc}}(\mathbb{R}^n)$ and φ is a continuous function with compact support. Prove that $\varphi * f$ is continuous.
3. If $f \in L^1_{\text{loc}}(\mathbb{R}^n)$ and
$$\int_{\mathbb{R}^n} f\varphi \, dx = 0$$
for every $\varphi \in C_c^\infty(\mathbb{R}^n)$, show that $f = 0$ almost everywhere. Hint: Use Theorem 10.1.

10.2. Basic Properties of Distributions

10.2. Definition. Let $\Omega \subset \mathbb{R}^n$ be an open set. A linear functional T on $\mathcal{D}(\Omega)$ is a **distribution** if and only if for every compact set $K \subset \Omega$, there exist constants C and N such that
$$|T(\varphi)| \leq C(K) \sup_{x \in K} \sum_{|\alpha| \leq N(K)} |D^\alpha \varphi(x)|$$
for all test functions $\varphi \in \mathcal{D}(\Omega)$ with support in K. If the integer N can be chosen independent of the compact set K, and N is the smallest possible choice, the distribution is said to be **of order** N. We use the notation
$$\|\varphi\|_{K;N} : = \sup_{x \in K} \sum_{|\alpha| \leq N} |D^\alpha \varphi(x)|.$$
Thus, in particular, $\|\varphi\|_{K;0}$ denotes the sup norm of φ on K.

Here are some examples of distributions. First, suppose μ is a signed Radon measure on Ω. Define the corresponding distribution by
$$T(\varphi) = \int_\Omega \varphi(x) d\mu(x)$$
for every test function φ. Note that the integral is finite, since μ is finite on compact sets, by definition. If $K \subset \Omega$ is compact and $C(K) = |\mu|(K)$ (recall the notation in Definition 6.39), then
$$\begin{aligned} |T(\varphi)| &\leq C(K) \|\varphi\|_{L^\infty} \\ &= C(K) \|\varphi\|_{K;0} \end{aligned}$$
for all test functions φ with support in K. Thus, the distribution T corresponding to the measure μ is of order 0. In the context of distribution theory, a Radon measure will be identified as a distribution in this way. In

particular, consider the Dirac measure δ, whose total mass is concentrated at the origin:

$$\delta(E) = \begin{cases} 1, & 0 \in E, \\ 0, & 0 \notin E. \end{cases}$$

The distribution identified with this measure is defined by

$$T(\varphi) = \varphi(0), \tag{10.13}$$

for every test function φ.

10.3. DEFINITION. Let $f \in L^1_{\text{loc}}(\Omega)$. The distribution corresponding to f is defined as

$$T(\varphi) = \int_\Omega \varphi(x) f(x)\, d\lambda(x).$$

Thus, a locally integrable function can be considered an absolutely continuous measure and is therefore identified with a distribution of order 0. The distribution corresponding to f is sometimes denoted by T_f.

We have just seen that a Radon measure is a distribution of order 0. The following result shows that we can actually identify measures and distributions of order 0.

10.4. THEOREM. *A distribution T is a Radon measure if and only if T is of order* 0.

PROOF. Assume that T is a distribution of order 0. We will show that T can be extended in a unique way to a linear functional T^* on the space $C_c(\Omega)$ such that

$$|T^*(\varphi)| \leq C(K)\, \|\varphi\|_{K;0}\,, \tag{10.14}$$

for each compact set K and φ supported on K. Then, by appealing to the Riesz representation theorem, it will follow that there is a (signed) Radon measure μ such that

$$T^*(\varphi) = \int_{\mathbb{R}^n} \varphi\, d\mu$$

for each $\varphi \in C_c(\Omega)$. In particular, we will have

$$T(\varphi) = T^*(\varphi) = \int_{\mathbb{R}^n} \varphi\, d\mu$$

whenever $\varphi \in \mathcal{D}(\Omega)$, thus establishing that μ is the measure identified with T.

In order to prove (10.14), select a continuous function φ with support in a compact set K. Using mollifiers and Theorem 10.1, it follows that there is a sequence of test functions $\{\varphi_i\} \in \mathcal{D}(\Omega)$ whose supports are contained in a fixed compact neighborhood of spt φ such that

$$\|\varphi_i - \varphi\|_{K;0} \to 0 \quad \text{as} \quad i \to \infty.$$

Now define

$$T^*(\varphi) = \lim_{i\to\infty} T(\varphi_i).$$

The limit exists, because when $i, j \to \infty$, we have

$$|T(\varphi_i) - T(\varphi_j)| = |T(\varphi_i - \varphi_j)| \le C \|\varphi_i - \varphi_j\|_{K;0} \to 0.$$

Similar reasoning shows that the limit is independent of the sequence chosen. Furthermore, since T is of order 0, it follows that

$$|T^*(\varphi)| \le C \|\varphi\|_{K;0},$$

which establishes (10.14). □

We conclude this section with a simple but very useful condition that ensures that a distribution is a measure.

10.5. DEFINITION. A distribution on Ω is **positive** if $T(\varphi) \ge 0$ for all test functions on Ω satisfying $\varphi \ge 0$.

10.6. THEOREM. *A distribution T on Ω is positive if and only if T is a positive measure.*

PROOF. From previous discussions, we know that a Radon measure is a distribution of order 0, so we need only consider the case that T is a positive distribution. Let $K \subset \Omega$ be a compact set. From Exercise 1, Section 10.1, there exists a function $\alpha \in C_c^\infty(\Omega)$ that equals 1 on a neighborhood of K. Now select a test function φ whose support is contained in K. Then we have

$$-\|\varphi\|_{K;0}\,\alpha(x) \le \varphi(x) \le \|\varphi\|_{K;0}\,\alpha(x) \quad \text{for all } x,$$

and therefore

$$-\|\varphi\|_{K;0}\,T(\alpha) \le T(\varphi) \le \|\varphi\|_{K;0}\,T(\alpha).$$

Thus, $|T(\varphi)| \le |T(\alpha)|\,\|\varphi\|_{K;0}$, which shows that T is of order 0 and is therefore a measure μ. The measure μ is clearly nonnegative. □

Exercises for Section 10.2

1. Use the Hahn–Banach theorem to provide an alternative proof of (10.14).
2. Show that the principal value integral

$$\text{p.v.} \int \frac{\phi(x)}{x}dx = \lim_{\varepsilon\to 0+}\left(\int_{-\infty}^{-\varepsilon} \frac{\phi(x)}{x}dx + \int_{\varepsilon}^{\infty} \frac{\phi(x)}{x}dx\right)$$

   exists for all $\phi \in \mathcal{D}(\mathbb{R})$ and is a distribution. What is its order?
3. Let T_j, $j = 1, 2, \ldots$, be a sequence of distributions. The sequence T_j is said to converge to the distribution T if

$$\lim T_j(\varphi) = T(\varphi), \text{ for all } \varphi \in \mathcal{D}(\Omega).$$

   Let $f_\varepsilon \in L^1_{\text{loc}}(\mathbb{R}^n)$ be a function that depends on a parameter $\varepsilon \in (0, 1)$ and is such that

(a) $\operatorname{supp} f_\varepsilon \subset \{|x| \le \varepsilon\}$;
(b) $\int_{\mathbb{R}^n} f_\varepsilon(x)dx = 1$;
(c) $\int_{\mathbb{R}^n} |f_\varepsilon(x)|dx \le \mu < \infty$, $0 < \varepsilon < 1$.

Show that $f_\varepsilon \to \delta$ as $\varepsilon \to 0$. Show also that if f_ε satisfies (a) and $f_\varepsilon \to \delta$ as $\varepsilon \to 0$, then (b) holds.

10.3. Differentiation of Distributions

One of the primary reasons why distributions were created was to provide a notion of differentiability for functions that are not differentiable in the classical sense. In this section, we define the derivative of a distribution and investigate some of its properties.

In order to motivate the definition of a distribution, consider the following simple case. Let Ω denote the open interval $(0,1)$ in $\mathbb{R}$, and let f denote an absolutely continuous function on $(0,1)$. If φ is a test function in Ω, we can integrate by parts (Exercise 7, Section 7.5) to obtain

$$\int_0^1 f'(x)\varphi(x)\,d\lambda(x) = -\int_0^1 f(x)\varphi'(x)\,d\lambda(x), \tag{10.15}$$

since by definition, $\varphi(1) = \varphi(0) = 0$. If we consider f a distribution T, we have

$$T(\varphi) = \int_0^1 f(x)\varphi(x)\,d\lambda(x)$$

for every test function φ in $(0,1)$. Now define a distribution S by

$$S(\varphi) = \int_0^1 f(x)\varphi'(x)\,d\lambda(x)$$

and observe that it is a distribution of order 1. From (10.15) we see that S can be identified with $-f'$. From this it is clear that the derivative T' should be defined as

$$T'(\varphi) = -T(\varphi')$$

for every test function φ. More generally, we have the following definition.

10.7. Definition. Let T be a distribution of order N defined on an open set $\Omega \subset \mathbb{R}^n$. The partial derivative of T with respect to the ith coordinate direction is defined by

$$\frac{\partial T}{\partial x_i}(\varphi) = -T(\frac{\partial \varphi}{\partial x_i}).$$

Observe that since the derivative of a test function is again a test function, the differentiated distribution is a linear functional on $\mathcal{D}(\Omega)$. It is in fact a distribution, since

$$\left|T\left(\frac{\partial \varphi}{\partial x_i}\right)\right| \le C\,\|\varphi\|_{N+1}$$

is valid whenever φ is a test function supported by a compact set K on which

$$|T(\varphi)| \le C\,\|\varphi\|_N.$$

Let α be any multi-index. More generally, the αth derivative of the distribution T is another distribution defined by

$$D^\alpha T(\varphi) = (-1)^{|\alpha|} T(D^\alpha \varphi).$$

Recall that a function $f \in L^1_{loc}(\Omega)$ is associated with the distribution T_f (see Definition 10.3). Thus, if $f, g \in L^1_{loc}(\Omega)$, we say that $D^\alpha f = g_\alpha$ in the sense of distributions if

$$\int_\Omega f D^\alpha \varphi d\lambda = (-1)^{|\alpha|} \int_\Omega \varphi g_\alpha d\lambda, \text{ for all } \varphi \in \mathcal{D}(\Omega).$$

Let us consider some examples. The first one has already been discussed above, but it is repeated for emphasis.

1. Let $\Omega = (a, b)$, and suppose f is an absolutely continuous function defined on $[a, b]$. If T is the distribution corresponding to f, we have

$$T(\varphi) = \int_a^b \varphi f \, d\lambda$$

for each test function φ. Since f is absolutely continuous and φ has compact support in $[a, b]$ (so that $\varphi(b) = \varphi(a) = 0$), we may employ integration by parts to conclude that

$$T'(\varphi) = -T(\varphi') = -\int_a^b \varphi' f \, d\lambda = \int_a^b \varphi f' \, d\lambda.$$

Thus, the distribution T' is identified with the function f'.

The next example shows how it is possible for a function to have a derivative in the sense of distributions but not be differentiable in the classical sense.

2. Let $\Omega = \mathbb{R}$ and define

$$f(x) = \begin{cases} 1, & x > 0, \\ 0, & x \leq 0. \end{cases}$$

With T defined as the distribution corresponding to f, we obtain

$$T'(\varphi) = -T(\varphi') = -\int_{\mathbb{R}} \varphi' f \, d\lambda = -\int_0^\infty \varphi' \, d\lambda = \varphi(0).$$

Thus, the derivative T' is equal to the Dirac measure; see (10.13).

3. We alter the function of Example 2 slightly:

$$f(x) = |x|.$$

Then

$$T(\varphi) = \int_0^\infty x\varphi(x) \, d\lambda(x) - \int_{-\infty}^0 x\varphi(x) \, d\lambda(x),$$

so that after integrating by parts, we obtain

$$\begin{aligned} T'(\varphi) &= \int_0^{\infty} \varphi(x)\,d\lambda(x) - \int_{-\infty}^{0} \varphi(x)\,d\lambda(x) \\ &= \int_{\mathbb{R}} \varphi(x)g(x)\,d\lambda(x), \end{aligned}$$

where

$$g(x) := \begin{cases} 1, & x > 0, \\ -1, & x \leq 0. \end{cases}$$

This shows that the derivative of f is g in the sense of distributions.

One would hope that the basic results of calculus carry over to the framework of distributions. The following is the first of many that do.

10.8. THEOREM. *If T is a distribution in $\mathbb{R}$ with $T' = 0$, then T is a constant. That is, T is the distribution that corresponds to a constant function.*

PROOF. Observe that $\varphi = \psi'$, where

$$\psi(x) := \int_{-\infty}^{x} \varphi(t)\,dt.$$

Since φ has compact support, it follows that ψ has compact support precisely when

$$\int_{-\infty}^{\infty} \varphi(t)\,dt = 0. \tag{10.16}$$

Thus, φ is the derivative of another test function if and only if (10.16) holds.

To prove the theorem, choose an arbitrary test function ψ with the property

$$\int \psi(x)\,d\lambda(x) = 1.$$

Then every test function φ can be written as

$$\varphi(x) = [\varphi(x) - a\psi(x)] + a\psi(x),$$

where

$$a = \int \varphi(t)\,d\lambda(t).$$

The function in brackets is a test function, call it α, whose integral is 0 and is therefore the derivative of another test function. Since $T' = 0$, it follows that $T(\alpha) = 0$, and therefore we obtain

$$T(\varphi) = aT(\psi) = \int T(\psi)\varphi(t)\,d\lambda(t),$$

which shows that T corresponds to the constant $T(\psi)$. □

This result along with Example 1 gives an interesting characterization of absolutely continuous functions.

10.9. THEOREM. *Suppose $f \in L^1_{\text{loc}}(a,b)$. Then f is equal almost everywhere to an absolutely continuous function on $[a,b]$ if and only if the derivative of the distribution corresponding to f is a function.*

PROOF. In Example 1 we have already seen that if f is absolutely continuous, then its derivative in the sense of distributions is again a function. Indeed, the function associated with the derivative of the distribution is f'.

Now suppose that T is the distribution associated with f and that $T' = g$. In order to show that f is equal almost everywhere to an absolutely continuous function, let

$$h(x) = \int_a^x g(t)\, d\lambda(t).$$

Observe that h is absolutely continuous and that $h' = g$ almost everywhere. Let S denote the distribution corresponding to h; that is,

$$S(\varphi) = \int h\varphi$$

for every test function φ. Then

$$S'(\varphi) = -\int h\varphi' = \int h'\varphi = \int g\varphi.$$

Thus, $S' = g$. Since $T' = g$ also, we have $T = S + k$ for some constant k by the previous theorem. This implies that

$$\begin{aligned} \int f\varphi\, d\lambda = T(\varphi) &= S(\varphi) + \int k\varphi \\ &= \int h\varphi + k\varphi \\ &= \int (h+k)\varphi \end{aligned}$$

for every test function φ. This implies that $f = h + k$ almost everywhere (see Exercise 3, Section 10.1). □

Since functions of bounded variation are closely related to absolutely continuous functions, it is natural to inquire whether they too can be characterized in terms of distributions.

10.10. THEOREM. *Suppose $f \in L^1_{\text{loc}}(a,b)$. Then f is equivalent to a function of bounded variation if and only if the derivative of the distribution corresponding to f is a signed measure whose total variation is finite.*

PROOF. Suppose $f \in L^1_{\text{loc}}(a,b)$ is a nondecreasing function and let T be the distribution corresponding to f. Then for every test function φ,

$$T(\varphi) = \int f\varphi\, d\lambda$$

and

$$T'(\varphi) = -T(\varphi') = -\int f\varphi'\,d\lambda\,.$$

Since f is nondecreasing, it generates a Lebesgue–Stieltjes measure λ_f. By the integration by parts formula for such measures (see Exercises 5 and 6 in Section 6.3), we have

$$-\int f\varphi'\,d\lambda = \int \varphi\,d\lambda_f,$$

which shows that T' corresponds to the measure λ_f. If f is of bounded variation on (a, b), we can write

$$f = f_1 - f_2,$$

where f_1 and f_2 are nondecreasing functions. Then the distribution T_f corresponding to f can be written as $T_f = T_{f_1} - T_{f_2}$, and therefore

$$T'_f = T'_{f_1} - T'_{f_2} = \lambda_{f_1} - \lambda_{f_2}.$$

Consequently, the signed measure $\lambda_{f_1} - \lambda_{f_2}$ corresponds to the distributional derivative of f.

Conversely, suppose the derivative of f is a signed measure μ. By the Jordan decomposition theorem, we can write μ as the difference of two nonnegative measures, $\mu = \mu_1 - \mu_2$. Let

$$f_i(x) = \mu_i((-\infty, x]), \quad i = 1, 2.$$

From Theorem 4.33, we have that μ_i agrees with the Lebesgue–Stieltjes measure λ_{f_i} on all Borel sets. Furthermore, utilizing the formula for integration by parts, we obtain

$$T'_{f_i}(\varphi) = -T_{f_i}(\varphi') = -\int f_i\varphi'\,d\lambda = \int \varphi\,d\lambda_{f_i} = \int \varphi\,d\mu_i.$$

Thus, with $g := f_1 - f_2$, we have that g is of bounded variation and that its distributional derivative is $\mu_1 - \mu_2 = \mu$. Since $f' = \mu$, we conclude from Theorem 10.8 that $f - g$ is a constant, and therefore that f is equivalent to a function of bounded variation. □

Exercises for Section 10.3

1. If a distribution T defined on $\mathbb{R}$ has the property that $T'' = 0$, what can be said about T?
2. Show that if T_k, $k = 1, 2, \ldots$, is a sequence of distributions that converges to a distribution T (see Exercise 3, Section 10.2), and α is a multi-index, then

$$D^\alpha T_k \to D^\alpha T.$$

3. Let $\Omega = \mathbb{R}$ and define

$$H(x) = \begin{cases} 1, & x > 0, \\ 0, & x \le 0. \end{cases}$$

Let T be the distribution corresponding to H and let $g \in C^\infty(\mathbb{R})$. Show that the function $x \to g(x)H(x)$ is locally integrable and so determines a distribution satisfying $D(gH) = g(0)\delta + HDg$.

4. Show that if T is a distribution on $\mathcal{D}(\mathbb{R})$, and $xDT + T = 0$, then $T = A(\frac{1}{x}) + B\delta$, where A and B are real numbers, and $\frac{1}{x}$ is the principal value distribution introduced in Exercise 2, Section 10.2.

10.4. Essential Variation

We have seen that a function of bounded variation defines a distribution whose derivative is a measure. Question: How is the variation of the function related to the variation of the measure? The notion of essential variation provides the answer.

A function f of bounded variation gives rise to a distribution whose derivative is a measure. Furthermore, every other function g agreeing with f almost everywhere defines the same distribution. Of course, g need not be of bounded variation. This raises the question of what condition on f is equivalent to f' being a measure. We will see that the needed ingredient is a concept of variation that remains unchanged when the function is altered on a set of measure zero.

10.11. DEFINITION. The **essential variation** of a function f defined on (a, b) is

$$\text{ess } V_a^b f = \sup\left\{\sum_{i=1}^{k} |f(t_{i+1}) - f(t_i)|\right\},$$

where the supremum is taken over all finite partitions $a < t_1 < \cdots < t_{k+1} < b$ such that each t_i is a point of approximate continuity of f.

Recall that the variation of a function is defined similarly, but without the restriction that f be approximately continuous at each t_i. Clearly, if $f = g$ a.e. on (a, b), then

$$\text{ess } V_a^b f = \text{ess } V_a^b g.$$

A signed Radon measure μ on (a, b) defines a linear functional on the space of continuous functions with compact support in (a, b) by

$$T_\mu(\varphi) = \int_a^b \varphi \, d\mu.$$

Recall that the **total variation** of μ on (a, b) is defined (see Definition 6.39) as the norm of T_μ; that is,

$$\|\mu\| = \sup\left\{\int_a^b \varphi \, d\mu : \varphi \in C_c(a, b), |\varphi| \leq 1\right\}. \tag{10.17}$$

Notice that the supremum could just as well be taken over C^∞ functions with compact support, since they are dense in $C_c(a, b)$ in the topology of uniform convergence.

10.12. THEOREM. *Suppose $f \in L^1(a,b)$. Then f' (in the sense of distributions) is a measure with finite total variation if and only if* ess $V_a^b f < \infty$. *Moreover, the total variation of the measure f' is given by $\|f'\| =$ ess $V_a^b f$.*

PROOF. First, under the assumption that ess $V_a^b f < \infty$, we will prove that f' is a measure and that

$$\|f'\| \leq \text{ess } V_a^b f. \tag{10.18}$$

For this purpose, choose $\varepsilon > 0$ and let $f_\varepsilon = \psi_\varepsilon * f$ denote the mollifier of f (see (10.9)). Consider an arbitrary partition with the property $a + \varepsilon < t_1 < \cdots < t_{m+1} < b - \varepsilon$. If a function g is defined for fixed t_i by $g(s) = f(t_i - s)$, then almost every s is a point of approximate continuity of g and therefore of $f(t_i - \cdot)$. Hence,

$$\begin{aligned}
\sum_{i=1}^{m} |f_\varepsilon(t_{i+1}) - f_\varepsilon(t_i)| &= \sum_{i=1}^{m} \left| \int_{-\varepsilon}^{\varepsilon} \psi_\varepsilon(s)(f(t_{i+1} - s) - f(t_i - s))\, d\lambda(s) \right| \\
&\leq \int_{-\varepsilon}^{\varepsilon} \psi_\varepsilon(s) \sum_{i=1}^{m} |f(t_{i+1} - s) - f(t_i - s)|\, d\lambda(s) \\
&\leq \int_{-\varepsilon}^{\varepsilon} \psi_\varepsilon(s) \text{ ess } V_a^b f\, d\lambda(s) \\
&\leq \text{ess } V_a^b f.
\end{aligned}$$

In obtaining the last inequality, we have used the fact that

$$\int_{-\varepsilon}^{\varepsilon} \psi_\varepsilon(s)\, d\lambda(s) = 1.$$

Now take the supremum of the left-hand side over all partitions and obtain

$$\int_{a+\varepsilon}^{b-\varepsilon} |(f_\varepsilon)'|\, d\lambda \leq \text{ess } V_a^b f.$$

Let $\varphi \in C_c^\infty(a,b)$ with $|\varphi| \leq 1$. Choosing $\varepsilon > 0$ such that spt$\varphi \subset (a+\varepsilon, b-\varepsilon)$, we obtain

$$\int_a^b f_\varepsilon \varphi'\, d\lambda = -\int_a^b (f_\varepsilon)'\varphi\, d\lambda \leq \int_{a+\varepsilon}^{b-\varepsilon} |(f_\varepsilon)'|\, d\lambda \leq \text{ess } V_a^b f.$$

Since f_ε converges to f in $L^1(a,b)$, it follows that

$$\int_a^b f\varphi'\, d\lambda \leq \text{ess } V_a^b f. \tag{10.19}$$

Thus, the distribution S defined for all test functions ψ by

$$S(\psi) = \int_a^b f\psi'\, d\lambda$$

is a distribution of order 0 and therefore a measure, since by (10.19),

$$S(\psi) = \int_a^b f\psi'\, d\lambda = \int_a^b f\varphi'\, \|\psi\|_{(a,b);0}\, d\lambda \leq \text{ess } V_a^b f\ \|\psi\|_{(a,b);0},$$

where we have taken

$$\varphi = \frac{\psi}{\|\psi\|_{(a,b);0}}.$$

Since $S = -f'$, we have that f' is a measure with

$$\|f'\| = \sup\left\{\int_a^b f\varphi'\, d\lambda : \varphi \in C_c^\infty(a,b), |\varphi| \leq 1\right\} \leq \text{ess } V_a^b f,$$

thus establishing (10.18), as desired.

Now for the opposite inequality. We assume that f' is a measure with finite total variation, and we will first show that $f \in L^\infty(a,b)$. For $0 < h < (b-a)/3$, let I_h denote the interval $(a+h, b-h)$ and let $\eta \in C_c^\infty(I_h)$ with $|\eta| \leq 1$. Then for all sufficiently small $\varepsilon > 0$, we have the mollifier $\eta_\varepsilon = \eta * \varphi_\varepsilon \in C_c^\infty(a,b)$. Thus, with the help of (10.10) and Fubini's theorem,

$$\begin{aligned}\int_{I_h} f_\varepsilon \eta'\, d\lambda &= \int_a^b f_\varepsilon \eta'\, d\lambda = \int_a^b (f * \varphi_\varepsilon)\eta'\, d\lambda \\ &= \int_a^b f(\varphi_\varepsilon * \eta)'\, d\lambda = \int_a^b f\eta_\varepsilon'\, d\lambda \leq \|f'\|,\end{aligned}$$

since $|\eta_\varepsilon| \leq 1$. Taking the supremum over all such η shows that

$$\|f_\varepsilon'\|_{L^1;I_h} \leq \|f'\|, \tag{10.20}$$

because

$$\int_{I_h} f_\varepsilon' \eta\, d\lambda = -\int_{I_h} f_\varepsilon \eta'\, d\lambda.$$

For arbitrary $y, z \in I_h$, we have

$$f_\varepsilon(z) = f_\varepsilon(y) + \int_y^z f_\varepsilon'\, d\lambda$$

and therefore, taking integral averages with respect to y,

$$\fint_{I_h} |f_\varepsilon(z)|\, d\lambda \leq \fint_{I_h} |f_\varepsilon|\, d\lambda + \int_a^b |f_\varepsilon'|\, d\lambda.$$

Thus, from Theorem 10.1 (iii) and (10.20), we have

$$|f_\varepsilon|(z) \leq \frac{3}{b-a}\|f\|_{L^1(a,b)} + \|f'\|$$

for $z \in I_h$. Since $f_\varepsilon \to f$ a.e. in (a,b) as $\varepsilon \to 0$, we conclude that $f \in L^\infty(a,b)$.

Now that we know that f is bounded on (a,b), we see that each point of approximate continuity of f is also a Lebesgue point (see Exercise 4, Section 7.9). Consequently, for each partition $a < t_1 < \cdots < t_{m+1} < b$, where each t_i is a point of approximate continuity of f, reference to Theorem

10.1 (ii) yields

$$\begin{aligned}\sum_{i=1}^{m}|f(t_{i+1})-f(t_i)| &= \lim_{\varepsilon\to 0}\sum_{i=1}^{m}|f_\varepsilon(t_{i+1})-f_\varepsilon(t_i)| \\ &\leq \limsup_{\varepsilon\to 0}\int_a^b |f'_\varepsilon|\, d\lambda \\ &\leq \|f\|\,(a,b). \qquad \text{by } (10.20)\end{aligned}$$

Now take the supremum over all such partitions to conclude that

$$\text{ess } V_a^b f \leq \|f'\|. \quad \square$$

Exercises for Section 10.4

1. Suppose f is an increasing function on $[a,b]$. Prove that we can write
$$f = f_A + f_C + f_J,$$
where each of the functions f_A, f_C, and f_J is increasing and
(a) f_A is absolutely continuous;
(b) f_C is continuous, but $f'_C(x) = 0$ for a.e. x;
(c) f_J is a jump function.
Show that each component f_A, f_C, f_J is uniquely determined up to an additive constant. Note: there is a similar decomposition for every f of bounded variation.
2. A bounded function f is said to be of bounded variation on $\mathbb{R}$ if f is of bounded variation on every finite subinterval $[a,b]$, and $\sup_{a,b} \text{ess} V_a^b f < \infty$. Prove that such an f has the following properties:
(a) $\int_{\mathbb{R}} |f(x+h) - f(x)| dx \leq A|h|$, for some constant A and all $h \in \mathbb{R}$.
(b) $\left|\int_{\mathbb{R}} f(x)\varphi'(x)dx\right| \leq A$, where φ is any C^1 function of bounded support with $\sup_{x\in\mathbb{R}} |\varphi(x)| \leq 1$.

CHAPTER 11

Functions of Several Variables

11.1. Differentiability

Because of the central role played by absolutely continuous functions and functions of bounded variation in the development of the fundamental theorem of calculus in $\mathbb{R}$, it is natural to ask whether they have analogues among functions of more than one variable. One of the main objectives of this chapter is to show that this is true. We have found that the BV functions in $\mathbb{R}$ constitute a large class of functions that are differentiable almost everywhere. Although there are functions on $\mathbb{R}^n$ that are analogous to BV functions, they are not differentiable almost everywhere. Among the functions that are often encountered in applications, Lipschitz functions on $\mathbb{R}^n$ form the largest class that are differentiable almost everywhere. This result is due to Rademacher and is one of the fundamental results in the analysis of functions of several variables.

The derivative of a function f at a point x_0 satisfies

$$\lim_{h\to 0}\left|\frac{f(x_0+h)-f(x_0)-f'(x_0)h}{h}\right| = 0.$$

This limit implies that

$$f(x_0+h)-f(x_0) = f'(x_0)h + r(h), \tag{11.1}$$

where the "remainder" $r(h)$ is small, in the sense that

$$\lim_{h\to 0}\frac{r(h)}{h} = 0.$$

Note that (11.1) expresses the difference $f(x_0+h)-f(x_0)$ as the sum of the linear function that takes h to $f'(x_0)h$, plus a small remainder. We can therefore regard the derivative of f at x_0 not as a real number, but as a linear function of $\mathbb{R}$ that takes h to $f'(x_0)h$. Observe that every real number α gives rise to the linear function $L:\mathbb{R}\to\mathbb{R}$, $L(h)=\alpha\cdot h$. Conversely, every linear function that carries $\mathbb{R}$ to $\mathbb{R}$ is multiplication by some real number. It is this natural one-to-one correspondence that motivates the definition of differentiability for functions of several variables. Thus, if $f\colon\mathbb{R}^n\to\mathbb{R}$, we say

W.P. Ziemer, *Modern Real Analysis*, Graduate Texts in Mathematics 278,
https://doi.org/10.1007/978-3-319-64629-9_11

that f is **differentiable** at $x_0 \in \mathbb{R}^n$ if there is a linear function $L\colon \mathbb{R}^n \to \mathbb{R}$ with the property that

$$\lim_{h\to 0} \frac{|f(x_0+h) - f(x_0) - L(h)|}{|h|} = 0. \tag{11.2}$$

The linear function L is called the **derivative** of f at x_0 and is denoted by $df(x_0)$. It is commonly accepted to use the term **differential** interchangeably with derivative.

Recall that the existence of partial derivatives at a point is not sufficient to ensure differentiability. For example, consider the following function of two variables:

$$f(x,y) = \begin{cases} \frac{xy^2+x^2y}{x^2+y^2} & \text{if } (x,y) \neq (0,0), \\ 0 & \text{if } (x,y) = (0,0). \end{cases}$$

Both partial derivatives are 0 at $(0,0)$, yet the function is not differentiable there. We leave it to the reader to verify this.

We now consider Lipschitz functions f on $\mathbb{R}^n$; see (3.10). Thus there is a constant C_f such that

$$|f(x) - f(y)| \leq C_f\,|x-y|$$

for all $x, y \in \mathbb{R}^n$. The next result is fundamental in the study of functions of several variables.

11.1. THEOREM. *If $f\colon \mathbb{R}^n \to \mathbb{R}$ is Lipschitz, then f is differentiable almost everywhere.*

PROOF. **Step 1:** Let $v \in \mathbb{R}^n$ with $|v| = 1$. We first show that $df_v(x)$ exists for λ-a.e. x, where $df_v(x)$ denotes the directional derivative of f at x in the direction of v.

Let

$$N_v := \mathbb{R}^n \cap \{x : df_v(x) \text{ fails to exist}\}.$$

For each $x \in \mathbb{R}^n$ define

$$\overline{df_v}(x) := \limsup_{t\to 0} \frac{f(x+tv) - f(x)}{t}$$

and

$$\underline{df_v}(x) := \liminf_{t\to 0} \frac{f(x+tv) - f(x)}{t}.$$

Notice that

$$N_v = \{x \in \mathbb{R}^n : \underline{df_v}(x) < \overline{df_v}(x)\}. \tag{11.3}$$

From Definition 3.63, it is clear that

$$\overline{df_v}(x) = \lim_{k\to\infty} \left(\sup_{\substack{0<|t|<1/k \\ t\in\mathbb{Q},\, k\in\mathbb{N}}} \frac{f(x+tv) - f(x)}{t} \right).$$

We now claim that

$$x \to \overline{df_v}(x) \text{ is a Borel measurable function of } x.$$

Indeed, the rational numbers $0 < |t| < \frac{1}{k}$ can be enumerated as $t_1^k, t_2^k, t_3^k, \ldots$. If we define for each $i = 1, 2, \ldots$ and fixed k the sequence $G_i^k(x) := \frac{f(x+t_i^k v)-f(x)}{t_i^k}$, then since f is continuous, the functions G_i^k are Borel measurable, and hence $F^k(x) := \sup_i\{G_i^k(x)\}$ is also Borel measurable. Note that

$$F^k(x) = \sup_{\substack{0<|t|<1/k \\ t\in\mathbb{Q}}} \frac{f(x+tv)-f(x)}{t}.$$

Since the pointwise limit of measurable functions is again measurable, it follows that

$$\overline{df_v}(x) = \lim_{k\to\infty} F^k(x) \tag{11.4}$$

is Borel measurable. Proceeding as before, we also have

$$x \to \underline{df_v}(x) \text{ is a Borel measurable function of } x.$$

Therefore, from (11.3) we conclude that

$$N_v \text{ is a Borel set.} \tag{11.5}$$

Now we proceed to show that

$$H^1(N_v \cap L) = 0, \text{ for each line } L \text{ parallel to } v. \tag{11.6}$$

In order to prove (11.6), we consider the line L_x that contains $x \in \mathbb{R}^n$ and is parallel to v. We now consider the restriction of f to L_x given by

$$\gamma : \mathbb{R} \to \mathbb{R},\ \gamma(t) = f(x+tv).$$

Note that γ is Lipschitz in $\mathbb{R}$, since f is Lipschitz in $\mathbb{R}^n$. Therefore, γ is absolutely continuous and hence differentiable at λ_1-a.e. t. Let $A_v = \{t \in \mathbb{R} : \gamma(t) \text{ is not differentiable}\}$ and $R_v = z(A_v)$, where $z : \mathbb{R} \to \mathbb{R}^n$ is given by $z(t) = x + tv$. Since z is Lipschitz, Exercise 10, Section 4.7, implies that $H^1(R_v) \le CH^1(A_v) = 0$, which gives

$$H^1(R_v) = 0. \tag{11.7}$$

We have that $t_0 \notin A_v$ if and only if $z(t_0) = x + t_0 v \notin N_v$. Indeed, for such t_0, $\gamma'(t_0)$ exists, and

$$\begin{aligned}
\gamma'(t_0) &= \lim_{t\to 0}\frac{\gamma(t)-\gamma(t_0)}{t-t_0} = \lim_{t\to 0}\frac{f(x+tv)-f(x+t_0v)}{t-t_0} \\
&= \lim_{t\to 0}\frac{f(x+t_0v+(t-t_0)v)-f(x+t_0v)}{t-t_0} \\
&= \lim_{h\to 0}\frac{f(x+t_0v+hv)-f(x+t_0v)}{h}; \text{ with } h = t-t_0 \\
&= \lim_{h\to 0}\frac{f(z(t_0)+hv)-f(z(t_0))}{h} = df_v(z(t_0)).
\end{aligned}$$

Hence the directional derivative of f exists at the point $z(t_0) = x + t_0 v$. It is now clear that

$$R_v = N_v \cap L_x,$$

and from (11.7) we conclude that

$$H^1(N_v \cap L_x) = 0. \tag{11.8}$$

Since (11.8) holds for arbitrary x, we have

$$H^1(N_v \cap L) = 0 \text{ for every line parallel to } v. \tag{11.9}$$

Therefore, since N_v is a Borel set, we can appeal to Fubini's theorem to conclude that

$$\lambda(N_v) = 0. \tag{11.10}$$

Step 2: Our next objective is to prove that

$$df_v(x) = \nabla f(x) \cdot v \tag{11.11}$$

for λ-almost all $x \in \mathbb{R}^n$. Here, ∇f denotes the gradient of f as defined in (10.6). Of course, this formula is valid for all x if $f \in C^\infty$. As a result of (11.10), we see that $\nabla f(x)$ exists for λ-almost all x. To establish (11.11), we begin with an observation that follows directly from Theorem 11.4 (change of variables formula), which will be established later:

$$\begin{aligned}\int_{\mathbb{R}^n} \left(\frac{f(x+tv) - f(x)}{t} \right) \varphi(x)\, d\lambda(x) \quad &= \\ &-\int_{\mathbb{R}^n} f(x) \left(\frac{\varphi(x) - \varphi(x - tv)}{t} \right) d\lambda(x)\end{aligned} \tag{11.12}$$

whenever $\varphi \in C_c^\infty(\mathbb{R}^n)$. Letting t assume the values $t = 1/k$ for all nonnegative integers k, we have

$$\left| \frac{f(x + \frac{1}{k}v) - f(x)}{\frac{1}{k}} \right| \leq C_f\, |v| = C_f. \tag{11.13}$$

From (11.12), (11.13), and the dominated convergence theorem, we have

$$\begin{aligned}
\int_{\mathbb{R}^n} df_v(x)\varphi(x)d\lambda(x) &= \int_{\mathbb{R}^n} \lim_{k\to\infty} \frac{f\left(x + \frac{1}{k}v\right) - f(x)}{\frac{1}{k}} \varphi(x)d\lambda(x) \\
&= \lim_{k\to\infty} \int_{\mathbb{R}^n} \frac{f\left(x + \frac{1}{k}v\right) - f(x)}{\frac{1}{k}} \varphi(x)d\lambda(x) \\
&= \lim_{k\to\infty} \int_{\mathbb{R}^n} \frac{\varphi\left(x - \frac{1}{k}v\right) - \varphi(x)}{\frac{1}{k}} f(x)d\lambda(x) \\
&= -\int_{\mathbb{R}^n} \lim_{k\to\infty} \frac{\varphi\left(x - \frac{1}{k}v\right) - \varphi(x)}{\frac{-1}{k}} f(x)d\lambda(x) \\
&= -\int_{\mathbb{R}^n} d\varphi_v(x) f(x)d\lambda(x).
\end{aligned}$$

We recall that f is absolutely continuous on lines, and therefore, using Fubini's theorem and integration by parts along lines, we compute

$$\begin{aligned}\int_{\mathbb{R}^n} df_v(x)\varphi(x)d\lambda(x) &= -\int_{\mathbb{R}^n} d\varphi_v(x)f(x)d\lambda(x)\\ &= -\int_{\mathbb{R}^n} f(x)\nabla\varphi(x)\cdot v d\lambda(x)\\ &= -\sum_{i=1}^n v_i\int_{\mathbb{R}^n} f(x)\frac{\partial\varphi}{\partial x_i}(x)d\lambda(x)\\ &= \sum_{i=1}^n v_i\int_{\mathbb{R}^n}\frac{\partial f}{\partial x_i}(x)\varphi(x)d\lambda(x)\\ &= \int_{\mathbb{R}^n}\nabla f(x)\cdot v\quad\varphi(x)d\lambda(x).\end{aligned}$$

Thus

$$\int_{\mathbb{R}^n} df_v(x)\varphi(x)d\lambda(x) = \int_{\mathbb{R}^n}\nabla f(x)\cdot v\varphi(x)d\lambda(x),\ \text{for all } \varphi\in C_c^\infty(\mathbb{R}^n).$$

Hence

$$df_v(x) = \nabla f(x)\cdot v,\ \lambda\text{-a.e. } x.$$

Now choose $\{v_k\}_{k=1}^\infty$ to be a countable dense subset of $\partial B(0,1)$. Observe that there is a set E, $\lambda(E)=0$, such that

$$df_v(x) = \nabla f(x)\cdot v_k,\ \text{for all } x\in\mathbb{R}^n\setminus E.$$

Step 3: We will now show that f is differentiable at each point $x\in\mathbb{R}^n\setminus E$. For $x\in\mathbb{R}^n\setminus E$ and $v\in\partial B(0,1)$ we define

$$Q(x,v,t) := \frac{f(x+tv)-f(x)}{t} - \nabla f(x)\cdot v,\quad t\neq 0.$$

For $v,v'\in\partial B(0,1)$ note that

$$\begin{aligned}|Q(x,v,t)-Q(x,v',t)| &\le \frac{|f(x+tv)-f(x+tv')|}{|t|} + |\nabla f(x)\cdot(v-v')|\\ &\le C_f|v-v'| + nC_f|v-v'|\\ &= C_f(n+1)|v-v'|.\end{aligned}$$

Hence

$$|Q(x,v,t)-Q(x,v',t)| \le C_f(n+1)|v-v'|,\quad v,v'\in\partial B(0,1). \tag{11.14}$$

Since $\{v_k\}$ is dense in the compact set $\partial B(0,1)$, it follows that for every $\varepsilon>0$ there exists N sufficiently large such that for every $v\in\partial B(0,1)$,

$$|v-v_k| < \frac{\varepsilon}{2(n+1)C_f},\ \text{for some } k\in\{1,\dots,N\}. \tag{11.15}$$

Thus, for every $v\in\partial B(0,1)$, there exists $k\in\{1,\dots,N\}$ such that

$$|Q(x,v,t)| \le |Q(x,v_k,t)| + |Q(x,v,t)-Q(x,v_k,t)|. \tag{11.16}$$

Since $df_{v_k}(x) = \nabla f(x) \cdot v_k$, it follows that for $0 < |t| \le \delta$ we have $|Q(x, v_k, t| \le \frac{\varepsilon}{2}$. Thus from (11.14), (11.15), (11.16), we have

$$|Q(x, v, t)| \le \frac{\varepsilon}{2} + \frac{\varepsilon}{2} = \varepsilon, \quad 0 \le |t| \le \delta.$$

We have shown that

$$\lim_{t \to 0} |Q(x, v, t)| = 0,$$

which is

$$\lim_{t \to 0} \left| \frac{f(x + tv) - f(x)}{t} - \nabla f(x) \cdot v \right| = 0. \tag{11.17}$$

The last step is to show that (11.17) implies that f is differentiable at every $x \in \mathbb{R}^n \setminus E$. Choose $y \in \mathbb{R}^n$, $y \ne x$. Let

$$v := \frac{y - x}{\|y - x\|}, \quad y = x + tv, \quad t = |y - x|.$$

From (11.17), we obtain

$$\lim_{|y - x| \to 0} \frac{|f(y) - f(x) - \nabla f(x) \cdot (y - x)|}{|y - x|} = 0, \tag{11.18}$$

or with $h := y - x$,

$$\lim_{|h| \to 0} \frac{|f(x + h) - f(x) - \nabla f(x) \cdot h|}{|h|} = 0, \tag{11.19}$$

which means that f is differentiable at x. Since $x \notin E$, we conclude that f is differentiable almost everywhere. □

The concept of differentiability for a transformation $T \colon \mathbb{R}^n \to \mathbb{R}^m$ is virtually the same as in (11.2). Thus, we say that T is **differentiable** at $x_0 \in \mathbb{R}^n$ if there is a linear mapping $L \colon \mathbb{R}^n \to \mathbb{R}^m$ such that

$$\lim_{h \to 0} \frac{|T(x_0 + h) - T(x_0) - L(h)|}{|h|} = 0. \tag{11.20}$$

As in the case of real-valued T, we call L the **derivative** of T at x_0. We will denote the linear function L by $dT(x_0)$. Thus, $dT(x_0)$ is a linear transformation, and when it is applied to a vector $v \in \mathbb{R}^n$, we will write $dT(x_0)(v)$. Writing T in terms of its coordinate functions, $T = (T^1, T^2, \ldots, T^m)$, it is easy to see that T is differentiable at a point x_0 if and only if each T^i is. Consequently, the following corollary is immediate.

11.2. COROLLARY. *If $T \colon \mathbb{R}^n \to \mathbb{R}^m$ is a Lipschitz transformation, then T is differentiable λ-almost everywhere.*

Exercises for Section 11.1

1. Prove that a linear map $L \colon \mathbb{R}^n \to \mathbb{R}^m$ is Lipschitz.
2. Show that a linear map $L \colon \mathbb{R}^n \to \mathbb{R}^n$ satisfies condition N and therefore leaves Lebesgue measurable sets invariant.

3. Use Fubini's theorem to prove that
$$\frac{\partial^2 f}{\partial x \partial y} = \frac{\partial^2 f}{\partial y \partial x}$$
if $f \in C^2(\mathbb{R}^2)$. Use this result to conclude that all second-order mixed partials are equal if $f \in C^2(\mathbb{R}^n)$.
4. Let $C^1[0,1]$ denote the space of functions on $[0,1]$ that have continuous derivatives on $[0,1]$, including one-sided derivatives at the endpoints. Define a norm on $C^1[0,1]$ by
$$\|f\| : = \sup_{x\in[0,1]} |f(x)| + \sup_{x\in[0,1]} f'(x).$$
Prove that $C^1[0,1]$ with this norm is a Banach space.

11.2. Change of Variables

We give a treatment of the behavior of the integral when the integrand is subjected to a change of variables by a Lipschitz transformation $T\colon \mathbb{R}^n \to \mathbb{R}^n$.

Consider $T\colon \mathbb{R}^n \to \mathbb{R}^n$ with $T = (T^1, T^2, \dots, T^n)$. If T is differentiable at x_0, then it follows immediately from the definitions that each of the partial derivatives
$$\frac{\partial T^i}{\partial x_j}, \; i, j = 1, 2, \dots, n,$$
exists at x_0. The linear mapping $L = dT(x_0)$ in (11.20) can be represented by the $n \times n$ matrix
$$dT(x_0) = \begin{bmatrix} \frac{\partial T^1}{\partial x_1} & \cdots & \frac{\partial T^1}{\partial x_n} \\ \vdots & & \vdots \\ \frac{\partial T^n}{\partial x_1} & \cdots & \frac{\partial T^n}{\partial x_n} \end{bmatrix},$$
where it is understood that each partial derivative is evaluated at x_0. The determinant of $[dT(x_0)]$ is called the **Jacobian** of T at x_0 and is denoted by $JT(x_0)$.

Recall from elementary linear algebra that a linear map $L\colon \mathbb{R}^n \to \mathbb{R}^n$ can be identified with the $n \times n$ matrix (L_{ij}), where
$$L_{ij} = L(e_i) \cdot e_j, \quad i, j = 1, 2, \dots, n,$$
and where $\{e_i\}$ is the standard basis for $\mathbb{R}^n$. The determinant of L_{ij} is denoted by $\det L$. If $M\colon \mathbb{R}^n \to \mathbb{R}^n$ is also a linear map, then

$$\det(M \circ L) = (\det M) \cdot (\det L). \tag{11.21}$$

Every nonsingular $n \times n$ matrix (L_{ij}) can be row-reduced to the identity matrix. That is, L can be written as the composition of finitely many linear transformations of the following three types:

(i) $L_1(x_1, \dots, x_i, \dots, x_n) = (x_1, \dots, cx_i, \dots, x_n), \quad 1 \le i \le n, c \ne 0.$
(ii) $L_2(x_1, \dots, x_i, \dots, x_n) = (x_1, \dots, x_i + cx_k, \dots, x_n), \quad 1 \le i \le n, k \ne i, c \ne 0.$

(iii) $L_3(x_1, \dots, x_i, \dots, x_j, \dots, x_n) = (x_1, \dots, x_j, \dots, x_i, \dots, x_n), \quad 1 \le i < j \le n.$

This leads to the following geometric interpretation of the determinant.

11.3. THEOREM. *If $L\colon \mathbb{R}^n \to \mathbb{R}^n$ is a nonsingular linear map, then $L(E)$ is Lebesgue measurable whenever E is Lebesgue measurable and*

$$\lambda[L(E)] = |\det L|\, \lambda(E). \tag{11.22}$$

PROOF. Exercise 2, Section 11.1, implies that $L(E)$ is Lebesgue measurable. In view of (11.21), it suffices to prove (11.22) when L is one of the three types mentioned above. In the case of L_3, we use Fubini's theorem and interchange the order of integration. In the case of L_1 or L_2, we integrate first with respect to x_i to arrive at formulas of the form

$$|c|\, \lambda_1(A) = \lambda_1(cA),$$
$$\lambda_1(A) = \lambda_1(c + A).$$

□

11.4. THEOREM. *If $f \in L^1(\mathbb{R}^n)$ and $L\colon \mathbb{R}^n \to \mathbb{R}^n$ is a nonsingular linear mapping, then*

$$\int_{\mathbb{R}^n} f \circ L\ |\det L|\ d\lambda = \int_{\mathbb{R}^n} f\ d\lambda.$$

PROOF. Assume first that $f \ge 0$. Note that

$$\{f > t\} = L(\{f \circ L > t\})$$

and therefore

$$\lambda(\{f > t\}) = \lambda(L(\{f \circ L > t\})) = |\det L|\, \lambda(\{f \circ L > t\})$$

for $t \ge 0$. Now apply Theorem 6.60 to obtain our desired result. The general case follows by writing $f = f^+ - f^-$. □

Our next result deals with an approximation property of Lipschitz transformations with nonvanishing Jacobians. First, we recall that a linear transformation $L\colon \mathbb{R}^n \to \mathbb{R}^n$ can be identified with its $n \times n$ matrix. Let $\mathcal{R}$ denote the family of $n \times n$ matrices whose entries are rational numbers. Clearly, $\mathcal{R}$ is countable. Furthermore, given an arbitrary nonsingular linear transformation L and $\varepsilon > 0$, there exists $R \in \mathcal{R}$ such that

$$\begin{aligned} |L(x) - R(x)| &< \varepsilon, \\ \left|L^{-1}(x) - R^{-1}(x)\right| &< \varepsilon, \end{aligned} \tag{11.23}$$

whenever $|x| \le 1$. Using linearity, this implies

$$|L(x) - R(x)| < \varepsilon\, |x|\,,$$
$$\left|L^{-1}(x) - R^{-1}(x)\right| < \varepsilon\, |x|\,,$$

for all $x \in \mathbb{R}^n$. Also, (11.23) implies

$$|L \circ R^{-1}(x-y)| \leq (1+\varepsilon)\,|x-y|$$

and

$$\left|R \circ L^{-1}(x-y)\right| \leq (1+\varepsilon)\,|x-y|$$

for all $x, y \in \mathbb{R}^n$. That is, the Lipschitz constants (see (3.10)) of $L \circ R^{-1}$ and $R \circ L^{-1}$ satisfy

$$C_{L \circ R^{-1}} < 1+\varepsilon \quad \text{and} \quad C_{R \circ L^{-1}} < 1+\varepsilon. \tag{11.24}$$

11.5. THEOREM. *Let $T\colon \mathbb{R}^n \to \mathbb{R}^n$ be a continuous transformation and set*

$$B = \{x : dT(x) \ \textit{exists}, JT(x) \neq 0\}.$$

Given $t > 1$, there exists a countable collection of Borel sets $\{B_k\}_{k=1}^{\infty}$ such that

(i) $B = \bigcup_{k=1}^{\infty} B_k$.

(ii) The restriction of T to B_k (denoted by T_k) is univalent.

(iii) For each positive integer k, there exists a nonsingular linear transformation $L_k\colon \mathbb{R}^n \to \mathbb{R}^n$ such that the Lipschitz constants of $T_k \circ L_k^{-1}$ and $L_k \circ T_k^{-1}$ satisfy

$$C_{T_k \circ L_k^{-1}} \leq t, \ C_{L_k \circ T_k^{-1}} \leq t$$

and

$$t^{-n}\,|\,det\, L_k| \leq |JT(x)| \leq t^n\,|\,det\, L_k|$$

for all $x \in B_k$.

PROOF. For fixed $t > 1$ choose $\varepsilon > 0$ such that

$$\frac{1}{t} + \varepsilon < 1 < t - \varepsilon.$$

Let C be a countable dense subset of B and let $\mathcal{R}$ (as introduced above) be the family of linear transformations whose matrices have rational entries. Now, for each $c \in C$, $R \in \mathcal{R}$, and each positive integer i, define $E(c, R, i)$ to be the set of all $b \in B \cap B(c, 1/i)$ that satisfy

$$\left(\frac{1}{t} + \varepsilon\right) |R(v)| \leq |dT(b)(v)| \leq (t-\varepsilon)\,|R(v)| \tag{11.25}$$

for all $v \in \mathbb{R}^n$ and

$$|T(a) - T(b) - dT(b)(a-b)| \leq \varepsilon\,|R(a-b)| \tag{11.26}$$

for all $a \in B(b, 2/i)$. Since T is continuous, each partial derivative of each coordinate function of T is a Borel function. Thus, it is an easy exercise to prove that each $E(c, R, i)$ is a Borel set. Observe that (11.25) and (11.26) imply

$$\frac{1}{t}\,|R(a-b)| \le |T(a)-T(b)| \le t\,|R(a-b)| \tag{11.27}$$

for all $b \in E(c, R, i)$ and $a \in B(b, 2/i)$.

Next, we will show that

$$\left(\frac{1}{t}+\varepsilon\right)^n |\det R| \le |JT(b)| \le (t-\varepsilon)^n\,|\det R| \tag{11.28}$$

for $b \in E(c, R, i)$. For the proof of this, let $L = dT(b)$. Then, from (11.25) we have

$$\left(\frac{1}{t}+\varepsilon\right)|R(v)| \le |L(v)| \le (t-\varepsilon)\,|R(v)| \tag{11.29}$$

whenever $v \in \mathbb{R}^n$, and therefore,

$$\left(\frac{1}{t}+\varepsilon\right)|v| \le \left|L \circ R^{-1}(v)\right| \le (t-\varepsilon)\,|v| \tag{11.30}$$

for $v \in \mathbb{R}^n$. This implies that

$$L \circ R^{-1}[B(0,1)] \subset B(0, t-\varepsilon),$$

and reference to Theorem 11.3 yields

$$\left|\det\left(L \circ R^{-1}\right)\right| \alpha(n) \le \lambda[B(0,t-\varepsilon)] = \alpha(n)(t-\varepsilon)^n,$$

where $\alpha(n)$ denotes the volume of the unit ball. Thus,

$$|\det L| \le (t-\varepsilon)^n\,|\det R|\,.$$

This proves one part of (11.28). The proof of the other part is similar.

We now are ready to define the Borel sets B_k appearing in the statement of our Theorem. Since the parameters c, R, and i used to define the sets $E(c, R, i)$ range over countable sets, the collection $\{E(c, R, i)\}$ is itself countable. We will relabel the sets $E(c, R, i)$ as $\{B_k\}_{k=1}^{\infty}$.

To show that property (i) holds, choose $b \in B$ and let $L = dT(b)$ as above. Now refer to (11.24) to find $R \in \mathcal{R}$ such that

$$C_{L \circ R^{-1}} < \left(\frac{1}{t}+\varepsilon\right)^{-1} \quad \text{and} \quad C_{R \circ L^{-1}} < t-\varepsilon.$$

Using the definition of the differentiability of T at b, select a nonnegative integer i such that

$$|T(a)-T(b)-dT(b)\cdot(a-b)| \le \frac{\varepsilon}{C_{R^{-1}}}\,|a-b| \le \varepsilon\,|R(a-b)|$$

for all $a \in B(b, 2/i)$. Now choose $c \in C$ such that $|b-c| < 1/i$ and conclude that $b \in E(c, R, i)$. Since this holds for all $b \in B$, property (i) holds.

To prove (ii), choose any set B_k. It is one of the sets of the form $E(c, R, i)$ for some $c \in C$, $R \in \mathcal{R}$, and some nonnegative integer i. According to (11.27),

$$\frac{1}{t}|R(a-b)| \leq |T(a)-T(b)| \leq t\,|R(a-b)|$$

for all $b \in B_k$, $a \in B(b, 2/i)$. Since $B_k \subset B(c, 1/i) \subset B(b, 2/i)$, we thus have

$$\frac{1}{t}|R(a-b)| \leq |T(a)-T(b)| \leq t\,|R(a-b)| \tag{11.31}$$

for all $a, b \in B_k$. Hence, T restricted to B_k is univalent.

With B_k of the form $E(c, R, i)$ as in the preceding paragraph, we define $L_k = R$. The proof of (iii) follows from (11.31) and (11.28), which imply

$$C_{T_k \circ L_k^{-1}} \leq t, \quad C_{L_k \circ T_k^{-1}} \leq t,$$

and

$$t^{-n}\,|\det L_k| \leq |JT_k| \leq t^n\,|\det L_k|\,,$$

since ε is arbitrary. □

We now proceed to develop the analogue of Banach's theorem (Theorem 7.33) for Lipschitz mappings $T\colon \mathbb{R}^n \to \mathbb{R}^n$. As in (7.40), for $E \subset \mathbb{R}^n$ we define

$$N(T, E, y)$$

as the (possibly infinite) number of points in $E \cap T^{-1}(y)$.

11.6. Lemma. *Let $T\colon \mathbb{R}^n \to \mathbb{R}^n$ be a Lipschitz transformation. If*

$$E := \{x \in \mathbb{R}^n : T \textit{ is differentiable at } x \textit{ and } JT(x) = 0\},$$

then

$$\lambda(T(E)) = 0.$$

Proof. By Rademacher's theorem, we know that T is differentiable almost everywhere. Since each entry of dT is a measurable function, it follows that the set E is measurable, and therefore so is $T(E)$, since T preserves sets of measure zero. It is sufficient to prove that $T(E_R)$ has measure zero for each $R > 0$, where $E_R := E \cap B(0, R)$. Let $\varepsilon \in (0, 1)$ and fix $x \in E_R$. Since T is differentiable at x,, there exists $0 < \delta_x < 1$ such that

$$|T(y)-T(x)-L(y-x)| \leq \varepsilon\,|y-x| \text{ for all } y \in B(x, \delta_x), \tag{11.32}$$

where $L := dT(x)$. We know that $JT(x) = 0$, and so L is represented by a singular matrix. Therefore, there is a linear subspace H of $\mathbb{R}^n$ of dimension $n-1$ such that L maps $\mathbb{R}^n$ into H. From (11.32) and the triangle inequality, we have

$$|T(y)-T(x)| \leq |L(y-x)| + \varepsilon\,|y-x|\,.$$

Let κ_L denote the Lipschitz constant of L: $|L(y)| \leq \kappa_L |y|$ for all $y \in \mathbb{R}^n$. Also, let κ_T denote the Lipschitz constant of T. Hence, for $x \in E_R$, $0 < r < \delta_x < 1$, and $y \in B(x, r)$ it follows that

$$\begin{aligned} |L(y) - L(x)| &\leq \kappa_L |y - x| \leq \kappa_L[|y| + R] \\ &= \kappa_L[|y - x + x| + R] \leq \kappa_L[r + 2R]. \end{aligned}$$

This implies

$$|L(y)| \leq |L(x)| + \kappa_L[r + 2R] \leq \kappa_L R + \kappa_L[r + 2R] = \kappa_L[r + 3R].$$

With $v := T(x) - L(x)$ we see that $|v| \leq |T(x)| + |L(x)| \leq \kappa_L |x| + \kappa_L |x| \leq R(\kappa_T + \kappa_L)$ and that (11.32) becomes

$$|T(y) - L(y) - v| \leq \varepsilon |y - x| ;$$

that is, $T(y)$ is within distance εr of $L(y)$ translated by the vector v. This means that the set $T(B(x, r))$ is contained within an εr-neighborhood of a bounded set in an affine space of dimension $n - 1$. The bound on this set depends only on r, R, κ_L and κ_T. Thus there is a constant $M = M(R, \kappa_L, \kappa_T)$ such that

$$\lambda(T(B(x, r))) \leq \varepsilon r M r^{n-1}.$$

The collection of balls $\{B(x, r)\}$ with $x \in E_R$ and $0 < r < \delta_x$ determines a Vitali covering of E_R, and accordingly there is a countable disjoint subcollection $B_i := B(x_i, r_i)$ whose union contains almost all of E_R:

$$\lambda(F) = 0 \text{ where } F := E_R \setminus \bigcup_{i=1}^{\infty} B_i.$$

Since T is Lipschitz, we know that $\lambda(T(F)) = 0$. Therefore, because

$$E_R \subset F \cup \bigcup_{i=1}^{\infty} B_i,$$

we have

$$T(E_R) \subset T(F) \cup \bigcup_{i=1}^{\infty} T(B_i),$$

and thus

$$\begin{aligned} \lambda(T(E_R)) &\leq \sum_{i=1}^{\infty} T(B_i) \\ &\leq \sum_{i=1}^{\infty} \varepsilon r_i M r_i^{n-1} \\ &\leq \varepsilon M \sum_{i=1}^{\infty} \lambda(B(x_i, r_i)). \end{aligned}$$

Since the balls $B(x_i, r_i)$ are disjoint and all are contained in $B(0, R + 1)$ and since $\varepsilon > 0$ was chosen arbitrarily, we conclude that $\lambda(T(E_R)) = 0$, as desired. □

11.7. THEOREM. **(Area formula)** *Let $T\colon \mathbb{R}^n \to \mathbb{R}^n$ be Lipschitz. Then for each Lebesgue measurable set $E \subset \mathbb{R}^n$,*

$$\int_E |JT(x)|\; d\lambda(x) = \int_{\mathbb{R}^n} N(T, E, y)\; d\lambda(y).$$

PROOF. Since T is Lipschitz, we know that T carries sets of measure zero into sets of measure zero. Thus, by Rademacher's theorem, we might as well assume that $dT(x)$ exists for all $x \in E$. Furthermore, it is easy to see that we may assume $\lambda(E) < \infty$.

In view of Lemma 11.6 it suffices to treat the case that $|JT| \neq 0$ on E. Fix $t > 1$ and let $\{B_k\}$ denote the sets provided by Theorem 11.5. By Lemma 4.7, we may assume that the sets $\{B_k\}$ are disjoint. For the moment, select some set B_k and set $B = B_k$. For each positive integer j, let $\mathcal{Q}_j$ denote a decomposition of $\mathbb{R}^n$ into disjoint "half-open" cubes with side length $1/j$ and of the form $[a_1, b_1) \times \cdots \times [a_n, b_n)$. Set

$$F_{i,j} = B \cap E \cap Q_i, \quad Q_i \in \mathcal{Q}_j.$$

Then for fixed j, the sets $F_{i,j}$ are disjoint, and

$$B \cap E = \bigcup_{i=1}^{\infty} F_{i,j}.$$

Furthermore, the sets $T(F_{i,j})$ are measurable, since T is Lipschitz. Thus, the functions g_j defined by

$$g_j = \sum_{i=1}^{\infty} \chi_{T(F_{i,j})}$$

are measurable. Now $g_j(y)$ is the number of sets $\{F_{i,j}\}$ such that $F_{i,j} \cap T^{-1}(y) \neq 0$ and

$$\lim_{j\to\infty} g_j(y) = N(T, B \cap E, y).$$

An application of the monotone convergence theorem yields

$$\lim_{j\to\infty} \sum_{i=1}^{\infty} \lambda[T(F_{i,j})] = \int_{\mathbb{R}^n} N(T, B \cap E, y)\; d\lambda(y). \tag{11.33}$$

Let L_k and T_k be as in Theorem 11.5. Then, recalling that $B = B_k$, we obtain

$$\begin{aligned} \lambda[T(F_{i,j})] &= \lambda[T_k(F_{i,j})] \\ &= \lambda[(T_k \circ L_k^{-1} \circ L_k)(F_{i,j})] \leq t^n \lambda[L_k(F_{i,j})], \end{aligned} \tag{11.34}$$

$$\begin{aligned} \lambda[L_k(F_{i,j})] &= \lambda[(L_k \circ T_k^{-1} \circ T_k)(F_{i,j})] \\ &\leq t^n \lambda[T_k(F_{i,j})] = t^n \lambda[T(F_{i,j})], \end{aligned} \tag{11.35}$$

and thus

$$\begin{aligned} t^{-2n}\lambda[T(F_{i,j})] &\le t^{-n}\lambda[L_k(F_{i,j})] && \text{by (11.34)} \\ &= t^{-n}\,|\det L_k|\,\lambda(F_{i,j}) && \text{by Theorem 11.3} \\ &\le \int_{F_{i,j}} |JT|\; d\lambda && \text{by Theorem 11.5} \\ &\le t^{n}\,|\det L_k|\,\lambda(F_{i,j}) && \text{by Theorem 11.5} \\ &= t^{n}\lambda[L_k(F_{i,j})] && \text{by Theorem 11.3} \\ &\le t^{2n}\lambda[T(F_{i,j})]. && \text{by (11.35)} \end{aligned}$$

With j fixed, sum on i and use the fact that the sets $F_{i,j}$ are disjoint:

$$t^{-2n}\sum_{i=1}^{\infty}\lambda[T(F_{i,j})] \le \int_{B\cap E} |JT|\; d\lambda \le t^{2n}\sum_{i=1}^{\infty}\lambda[T(F_{i,j})]. \tag{11.36}$$

Now let $j \to \infty$ and recall (11.33):

$$\begin{aligned} t^{-2n}\int_{\mathbb{R}^n} N(T, B\cap E, y)\, d\lambda(y) &\le \int_{B\cap E} |JT|\; d\lambda \\ &\le t^{2n}\int_{\mathbb{R}^n} N(T, B\cap E, y)\, d\lambda(y). \end{aligned}$$

Since t was initially chosen as any number larger than 1, we conclude that

$$\int_{B\cap E} |JT|\; d\lambda = \int_{\mathbb{R}^n} N(T, B\cap E, y)\, d\lambda(y). \tag{11.37}$$

Now B was defined as an arbitrary set of the sequence $\{B_k\}$ that was provided by Theorem 11.5. Thus (11.37) holds with B replaced by an arbitrary B_k. Since the sets $\{B_k\}$ are disjoint and both sides of (11.37) are additive relative to $\cup B_k$, the monotone convergence theorem implies

$$\int_E |JT|\; d\lambda = \int_{\mathbb{R}^n} N(T, E, y)\, d\lambda(y).$$

We have reached this conclusion under the assumption that $|JT| \neq 0$ on E. The proof will be concluded if we can show that

$$\lambda[T(E_0)] = 0,$$

where $E_0 = E \cap \{x : JT(x) = 0\}$. Note that E_0 is a Borel set. Note also that nothing is lost if we assume that E_0 is bounded. Choose $\varepsilon > 0$ and let $U \supset E_0$ be a bounded open set with $\lambda(U - E_0) < \varepsilon$. For each $x \in E_0$, there exists $\delta_x > 0$ such that $B(x, \delta_x) \subset U$ and

$$|T(x+h) - T(x)| < \varepsilon\,|h|$$

for all $h \in \mathbb{R}^n$ with $|h| < \delta_x$. In other words,

$$T[B(x, r)] \subset B(T(x), \varepsilon r) \tag{11.38}$$

whenever $r < \delta_x$. The collection of all balls $B(x, r)$, where $x \in E_0$ and $r < \delta_x$, provides a Vitali covering of E_0 in the sense of Definition 7.4. Thus by Theorem 7.7, there exists a disjoint countable subcollection $\{B_i\}$ such that

$$\lambda\left(E_0 - \bigcup_{i=1}^{\infty} B_i\right) = 0.$$

Note that $\cup B_i \subset U$. Let us suppose that each B_i is of the form $B_i = B(x_i, r_i)$. Then, using the fact that T carries sets of measure zero into sets of measure zero,

$$\begin{aligned}
\lambda[T(E_0)] &\leq \sum_{i=1}^{\infty} \lambda[T(B_i)] \\
&\leq \sum_{i=1}^{\infty} \lambda[B(T(x_i), \varepsilon r_i)] && \text{by (11.38)} \\
&= \sum_{i=1}^{\infty} \alpha(n)(\varepsilon r_i)^n \\
&= \varepsilon^n \sum_{i=1}^{\infty} \alpha(n) r_i^n \\
&= \varepsilon^n \sum_{i=1}^{\infty} \lambda(B_i) \\
&\leq \varepsilon^n \lambda(U). && \text{since the } \{B_i\} \text{ are disjoint}
\end{aligned}$$

Since $\lambda(U) < \infty$ (because U is bounded) and ε is arbitrary, we have $\lambda[T(E_0)] = 0$. □

11.8. COROLLARY. *If $T\colon \mathbb{R}^n \to \mathbb{R}^n$ is Lipschitz and univalent, then*

$$\int_E |JT(x)| \; d\lambda(x) = \lambda[T(E)]$$

whenever $E \subset \mathbb{R}^n$ is measurable.

PROOF. Since T is univalent, we have $N(T, E, y) = 1$ on $T(E)$ and $N(T, E, y) = 0$ in $\mathbb{R}^n \setminus T(E)$. Therefore, the result is clear from Theorem 11.7. See also Theorem 11.3. □

11.9. THEOREM. *Let $T\colon \mathbb{R}^n \to \mathbb{R}^n$ be a Lipschitz transformation and suppose $f \in L^1(\mathbb{R}^n)$. Then*

$$\int_{\mathbb{R}^n} f \circ T(x) \, |JT(x)| \; d\lambda(x) = \int_{\mathbb{R}^n} f(y) N(T, \mathbb{R}^n, y) \; d\lambda(y).$$

PROOF. First, suppose f is the characteristic function of a measurable set $A \subset \mathbb{R}^n$. Then $f \circ T = \chi_{T^{-1}(A)}$ and

$$\begin{aligned}\int_{\mathbb{R}^n} f \circ T\,|JT|\; d\lambda = \int_{T^{-1}(A)} |JT|\; d\lambda &= \int_{\mathbb{R}^n} N(T, T^{-1}(A), y)\; d\lambda(y)\\ &= \int_{\mathbb{R}^n} \chi_A(y) N(T, \mathbb{R}^n, y)\; d\lambda(y).\\ &= \int_{\mathbb{R}^n} f(y) N(T, \mathbb{R}^n, y)\; d\lambda(y).\end{aligned}$$

Clearly, this also holds whenever f is a simple function. Reference to Theorem 5.27 and the monotone convergence theorem shows that it then holds whenever f is nonnegative. Finally, writing $f = f^+ - f^-$ yields the final result. □

11.10. COROLLARY. *If* $T\colon \mathbb{R}^n \to \mathbb{R}^n$ *is Lipschitz and univalent, then*

$$\int_{\mathbb{R}^n} f \circ T(x)\,|JT(x)|\; d\lambda(x) = \int_{\mathbb{R}^n} f(y)\; d\lambda(y).$$

PROOF. This is immediate from Theorem 11.9, since in this case, $N(T, \mathbb{R}^n, y) \equiv 1$. □

11.11. COROLLARY. **(Change of Variables Formula)** *Let* $T\colon\mathbb{R}^n \to \mathbb{R}^n$ *be a Lipschitz map and* $f \in L^1(\mathbb{R}^n)$*. If* $E \subset \mathbb{R}^n$ *is Lebesgue measurable and* T *is injective on* E*, then*

$$\int_{T(E)} f(y) d\lambda(y) = \int_E f \circ T(x) |JT(x)| d\lambda(x).$$

PROOF. We apply Theorem 11.9 with $\chi_{T(E)} f$ instead of f. □

11.12. REMARK. This result provides a geometric interpretation of $JT(x)$. If T is linear, Theorem 11.3 states that $|JT|$ is given by

$$\frac{\lambda[T(E)]}{\lambda(E)} \tag{11.39}$$

for every measurable set E. Roughly speaking, the same is true locally when T is Lipschitz and univalent, because Corollary 11.8 implies

$$\int_{B(x_0,r)} |JT(x)|\; d\lambda(x) = \lambda[T(B(x_0, r))]$$

for an arbitrary ball $B(x_0, r)$. Now use the result on Lebesgue points (Theorem 7.11) to conclude that

$$|JT(x_0)| = \lim_{r\to 0} \frac{\lambda[T(B(x_0, r))]}{\lambda(B(x_0, r))}$$

for λ-almost all x_0, which is the infinitesimal analogue of (11.39).

We close this section with a discussion of spherical coordinates in $\mathbb{R}^n$. First, consider $\mathbb{R}^2$ with points designated by (x_1, x_2). Let Ω denote $\mathbb{R}^2$ with the set $N := \{(x_1, 0) : x_1 \geq 0\}$ removed. Let $r = \sqrt{x_1^2 + x_2^2}$ and let θ be the angle from N to the ray emanating from $(0,0)$ passing through (x_1, x_2) with $0 < \theta(x_1, x_2) < 2\pi$. Then (r, θ) are the coordinates of (x_1, x_2), and

$$x_1 = r\cos\theta \colon = T^1(r,\theta), \qquad x_2 = r\sin\theta \colon = T^2(r,\theta).$$

The transformation $T = (T^1, T^2)$ is a C^∞ bijection of $(0,\infty) \times (0, 2\pi)$ onto $\mathbb{R}^2 - N$. Furthermore, since $JT(r,\theta) = r$, we obtain from Corollary 11.11 that

$$\int_A f \circ T(r,\theta) r \, dr \, d\theta = \int_B f \, d\lambda,$$

where $T(A) = B - N$. Of course, $\lambda(N) = 0$, and therefore the integral over B is the same as the integral over $T(A)$.

Now we consider the case $n > 2$ and proceed inductively. For $x = (x_1, x_2, \dots, x_n) \in \mathbb{R}^n$, Let $r = |x|$ and let $\theta_1 = \cos^{-1}(x_1/r)$, $0 < \theta_1 < \pi$. Also, let $(\rho, \theta_2, \dots, \theta_{n-1})$ be spherical coordinates for $x' = (x_2, \dots, x_n)$, where $\rho = |x'| = r \sin\theta_1$. The coordinates of x are

$$\begin{aligned}
x_1 &= r\cos\theta_1 \colon = T^1(r, \theta_1, \dots, \theta_n),\\
x_2 &= r\sin\theta_1\cos\theta_2 \colon = T^2(r, \theta_1, \dots, \theta_n),\\
&\ \vdots\\
x_{n-1} &= r\sin\theta_1\sin\theta_2 \dots \sin\theta_{n-2}\cos\theta_{n-1} \colon = T^{n-1}(r, \theta_1, \dots, \theta_n),\\
x_n &= r\sin\theta_1\sin\theta_2 \dots \sin\theta_{n-2}\sin\theta_{n-1} \colon = T^n(r, \theta_1, \dots, \theta_n).
\end{aligned}$$

The mapping $T = (T^1, T^2, \dots, T^n)$ is a bijection of

$$(0,\infty) \times (0,\pi)^{n-2} \times (0, 2\pi)$$

onto

$$\mathbb{R}^n - (\mathbb{R}^{n-2} \times [0,\infty) \times \{0\}).$$

A straightforward calculation shows that the Jacobian is

$$JT(r, \theta_1, \dots, \theta_{n-1}) = r^{n-1} \sin^{n-2}\theta_1 \sin^{n-3}\theta_2 \dots \sin\theta_{n-2}.$$

Hence, with $\theta = (\theta_1, \dots, \theta_{n-1})$, we obtain

$$\begin{aligned}
&\int_A f \circ T(r,\theta) r^{n-1} \sin^{n-2}\theta_1 \sin^{n-3}\theta_2 \dots \sin\theta_{n-2} \, dr \, d\theta_1 \dots \theta_{n-1}\\
&\qquad = \int_{T(A)} f \, d\lambda.
\end{aligned} \tag{11.40}$$

Exercises for Section 11.2

1. Prove that the sets $E(c, R, i)$ defined by (11.25) and (11.26) are Borel sets.
2. Give another proof of Theorem 11.4.

11.3. Sobolev Functions

It is not obvious that there is a class of functions defined on $\mathbb{R}^n$ that is analogous to the absolutely continuous functions on $\mathbb{R}$. However, it is tempting to employ the multidimensional analogue of Theorem 10.9, which states roughly that a function is absolutely continuous if and only if its derivative in the sense of distributions is a function. We will follow this direction by considering functions whose partial derivatives are functions and show that this definition leads to a fruitful development.

11.13. DEFINITION. Let $\Omega \subset \mathbb{R}^n$ be an open set and let $f \in L^1_{\text{loc}}(\Omega)$. We use Definition 10.7 to define the partial derivatives of f in the sense of distributions. Thus, for $1 \leq i \leq n$, we say that a function $g_i \in L^1_{\text{loc}}(\Omega)$ is the ith **partial derivative of f in the sense of distributions** if

$$\int_\Omega f \frac{\partial \varphi}{\partial x_i}\, d\lambda = -\int_\Omega g_i \varphi \, d\lambda \tag{11.41}$$

for every test function $\varphi \in C_c^\infty(\Omega)$. We will write

$$\frac{\partial f}{\partial x_i} = g_i;$$

thus, $\frac{\partial f}{\partial x_i}$ is defined to be merely the function g_i.

At this time we cannot assume that the partial derivative $\frac{\partial f}{\partial x_i}$ exists in the classical sense for the Sobolev function f. This existence will be discussed later after Theorem 11.19 has been proved. Consistent with the notation introduced in (10.5), we will sometimes write $D_i f$ to denote $\frac{\partial f}{\partial x_i}$.

The definition in (11.41) is a restatement of Definition 10.7 with the requirement that the derivative of f (in the sense of distributions) be again a function and not merely a distribution. This requirement imposes a condition on f, and the purpose of this section is to see what properties f must possess in order to satisfy this condition.

In general, we recall what it means for a higher-order derivative of f to be a function (see Definition 10.7). If α is a multi-index, then $g_\alpha \in L^1_{\text{loc}}(\Omega)$ is called the αth **distributional derivative** of f if

$$\int_\Omega f D^\alpha \varphi \, d\lambda = (-1)^{|\alpha|} \int_\Omega \varphi g_\alpha \, d\lambda$$

for every test function $\varphi \in C_c^\infty(\Omega)$. We write $D^\alpha f := g_\alpha$. For $1 \leq p \leq \infty$ and k a nonnegative integer, we say that f belongs to the **Sobolev space**

$$W^{k,p}(\Omega)$$

if $D^\alpha f \in L^p(\Omega)$ for each multi-index α with $|\alpha| \leq k$. In particular, this implies that $f \in L^p(\Omega)$. Similarly, the space

$$W^{k,p}_{\text{loc}}(\Omega)$$

consists of all f with $D^\alpha f \in L^p_{\text{loc}}(\Omega)$ for $|\alpha| \leq k$.

In order to motivate the definition of the distributional partial derivative, we recall the classical Gauss–Green theorem.

11.14. THEOREM. *Suppose that U is an open set with smooth boundary and let $\nu(x)$ denote the unit exterior normal to U at $x \in \partial U$. If $V \colon \mathbb{R}^n \to \mathbb{R}^n$ is a transformation of class C^1, then*

$$\int_U \operatorname{div} V \, d\lambda = \int_{\partial U} V(x) \cdot \nu(x) \, dH^{n-1}(x), \tag{11.42}$$

where div$V$*, the divergence of $V = (V^1, \ldots, V^n)$, is defined by*

$$\operatorname{div} V = \sum_{i=1}^{n} \frac{\partial V^i}{\partial x_i}.$$

Now suppose that $f \colon \Omega \to \mathbb{R}$ is of class C^1 and $\varphi \in C_c^\infty(\Omega)$. Define $V \colon \mathbb{R}^n \to \mathbb{R}^n$ to be the transformation whose coordinate functions are all 0 except for the ith one, which is $f\varphi$. Then

$$\operatorname{div} V = \frac{\partial(f\varphi)}{\partial x_i} = f\frac{\partial \varphi}{\partial x_i} + \frac{\partial f}{\partial x_i}\varphi.$$

Since the support of φ is a compact set contained in Ω, it is possible to find an open set U with smooth boundary containing the support of φ such that $\overline{U} \subset \Omega$. Then,

$$\int_\Omega \operatorname{div} V \, d\lambda = \int_U \operatorname{div} V \, d\lambda = \int_{\partial U} V \cdot \nu \, dH^{n-1} = 0.$$

Thus,

$$\int_\Omega f \frac{\partial \varphi}{\partial x_i} \, d\lambda = -\int_\Omega \frac{\partial f}{\partial x_i} \varphi \, d\lambda, \tag{11.43}$$

which is precisely (11.41) when $f \in C^1(\Omega)$. Note that for a Sobolev function f, the formula (11.43) is valid for all test functions $\varphi \in C_c^\infty(\Omega)$ by the definition of the distributional derivative. The **Sobolev norm** of $f \in W^{1,p}(\Omega)$ is defined by

$$\|f\|_{1,p;\Omega} : = \|f\|_{p;\Omega} + \sum_{i=1}^{n} \|D_i f\|_{p;\Omega}, \tag{11.44}$$

for $1 \le p < \infty$ and

$$\|f\|_{1,\infty;\Omega} : = \operatorname*{ess\,sup}_\Omega (|f| + \sum_{i=1}^{n} |D_i f|).$$

One can readily verify that $W^{1,p}(\Omega)$ becomes a Banach space when it is endowed with the above norm (Exercise 1, Section 11.3).

11.15. REMARK. When $1 \le p < \infty$, it can be shown (Exercise 2, Section 11.3) that the norm

$$\|f\| := \|f\|_{p;\Omega} + \left(\sum_{i=1}^{n} \|D_i f\|_{p;\Omega}^{p} \right)^{1/p} \tag{11.45}$$

is equivalent to (11.44). Also, it is sometimes convenient to regard $W^{1,p}(\Omega)$ as a subspace of the Cartesian product of spaces $L^p(\Omega)$. The identification is made by defining $P\colon W^{1,p}(\Omega) \to \prod_{i=1}^{n+1} L^p(\Omega)$ as

$$P(f) = (f, D_1 f, \ldots, D_n f) \quad \text{for} \quad f \in W^{1,p}(\Omega).$$

In view of Exercise 2, Section 8.1, it follows that P is an isometric isomorphism of $W^{1,p}(\Omega)$ onto a subspace W of this Cartesian product. Since $W^{1,p}(\Omega)$ is a Banach space, W is a closed subspace. By Exercise 10, Section 8.4, W is reflexive, and therefore so is $W^{1,p}(\Omega)$.

11.16. REMARK. Observe that $f \in W^{1,p}(\Omega)$ is determined only up to a set of Lebesgue measure zero. We agree to call the Sobolev function f continuous, bounded, etc. if there is a function $\overline{f}$ with these respective properties such that $\overline{f} = f$ almost everywhere.

We will show that elements in $W^{1,p}(\Omega)$ have representatives that permit us to regard them as generalizations of absolutely continuous functions on $\mathbb{R}^1$. First, we prove an important result concerning the convergence of regularizers of Sobolev functions.

11.17. NOTATION. If Ω' and Ω are open sets of $\mathbb{R}^n$, we will write $\Omega' \subset\subset \Omega$ to signify that the closure of Ω' is compact and that the closure of Ω' is a subset of Ω. Also, we will frequently use dx instead of $d\lambda(x)$ to denote integration with respect to Lebesgue measure.

11.18. LEMMA. *Suppose $f \in W^{1,p}(\Omega)$, $1 \le p < \infty$. Then the mollifiers, f_ε, of f* (see (10.9)) *satisfy*

$$\lim_{\varepsilon \to 0} \|f_\varepsilon - f\|_{1,p;\Omega'} = 0$$

whenever $\Omega' \subset\subset \Omega$.

PROOF. Since Ω' is a bounded domain, there exists $\varepsilon_0 > 0$ such that $\varepsilon_0 <$ dist $(\Omega', \partial\Omega)$. For $\varepsilon < \varepsilon_0$, we may differentiate under the integral sign (see the proof of (10.10)) to obtain, for $x \in \Omega'$ and $1 \le i \le n$,

$$\begin{aligned}\frac{\partial f_\varepsilon}{\partial x_i}(x) &= \varepsilon^{-n}\int_\Omega \frac{\partial\varphi}{\partial x_i}\left(\frac{x-y}{\varepsilon}\right) f(y)\,d\lambda(y)\\ &= -\varepsilon^{-n}\int_\Omega \frac{\partial\varphi}{\partial y_i}\left(\frac{x-y}{\varepsilon}\right) f(y)\,d\lambda(y)\\ &= \varepsilon^{-n}\int_\Omega \varphi\left(\frac{x-y}{\varepsilon}\right)\frac{\partial f}{\partial y_i}\,d\lambda(y) \qquad \text{by(11.41)}\\ &= \left(\frac{\partial f}{\partial x_i}\right)_\varepsilon (x).\end{aligned}$$

Our result now follows from Theorem 10.1. □

Since the definition of Sobolev functions requires that their distributional derivatives belong to L^p, it is natural to inquire whether they have partial derivatives in the classical sense. To this end, we begin by showing that their partial derivatives exist almost everywhere. That is, in keeping with Remark 11.16, we will show that there is a function f^* such that $f^* = f$ a.e. and that the partial derivatives of f^* exist almost everywhere.

In the next theorem, the set R is an interval of the form

$$R\colon = (a_1, b_1)\times\cdots\times(a_n, b_n). \tag{11.46}$$

11.19. THEOREM. *Suppose $f \in W^{1,p}(\Omega)$, $1 \le p < \infty$. Let $R \subset\subset \Omega$. Then f has a representative f^* that is absolutely continuous on almost all line segments of R that are parallel to the coordinate axes, and the classical partial derivatives of f^* agree almost everywhere with the distributional derivatives of f. Conversely, if $f = f^*$ almost everywhere for some function f^* that is absolutely continuous on almost all line segments of R that are parallel to the coordinate axes and if all partial derivatives of f^* belong to $L^p(R)$, then $f \in W^{1,p}(R)$.*

PROOF. First, suppose $f \in W^{1,p}(\Omega)$ and fix i with $1 \le i \le n$. Since $R \subset\subset \Omega$, the mollifiers f_ε are defined for all $x \in R$ for ε sufficiently small. Throughout the proof, only such mollifiers will be considered. We know from Lemma 11.18 that $\|f_\varepsilon - f\|_{1,p;R} \to 0$ as $\varepsilon \to 0$. Choose a sequence $\varepsilon_k \to 0$ and let $f_k\colon = f_{\varepsilon_k}$. Also write $x \in R$ as $x = (x', t)$, where

$$x' \in R_i\colon = (a_1, b_1)\times\cdots\times\underbrace{(a_i, b_i)}_{\text{omitted}}\times\cdots\times(a_n, b_n)$$

and $t \in (a_i, b_i)$, $1 \le i \le n$. Then it follows that

$$\lim_{k\to\infty}\int_{R_i}\int_{a_i}^{b_i} |f_k(x', t) - f(x', t)|^p + |\nabla f_k(x', t) - \nabla f(x', t)|^p\,dt\,d\lambda(x') = 0,$$

which can be rewritten as

$$\lim_{k\to\infty}\int_{R_i} F_k(x')d\lambda(x') = 0, \tag{11.47}$$

where

$$F_k(x') = \int_{a_i}^{b_i} |f_k(x',t) - f(x',t)|^p + |\nabla f_k(x',t) - \nabla f(x',t)|^p dt.$$

As in the classical setting, we use ∇f to denote $(D_1 f, \dots, D_n f)$, which is defined in terms of the distributional derivatives of f. By Vitali's convergence theorem (see Theorem 6.30), and since (11.47) means $F_k \to 0$ in $L^1(R_i)$, there exists a subsequence (which will still be denoted the same as the full sequence) such that $F_k(x') \to 0$ for λ_{n-1}-a.e. x'. That is,

$$\lim_{k\to\infty} \int_{a_i}^{b_i} |f_k(x',t) - f(x',t)|^p + |\nabla f_k(x',t) - \nabla f(x',t)|^p \, dt = 0 \tag{11.48}$$

for λ_{n-1}-almost all $x' \in R_i$. From Hölder's inequality we have

$$\begin{aligned} &\int_{a_i}^{b_i} |\nabla f_k(x',t) - \nabla f(x',t)| \, dt \\ &\qquad \le (b_i - a_i)^{1/p'} \left(\int_{a_i}^{b_i} |\nabla f_k(x',t) - \nabla f(x',t)|^p \, dt \right)^{1/p}. \end{aligned} \tag{11.49}$$

The fundamental theorem of calculus implies that for all $[a,b] \subset [a_i,b_i]$,

$$\begin{aligned} |f_k(x',b) - f_k(x',a)| &= \left| \int_a^b \frac{\partial f_k}{\partial x_i}(x',t) dt \right| \\ &\le \int_a^b \left| \frac{\partial f_k}{\partial x_i}(x',t) \right| dt \\ &\le \int_a^b |\nabla f_k(x',t)| \, dt \\ &\le \int_a^b |\nabla f_k(x',t) - \nabla f(x',t)| dt + \int_a^b |\nabla f(x',t)| \, dt. \end{aligned} \tag{11.50}$$

Consequently, it follows from (11.48), (11.49), and (11.50) that there is a constant $M_{x'}$ such that

$$|f_k(x',b) - f_k(x',a)| \le \int_{a_i}^{b_i} |\nabla f(x',t)| \, dt + 1 \tag{11.51}$$

whenever $k > M_{x'}$ and $a, b \in [a_i, b_i]$. Note that (11.48) implies that there exists a subsequence of $\{f_k(x',\cdot)\}$, which again is denoted the same as the full sequence, such that

$$f_k(x',t) \to f(x',t), \quad \lambda\text{-a.e. } t. \tag{11.52}$$

Fix t_0 for which $f_k(x',t_0) \to f_k(x',t_0)$, $t_0 < b$. Then (using this further subsequence) (11.51) with $a = t_0$ yields

$$|f_k(x',b)| \le 1 + \int_{a_i}^{b_i} |\nabla f(x',t)| dt + |f_k(x',t_0)|. \tag{11.53}$$

Thus for k large enough (depending on x'), we have

$$|f_k(x',b)| \leq 1 + \int_{a_i}^{b_i} |\nabla f(x',t)|dt + |f(x',t_0)| + 1, \tag{11.54}$$

which proves that the sequence of functions $\{f_k(x',\cdot\}$ is pointwise bounded for almost every x'.

We now note that for each x' under consideration, the functions $f_k(x',\cdot)$, as functions of t, are absolutely continuous (see Theorem 7.36). Moreover, the functions are absolutely continuous *uniformly in* k. Indeed, it is easy to check, using (11.50), that for each $\varepsilon > 0$ there exists $\delta > 0$ such that for every finite collection $\mathcal{F}$ of nonoverlapping intervals in $[a_i, b_i]$ with $\sum_{I\in\mathcal{F}} |b_I - a_I| < \delta$, one has

$$\sum_{I\in\mathcal{F}} |f_k(x',b_I) - f_k(x',a_I)| < \varepsilon$$

for all positive integers k. Here, as in Definition 7.23, the endpoints of the interval I are denoted by a_I, b_I. In particular, the sequence is equicontinuous on $[a_i, b_i]$. Since the sequence $\{f_k(x',\cdot)\}$ is pointwise bounded and equicontinuous, we now use the Arzelà–Ascoli theorem, and we find that there is a subsequence that converges uniformly to a function on $[a_i, b_i]$; call it $f_i^*(x',\cdot)$. The uniform absolute continuity of this subsequence implies that $f_i^*(x',\cdot)$ is absolutely continuous. We now recall (11.52) and conclude that

$$f(x',t) = f_i^*(x',t) \text{ for } \lambda\text{-a.e. } t \in [a_i, b_i].$$

To summarize what has been done so far, recall that for each interval $R \subset \Omega$ of the form (11.46) and each $1 \leq i \leq n$, there is a representative f_i^* of f that is absolutely continuous in t for λ_{n-1}-almost all $x' \in R_i$. This representative was obtained as the pointwise a.e. limit of a subsequence of mollifiers of f. Observe that there exist a single subsequence and a single representative that can be found for all i simultaneously. This can be accomplished by first obtaining a subsequence and a representative for $i = 1$. Then a subsequence can be extracted from this one that can be used to define a representative for $i = 1$ and 2. Continuing in this way, the desired subsequence and representative are obtained. Thus, there exist a sequence of mollifiers, denoted by $\{f_k\}$, and a function f^* such that for each $1 \leq i \leq n$, $f_k(x',\cdot)$ converges uniformly to $f^*(x',\cdot)$ for λ_{n-1}-almost all x'. Furthermore, $f^*(x',t)$ is absolutely continuous in t for λ_{n-1}-almost all x'.

The proof that the classical partial derivatives of f^* agree with the distributional derivatives almost everywhere is not difficult. Consider any partial derivative, say the ith one, and recall that $D_i = \frac{\partial}{\partial x_i}$. The distributional derivative is defined by

$$D_i f(\varphi) = -\int_R f D_i\varphi \, dx$$

for a test function $\varphi \in C_c^\infty(R)$. Since f^* is absolutely continuous on almost all line segments parallel to the ith axis, we have

$$\begin{aligned}\int_R f^* D_i\varphi \, dx &= \int_{R_i} \int_{a_i}^{b_i} f^* D_i\varphi \, dx_i dx' \\ &= -\int_{R_i} \int_{a_i}^{b_i} (D_i f^*)\varphi \, dx_i dx' \\ &= -\int_R (D_i f^*)\varphi \, dx.\end{aligned}$$

Since $f = f^*$ almost everywhere, we have

$$\int_R f D_i\varphi \, dx = \int_R f^* D_i\varphi \, dx,$$

and therefore

$$\int_R D_i f \varphi \, dx = \int_R D_i f^* \varphi \, dx,$$

for every $\varphi \in C_c^\infty(R)$. This implies that the partial derivatives of f^* agree with the distributional derivatives almost everywhere (see Exercise 3, Section 10.1). To prove the converse, suppose that f has a representative f^* as in the statement of the theorem. Then, for $\varphi \in C_c^\infty(R)$, $f^*\varphi$ has the same absolutely continuous properties as does f^*. Thus, for $1 \le i \le n$, we can apply the fundamental theorem of calculus to obtain

$$\int_{a_i}^{b_i} \frac{\partial(f^*\varphi)}{\partial x_i}(x', t) \, dt = 0$$

for λ_{n-1}-almost all $x' \in R_i$, and therefore (see problem 10, Section 7.5),

$$\int_{a_i}^{b_i} f^*(x', t) \frac{\partial \varphi}{\partial x_i}(x', t) \, dt = -\int_{a_i}^{b_i} \frac{\partial f^*}{\partial x_i}(x', t) \, \varphi(x', t) \, dt.$$

Fubini's theorem implies

$$-\int_R f^* \frac{\partial \varphi}{\partial x_i} \, d\lambda = \int_R \frac{\partial f^*}{\partial x_i} \varphi \, d\lambda$$

for all $\varphi \in C_c^\infty(R)$. Recall that the distributional derivative $\frac{\partial f}{\partial x_i}$ is defined as

$$\frac{\partial f}{\partial x_i}(\varphi) = -\int_R f \frac{\partial \varphi}{\partial x_i} d\lambda, \quad \varphi \in C_c^\infty(R).$$

Hence, since $f = f^*$ a.e., we have

$$\frac{\partial f}{\partial x_i}(\varphi) = \int_R \frac{\partial f^*}{\partial x_i} \varphi \, d\lambda, \qquad \text{for all } C_c^\infty(R). \tag{11.55}$$

From (11.55), it follows that the distribution $\frac{\partial f}{\partial x_i}$ is the function $\frac{\partial f^*}{\partial x_i} \in L^p(R)$. Therefore, $\frac{\partial f}{\partial x_i} \in L^p(R)$, and we conclude that $f \in W^{1,p}(R)$. □

Exercises for Section 11.3

1. Prove that $W^{1,p}(\Omega)$ endowed with the norm (11.44) is a Banach space.
2. Prove that the norm

$$\|f\| : = \|f\|_{p;\Omega} + \left(\sum_{i=1}^{n} \|D_i f\|_{p;\Omega}^{p} \right)^{1/p}$$

is equivalent to (11.44).

3. With the help of Exercise 2, Section 8.1, show that $W^{1,p}(\Omega)$ can be regarded as a closed subspace of the Cartesian product of $L^p(\Omega)$ spaces. Referring now to Exercise 10, Section 8.4, show that this subspace is reflexive and hence $W^{1,p}(\Omega)$ is reflexive if $1 < p < \infty$.
4. Suppose $f \in W^{1,p}(\mathbb{R}^n)$. Prove that f^+ and f^- are in $W^{1,p}(\mathbb{R}^n)$.

11.4. Approximating Sobolev Functions

We will show that the Sobolev space $W^{1,p}(\Omega)$ can be characterized as the closure of $C^\infty(\Omega)$ in the Sobolev norm. This is a very useful result and is employed frequently in applications. In the next section we will demonstrate its utility in proving the Sobolev inequality, which implies that $W^{1,p}(\Omega) \subset L^{p^*}(\Omega)$, where $p^* = np/(n-p)$.

We begin with a smooth version of Theorem 9.8.

11.20. LEMMA. *Let $\mathcal{G}$ be an open cover of a set $E \subset \mathbb{R}^n$. Then there exists a family $\mathcal{F}$ of functions $f \in C_c^\infty(\mathbb{R}^n)$ such that $0 \le f \le 1$ and the following hold:*

(i) For each $f \in \mathcal{F}$, there exists $U \in \mathcal{G}$ such that $\operatorname{spt} f \subset U$.

(ii) If $K \subset E$ is compact, then $\operatorname{spt} f \cap K \neq 0$ for only finitely many $f \in \mathcal{F}$.

(iii) $\sum_{f \in \mathcal{F}} f(x) = 1$ for each $x \in E$. The family $\mathcal{F}$ is called a **smooth partition of unity** *of E subordinate to the open covering $\mathcal{G}$.*

PROOF. If E is compact, our desired result follows from the proof of Theorem 9.8 and Exercise 1, Section 10.1.

Now assume that E is open, and for each positive integer i define

$$E_i = E \cap \overline{B}(0,i) \cap \{x : \operatorname{dist}(x, \partial E) \ge \frac{1}{i}\}.$$

Thus, E_i is compact, $E_i \subset \operatorname{int} E_{i+1}$, and $E = \cup_{i=1}^\infty E_i$. Let $\mathcal{G}_i$ denote the collection of all open sets of the form

$$U \cap \{\operatorname{int} E_{i+1} - E_{i-2}\},$$

where $U \in \mathcal{G}$ and where we take $E_0 = E_{-1} = \emptyset$. The family $\mathcal{G}_i$ forms an open covering of the compact set $E_i - \operatorname{int} E_{i-1}$. Therefore, our first case applies, and we obtain a smooth partition of unity, $\mathcal{F}_i$, subordinate to $\mathcal{G}_i$, which consists of finitely many elements. For each $x \in \mathbb{R}^n$, define

$$s(x) = \sum_{i=1}^{\infty} \sum_{g \in \mathcal{F}_i} g(x).$$

The sum is well defined, since only a finite number of terms are nonzero for each x. Note that $s(x) > 0$ for $x \in E$. Corresponding to each positive integer i and to each function $g \in \mathcal{G}_i$, define a function f by

$$f(x) = \begin{cases} \frac{g(x)}{s(x)} & \text{if } x \in E, \\ 0 & \text{if } x \notin E. \end{cases}$$

The partition of unity $\mathcal{F}$ that we want comprises all such functions f.

If E is arbitrary, then a partition of unity for the open set $\cup\{U : U \in \mathcal{G}\}$ is also one for E. $\square$

Clearly, the set

$$S = C^{\infty}(\Omega) \cap \{f : \|f\|_{1,p;\Omega} < \infty\}$$

is contained in $W^{1,p}(\Omega)$. Moreover, the same is true of the closure of S in the Sobolev norm, since $W^{1,p}(\Omega)$ is complete. The next result shows that $\overline{S} = W^{1,p}(\Omega)$.

11.21. Theorem. *If $1 \le p < \infty$, then the space*

$$C^{\infty}(\Omega) \cap \{f : \|f\|_{1,p;\Omega} < \infty\}$$

is dense in $W^{1,p}(\Omega)$.

Proof. Let Ω_i be subdomains of Ω such that $\overline{\Omega_i} \subset \Omega_{i+1}$ and $\cup_{i=1}^{\infty}\Omega_i = \Omega$. Let $\mathcal{F}$ be a partition of unity of Ω subordinate to the covering $\{\Omega_{i+1} - \overline{\Omega_{i-1}}\}$, where Ω_{-1} is defined as the null set. Let f_i denote the sum of the finitely many $f \in \mathcal{F}$ with $\operatorname{spt} f \subset \Omega_{i+1} - \overline{\Omega_{i-1}}$. Then $f_i \in C_c^{\infty}(\Omega_{i+1} - \overline{\Omega_{i-1}})$ and

$$\sum_{i=1} f_i \equiv 1 \quad \text{on} \quad \Omega. \tag{11.56}$$

Choose $\varepsilon > 0$. For $f \in W^{1,p}(\Omega)$, refer to Lemma 11.18 to obtain $\varepsilon_i > 0$ such that

$$\begin{aligned} &\operatorname{spt}(f_i f)_{\varepsilon_i} \subset \Omega_{i+1} - \overline{\Omega_{i-1}}, \\ &\|(f_i f)_{\varepsilon_i} - f_i f\|_{1,p;\Omega} < \varepsilon 2^{-i}. \end{aligned} \tag{11.57}$$

With $g_i \colon = (f_i f)_{\varepsilon_i}$, (11.57) implies that only finitely many of the g_i can fail to vanish on any given $\Omega' \subset\subset \Omega$. Therefore, $g \colon = \sum_{i=1}^{\infty} g_i$ is defined and is an element of $C^{\infty}(\Omega)$. For $x \in \Omega_i$, we have

$$f(x) = \sum_{j=1}^{i} f_j(x) f(x),$$

and by (11.57),

$$g(x) = \sum_{j=1}^{i} (f_j f)_{\varepsilon_j}(x).$$

Consequently,

$$\|f - g\|_{1,p;\Omega_i} \leq \sum_{j=1}^{i} \|(f_j f)_{\varepsilon_j} - f_j f\|_{1,p;\Omega} < \varepsilon.$$

Now an application of the monotone convergence theorem establishes our desired result. □

The previous result holds in particular when $\Omega = \mathbb{R}^n$, in which case we get the following apparently stronger result.

11.22. COROLLARY. *If $1 \leq p < \infty$, the space $C_c^\infty(\mathbb{R}^n)$ is dense in $W^{1,p}(\mathbb{R}^n)$.*

PROOF. This follows from the previous result and the fact that $C_c^\infty(\mathbb{R}^n)$ is dense in

$$C^\infty(\mathbb{R}^n) \cap \{f : \|f\|_{1,p;\mathbb{R}^n} < \infty\}$$

relative to the Sobolev norm (see Exercise 1, Section 11.4). □

Recall that if $f \in L^p(\mathbb{R}^n)$, then $\|f(x+h) - f(x)\|_p \to 0$ as $h \to 0$. A similar result provides a very useful characterization of $W^{1,p}$.

11.23. THEOREM. *Let $1 < p < \infty$ and suppose $\Omega \subset \mathbb{R}^n$ is an open set. If $f \in W^{1,p}(\Omega)$ and $\Omega' \subset\subset \Omega$, then $|h^{-1}| \, \|f(x+h) - f(x)\|_{p;\Omega'}$ remains bounded for all sufficiently small h. Conversely, if $f \in L^p(\Omega)$ and*

$$|h^{-1}| \, \|f(x+h) - f(x)\|_{p;\Omega'}$$

remains bounded for all sufficiently small h, then $f \in W^{1,p}(\Omega')$.

PROOF. Assume $f \in W^{1,p}(\Omega)$ and let $\Omega' \subset\subset \Omega$. By Theorem 11.21, there exists a sequence of $C^\infty(\Omega)$ functions $\{f_k\}$ such that $\|f_k - f\|_{1,p;\Omega} \to 0$ as $k \to \infty$. For each $g \in C^\infty(\Omega)$, we have

$$\frac{g(x+h) - g(x)}{|h|} = \frac{1}{|h|} \int_0^{|h|} \nabla g\left(x + t\frac{h}{|h|}\right) \cdot \frac{h}{|h|} \, dt,$$

so by Jensen's inequality (Exercise 10, Section 6.5),

$$\left|\frac{g(x+h) - g(x)}{h}\right|^p \leq \frac{1}{|h|} \int_0^{|h|} \left|\nabla g\left(x + t\frac{h}{|h|}\right)\right|^p \, dt$$

whenever $x \in \Omega'$ and $h < \delta := \operatorname{dist}(\partial\Omega', \partial\Omega)$. Therefore,

$$\begin{aligned}\|g(x+h) - g(x)\|_{p;\Omega'}^p &\le |h|^p \frac{1}{|h|} \int_0^{|h|} \int_{\Omega'} \left| \nabla g \left(x + t\frac{h}{|h|} \right) \right|^p dx\, dt \\ &\le |h|^{p-1} \int_0^{|h|} \int_{\Omega} |\nabla g(x)|^p \, dx\, dt,\end{aligned}$$

or

$$\|g(x+h) - g(x)\|_{p;\Omega'} \le |h| \, \|\nabla g\|_{p;\Omega}$$

for all $h < \delta$. Since this inequality holds for each f_k, it also holds for f.

For the proof of the converse, let e_i denote the ith unit basis vector. By assumption, the sequence

$$\left\{ \frac{f(x + e_i/k) - f(x)}{1/k} \right\}$$

is bounded in $L^p(\Omega')$ for all large k. Therefore by Alaoglu's theorem (Theorem 8.40), there exist a subsequence (denoted by the full sequence) and $f_i \in L^p(\Omega')$ such that

$$\frac{f(x + e_i/k) - f(x)}{1/k} \to f_i$$

weakly in $L^p(\Omega')$. Thus, for $\varphi \in C_0^\infty(\Omega')$, we have

$$\begin{aligned}\int_{\Omega'} f_i \varphi \, dx &= \lim_{k\to\infty} \int_{\Omega'} \left[\frac{f(x + e_i/k) - f(x)}{1/k} \right] \varphi(x) dx \\ &= \lim_{k\to\infty} \int_{\Omega'} f(x) \left[\frac{\varphi(x - e_i/k) - \varphi(x)}{1/k} \right] dx \\ &= -\int_{\Omega'} f D_i \varphi \, dx.\end{aligned}$$

This shows that $D_i f = f_i$ in the sense of distributions. Hence, $f \in W^{1,p}(\Omega')$. □

Exercises for Section 11.4

1. Relative to the Sobolev norm, prove that $C_c^\infty(\mathbb{R}^n)$ is dense in
$$C^\infty(\mathbb{R}^n) \cap W^{1,p}(\mathbb{R}^n).$$
2. Let $f \in C^2(\Omega)$. Show that f is harmonic in Ω if and only if f is weakly harmonic in Ω (i.e., $\int_\Omega f \Delta\varphi \, dx = 0$ for every $\varphi \in C_c^\infty(\Omega)$). Show also that f is weakly harmonic in Ω if and only if $\int_\Omega \nabla\varphi \cdot \nabla f \, dx = 0$ for every $\varphi \in C_c^\infty(\Omega)$.
3. Let $f \in W^{1,2}(\Omega)$. Show that f is weakly harmonic (i.e.,
$$\int_\Omega f \Delta\varphi \, dx = 0$$

for every $\varphi \in C_c^\infty(\Omega)$) if and only if

$$\int_\Omega \nabla f \cdot \nabla \varphi \, dx = 0$$

for every test function φ. Hint: Use Theorem 11.21 along with (11.42).

4. Let $\Omega \subset \mathbb{R}^n$ be an open connected set and suppose $f \in W^{1,p}(\Omega)$ has the property that $\nabla f = 0$ almost everywhere in Ω. Prove that f is constant on Ω.

11.5. Sobolev Embedding Theorem

One of the most useful estimates in the theory of Sobolev functions is the Sobolev inequality. It implies that a Sobolev function is in a higher Lebesgue class than the one in which it was originally defined. In fact, for $1 \le p < n$, one has $W^{1,p}(\Omega) \subset L^{p^*}(\Omega)$, where $p^* = np/(n-p)$.

We use the result of the previous section to set the stage for the next definition. First, consider a bounded open set $\Omega \subset \mathbb{R}^n$ whose boundary has Lebesgue measure zero. Recall that Sobolev functions are defined only almost everywhere. Consequently, it is not possible in our present state of development to define what it means for a Sobolev function to be zero (pointwise) on the boundary of a domain Ω (see [25, 26, 51] for further reading on traces of Sobolev functions). Instead, we define what it means for a Sobolev function to be zero on $\partial\Omega$ in a global sense.

11.24. DEFINITION. Let $\Omega \subset \mathbb{R}^n$ be an arbitrary open set. The space $W_0^{1,p}(\Omega)$ is defined as the closure of $C_c^\infty(\Omega)$ relative to the Sobolev norm. Thus, $f \in W_0^{1,p}(\Omega)$ if and only if there is a sequence of functions $f_k \in C_c^\infty(\Omega)$ such that

$$\lim_{k\to\infty} \|f_k - f\|_{1,p;\Omega} = 0.$$

11.25. REMARK. If $\Omega = \mathbb{R}^n$, then Exercise 1 in Section 11.4 implies that $W_0^{1,p}(\mathbb{R}^n) = W^{1,p}(\mathbb{R}^n)$.

11.26. THEOREM. *Let $1 \le p < n$ and let $\Omega \subset \mathbb{R}^n$ be an open set. Then there is a constant $C = C(n,p)$ such that for $f \in W_0^{1,p}(\Omega)$,*

$$\|f\|_{p^*;\Omega} \le C \|\nabla f\|_{p;\Omega} \, .$$

PROOF. **Step 1:** Assume first that $p = 1$ and $f \in C_c^\infty(\mathbb{R}^n)$. Appealing to the fundamental theorem of calculus and using the fact that f has compact support, it follows that for each integer i, $1 \le i \le n$, one has

$$f(x_1, \ldots, x_i, \ldots, x_n) = \int_{-\infty}^{x_i} \frac{\partial f}{\partial x_i}(x_1, \ldots, t_i, \ldots, x_n) \, dt_i,$$

and therefore,

$$\begin{aligned}|f(x)| &\leq \int_{-\infty}^{\infty} \left| \frac{\partial f}{\partial x_i}(x_1, \ldots, t_i, \ldots, x_n) \right| dt_i \\ &\leq \int_{-\infty}^{\infty} |\nabla f(x_1, \ldots, t_i, \ldots, x_n)|\, dt_i, \quad 1 \leq i \leq n.\end{aligned}$$

Consequently,

$$|f(x)|^{\frac{n}{n-1}} \leq \prod_{i=1}^{n} \left(\int_{-\infty}^{\infty} |\nabla f(x_1, \ldots, t_i, \ldots, x_n)|\; dt_i \right)^{\frac{1}{n-1}}.$$

This can be rewritten as

$$\begin{aligned}&|f(x)|^{\frac{n}{n-1}} \\ &\leq \left(\int_{-\infty}^{\infty} |\nabla f(x)|\, dt_1 \right)^{\frac{1}{n-1}} \cdot \prod_{i=2}^{n} \left(\int_{-\infty}^{\infty} |\nabla f(x_1, \ldots, t_i, \ldots, x_n)|\, dt_i \right)^{\frac{1}{n-1}}.\end{aligned}$$

Only the first factor on the right is independent of x_1. Thus, when the inequality is integrated with respect to x_1 we obtain, with the help of the generalized Hölder inequality (see Exercise 14, Section 6.5),

$$\begin{aligned}&\int_{-\infty}^{\infty} |f|^{\frac{n}{n-1}}\; dx_1 \\ &\quad\leq \left(\int_{-\infty}^{\infty} |\nabla f|\; dt_1 \right)^{\frac{1}{n-1}} \int_{-\infty}^{\infty} \prod_{i=2}^{n} \left(\int_{-\infty}^{\infty} |\nabla f|\; dt_i \right)^{\frac{1}{n-1}} dx_1 \\ &\quad\leq \left(\int_{-\infty}^{\infty} |\nabla f|\; dt_1 \right)^{\frac{1}{n-1}} \left(\prod_{i=2}^{n} \int_{-\infty}^{\infty} \int_{-\infty}^{\infty} |\nabla f|\; dx_1 dt_i \right)^{\frac{1}{n-1}}.\end{aligned}$$

Similarly, integration with respect to x_2 yields

$$\begin{aligned}&\int_{-\infty}^{\infty} \int_{-\infty}^{\infty} |f|^{\frac{n}{n-1}}\; dx_1 dx_2 \\ &\quad\leq \left(\int_{-\infty}^{\infty} \int_{-\infty}^{\infty} |\nabla f|\; dx_1 dt_2 \right)^{\frac{1}{n-1}} \int_{-\infty}^{\infty} \prod_{i=1, i\neq 2}^{n} I_i^{\frac{1}{n-1}} dx_2,\end{aligned}$$

$$I_1 := \int_{-\infty}^{\infty} |\nabla f| dt_1 \quad I_i = \int_{-\infty}^{\infty} \int_{-\infty}^{\infty} |\nabla f| dx_1\, dt_i, \quad i = 3, \ldots, n.$$

Applying once more the generalized Hölder inequality, we obtain

$$\begin{aligned}&\int_{-\infty}^{\infty}\int_{-\infty}^{\infty}|f|^{\frac{n}{n-1}}\,dx_1dx_2\\&\leq\left(\int_{-\infty}^{\infty}\int_{-\infty}^{\infty}|\nabla f|\,dx_1dt_2\right)^{\frac{1}{n-1}}\left(\int_{-\infty}^{\infty}\int_{-\infty}^{\infty}|\nabla f|\,dt_1dx_2\right)^{\frac{1}{n-1}}\\&\quad\times\prod_{i=3}^{n}\left(\int_{-\infty}^{\infty}\int_{-\infty}^{\infty}\int_{-\infty}^{\infty}|\nabla f|\,dx_1dx_2dt_i\right)^{\frac{1}{n-1}}.\end{aligned}$$

Continuing in this way for the remaining $n-2$ steps, we finally arrive at

$$\int_{\mathbb{R}^n}|f|^{\frac{n}{n-1}}\,dx\leq\prod_{i=1}^{n}\left(\int_{\mathbb{R}^n}|\nabla f|\,dx\right)^{\frac{1}{n-1}},$$

or

$$\|f\|_{\frac{n}{n-1}}\leq\int_{\mathbb{R}^n}|\nabla f|\,dx,\quad f\in C_c^{\infty}(\mathbb{R}^n),\tag{11.58}$$

which is the desired result in the case $p=1$ and $f\in C_c^{\infty}(\mathbb{R}^n)$.

Step 2: Assume now that $1\leq p<n$ and $f\in C_c^{\infty}(\mathbb{R}^n)$. This case is treated by replacing f by positive powers of $|f|$. Thus, for q to be determined later, apply our previous step to $g\colon=|f|^q$. Technically, the previous step requires $g\in C_c^{\infty}(\Omega)$. However, a close examination of the proof reveals that we need g to be an absolutely continuous function only in each variable separately. Then,

$$\begin{aligned}\|f^q\|_{n/(n-1)}&\leq\int_{\mathbb{R}^n}|\nabla|f|^q|\,dx\\&=\int_{\mathbb{R}^n}q\,|f|^{q-1}\,|\nabla f|\,dx\\&\leq q\left\|f^{q-1}\right\|_{p'}\|\nabla f\|_p,\end{aligned}$$

where we have used Hölder's inequality in the last inequality. Requesting $(q-1)p'=q\frac{n}{n-1}$, with $\frac{1}{p'}=1-\frac{1}{p}$, we obtain $q=(n-1)p/(n-p)$. With this q we have $q\frac{n}{n-1}=\frac{np}{n-p}$ and hence

$$\left(\int_{\mathbb{R}^n}|f|^{\frac{np}{n-p}}\right)^{\frac{n-1}{n}}\leq q\left(\int_{\mathbb{R}^n}|f|^{\frac{np}{n-p}}\right)^{\frac{1}{p'}}\|\nabla f\|_{p;\mathbb{R}^n}.$$

Since $\frac{n-1}{n}-\frac{1}{p'}=\frac{n-p}{np}$, we obtain

$$\left(\int_{\mathbb{R}^n}|f|^{\frac{np}{n-p}}\right)^{\frac{n-p}{np}}\leq q\,\|\nabla f\|_{p;\mathbb{R}^n},$$

which is

$$\|f\|_{\frac{np}{n-p};\mathbb{R}^n}\leq\frac{(n-1)p}{n-p}\,\|\nabla f\|_{p;\mathbb{R}^n},\ f\in C_c^{\infty}(\mathbb{R}^n).\tag{11.59}$$

Step 3:

Let $f \in W_0^{1,p}(\Omega)$. Let $\{f_i\}$ be a sequence of functions in $C_c^\infty(\Omega)$ converging to f in the Sobolev norm. We have

$$\|f_i - f_j\|_{1,p;\Omega} \to 0. \tag{11.60}$$

Applying (11.59) to $\|f_i - f_j\|$, we obtain

$$\|f_i - f_j\|_{p^*;\mathbb{R}^n} \leq C\,\|f_i - f_j\|_{1,p;\mathbb{R}^n},$$

where we have extended the functions f_i by zero outside Ω. Therefore, we have

$$\|f_i - f_j\|_{p^*;\Omega} \leq C\,\|f_i - f_j\|_{1,p;\Omega}. \tag{11.61}$$

From (11.60) and (11.61) it follows that $\{f_i\}$ is Cauchy in $L^{p^*}(\Omega)$, and hence there exists $g \in L^{p^*}(\Omega)$ such that

$$f_i \to g \text{ in } L^{p^*}(\Omega).$$

Therefore, there exists a subsequence f_{i_k} such that $f_{i_k} \to g$ pointwise. Since $f_{i_k} \to f$ in $L^p(\Omega)$, it follows that up to a further subsequence, $f_{i_k} \to f$ pointwise. By uniqueness of the limit, we conclude that $f = g$ almost everywhere. That is,

$$f_i \to f \text{ in } L^{p^*}(\Omega). \tag{11.62}$$

Using that $|\nabla f_i| \to |\nabla f|$ in $L^p(\Omega)$ and (11.62), we can let $i \to \infty$ in

$$\|f_i\|_{p^*;\Omega} \leq C\,\|\nabla f_i\|_{p;\Omega}$$

to obtain

$$\|f\|_{p^*;\Omega} \leq C\,\|\nabla f\|_{p;\Omega}.$$

□

11.27. Remark. Let Ω be a bounded open subset of $\mathbb{R}^n$ and assume that ∂U is C^1. Assume $1 \leq p < n$ and $f \in W^{1,p}(\Omega)$. Then $f \in L^{p^*}(\Omega)$, with the estimate

$$\|f\|_{p^*;\Omega} \leq C\,\|f\|_{1,p;\Omega},$$

where the constant C depends only on p, n, and Ω. This inequality can be proven by extending f to a Sobolev function in the whole space $\mathbb{R}^n$ (see [26, Theorem 1, Section 5.4]) and then using the approximation by smooth functions in Corollary 11.22 and the Sobolev embedding theorem, Theorem 11.26.

Exercise for Section 11.5

1. Suppose $\|f_i - f_j\|_{q;\Omega} \to 0$ and $\|f_i - f\|_{p;\Omega} \to 0$, where $\Omega \subset \mathbb{R}^n$ and $q > p$. Prove that $\|f_i - f\|_{q;\Omega} \to 0$.

11.6. Applications

A basic problem in the calculus of variations is to find a harmonic function in a domain that assumes values that have been prescribed on the boundary. Using results of the previous sections, we will discuss a solution to this problem. Our first result shows that this problem has a "weak solution," that is, a solution in the sense of distributions.

11.28. DEFINITION. A function $f \in C^2(\Omega)$ is called **harmonic** in Ω if

$$\frac{\partial^2 f}{\partial x_1^2}(x) + \frac{\partial^2 f}{\partial x_2^2}(x) + \cdots + \frac{\partial^2 f}{\partial x_n^2}(x) = 0$$

for each $x \in \Omega$.

A straightforward calculation shows that this is equivalent to $\operatorname{div}(\nabla f)(x) = 0$. In general, if we let Δ denote the operator

$$\Delta = \frac{\partial^2}{\partial x_1^2} + \frac{\partial^2}{\partial x_2^2} + \cdots + \frac{\partial^2}{\partial x_n^2},$$

then taking $V = \varphi \nabla f$ in (11.42), we have

$$0 = \int_\Omega \operatorname{div} V = \int_\Omega \varphi \Delta f \, dx + \int_\Omega \nabla\varphi \cdot \nabla f \, dx \tag{11.63}$$

whenever $f \in C^2(\Omega)$ and $\varphi \in C_c^\infty(\Omega)$, and hence if f is harmonic in Ω, then

$$\int_\Omega \nabla f \cdot \nabla \varphi \, dx = 0$$

whenever $\varphi \in C_c^\infty(\Omega)$.

Returning to the context of distributions, recall that a distribution can be differentiated any number of times. In particular, it is possible to define ΔT whenever T is a distribution. If T is taken as an integrable function f, we have

$$\Delta f(\varphi) = \int_\Omega f \Delta\varphi \, dx$$

for all $\varphi \in C_c^\infty(\Omega)$.

11.29. DEFINITION. We say that f **is harmonic in the sense of distributions** or **weakly harmonic** if $\Delta f(\varphi) = 0$ for every test function φ.

Therefore (see Exercise 1, Section 11.4), $f \in C^2(\Omega)$ is harmonic in Ω if and only if f is weakly harmonic in Ω. Also, $f \in C^2(\Omega)$ is weakly harmonic in Ω if and only if $\int_\Omega \nabla\varphi \cdot \nabla f \, dx = 0$.

If $f \in W^{1,2}(\Omega)$ is weakly harmonic, then (11.63) implies (with f and φ interchanged) that

$$\int_\Omega \nabla f \cdot \nabla \varphi \, dx = 0 \tag{11.64}$$

for every test function $\varphi \in C_c^\infty(\Omega)$ (see Exercise 2, Section 11.6). Since $f \in W^{1,2}(\Omega)$, note that (11.64) remains valid with $\varphi \in W_0^{1,2}(\Omega)$.

11.30. THEOREM. *Suppose* $\Omega \subset \mathbb{R}^n$ *is a bounded open set and let* $\psi \in W^{1,2}(\Omega)$. *Then there exists a weakly harmonic function* $f \in W^{1,2}(\Omega)$ *such that* $f - \psi \in W_0^{1,2}(\Omega)$.

The theorem states that for a given function $\psi \in W^{1,2}(\Omega)$, there exists a weakly harmonic function f that assumes the same values (in a weak sense) on $\partial\Omega$ as does ψ. That is, f and ψ have the same boundary values and

$$\int_\Omega \nabla f \cdot \nabla \varphi \, dx = 0$$

for every test function $\varphi \in C_c^\infty(\Omega)$. Later we will show that f is, in fact, harmonic in the sense of Definition 11.28. In particular, it will be shown that $f \in C^\infty(\Omega)$.

PROOF. Let

$$m: = \inf \left\{ \int_\Omega |\nabla f|^2 \, dx : f - \psi \in W_0^{1,2}(\Omega) \right\}. \tag{11.65}$$

This definition requires $f - \psi \in W_0^{1,2}(\Omega)$. Since $\psi \in W^{1,2}(\Omega)$, note that f must be an element of $W^{1,2}(\Omega)$ and therefore that $|\nabla f| \in L^2(\Omega)$.

Our first objective is to prove that the infimum is attained. For this purpose, let f_i be a sequence such that $f_i - \psi \in W_0^{1,2}(\Omega)$ and

$$\int_\Omega |\nabla f_i|^2 \, dx \to m \quad \text{as} \quad i \to \infty. \tag{11.66}$$

Now apply both Hölder's and Sobolev's inequalities (Theorem 11.26) to obtain

$$\|f_i - \psi\|_{2;\Omega} \le \lambda(\Omega)^{(\frac{1}{2} - \frac{1}{2^*})} \|f_i - \psi\|_{2^*;\Omega} \le C \|\nabla(f_i - \psi)\|_{2;\Omega} .$$

This along with (11.66) shows that $\{\|f_i\|_{1,2;\Omega}\}_{i=1}^\infty$ is a bounded sequence. Referring to Remark 11.15, we know that $W^{1,2}(\Omega)$ is a reflexive Banach space, and therefore Theorem 8.37 implies that there exist a subsequence (denoted the same as the full sequence) and $f \in W^{1,2}(\Omega)$ such that $f_i \to f$ weakly in $W^{1,2}(\Omega)$. This is equivalent to

$$f_i \to f \text{ weakly in } L^2(\Omega) \tag{11.67}$$

$$\nabla f_i \to \nabla f \text{ weakly in } L^2(\Omega). \tag{11.68}$$

Furthermore, it follows from the lower semicontinuity of the norm (note that Theorem 8.35 (iii) is also true with $\|x\|^2 \le \lim_{k\to\infty} \|x_k\|^2$) and (11.68) that

$$\int_\Omega |\nabla f|^2 \, dx \le \liminf_{i\to\infty} \int_\Omega |\nabla f_i|^2 \, dx.$$

To show that f is a valid competitor in (11.65) we need to establish that $f - \psi \in W_0^{1,2}(\Omega)$, for then we will have

$$\int_\Omega |\nabla f|^2 \, dx = m,$$

thus establishing that the infimum in (11.65) is attained. To show that f assumes the correct boundary values, note that (for a subsequence) $f_i - \psi \to g$ weakly for some $g \in W_0^{1,2}(\Omega)$, since $\{\|f_i - \psi\|_{1,2;\Omega}\}$ is a bounded sequence. But $f_i - \psi \to f - \psi$ weakly in $W^{1,2}(\Omega)$, and therefore $f - \psi = g \in W_0^{1,2}(\Omega)$.

The next step is to show that f is weakly harmonic. Choose $\varphi \in C_c^\infty(\Omega)$ and for each real number t, let

$$\begin{aligned}\alpha(t) &= \int_\Omega |\nabla(f + t\varphi)|^2 \, dx \\ &= \int_\Omega |\nabla f|^2 + 2t\nabla f \cdot \nabla\varphi + t^2 |\nabla\varphi|^2 \, dx.\end{aligned}$$

Note that α has a local minimum at $t = 0$. Furthermore, referring to Exercise 7, Section 6.2, we see that it is permissible to differentiate under the integral sign to compute $\alpha'(t)$. Thus, it follows that

$$0 = \alpha'(0) = 2\int_\Omega \nabla f \cdot \nabla\varphi \, dx,$$

which shows that f is weakly harmonic. □

Exercises for Section 11.6

1. Let $f \in C^2(\Omega)$. Show that f is harmonic in Ω if and only if f is weakly harmonic in Ω (i.e., $\int_\Omega f\Delta\varphi \, dx = 0$ for every $\varphi \in C_c^\infty(\Omega)$). Show also that f is weakly harmonic in Ω if and only if $\int_\Omega \nabla\varphi \cdot \nabla f \, dx = 0$ for every $\varphi \in C_c^\infty(\Omega)$.
2. Let $f \in W^{1,2}(\Omega)$. Show that f is weakly harmonic (i.e.,
$$\int_\Omega f\Delta\varphi \, dx = 0$$
for every $\varphi \in C_c^\infty(\Omega)$) if and only if
$$\int_\Omega \nabla f \cdot \nabla\varphi \, dx = 0$$
for every test function φ. Hint: Use Theorem 11.21 along with (11.42).

11.7. Regularity of Weakly Harmonic Functions

We will now show that the weak solution found in the previous section is actually a classical C^∞ solution.

11.31. Theorem. *If $f \in W_{loc}^{1,2}(\Omega)$ is weakly harmonic, then f is continuous in Ω and*

$$f(x_0) = \fint_{B(x_0,r)} f(y) \, dy$$

whenever $\overline{B}(x_0, r) \subset \Omega$.

PROOF. **Step 1:** We will prove first that for every $x_0 \in \Omega$, the function

$$F(r) = \int_{\partial B(x_0,1)} f(r,z)d\mathcal{H}^{n-1}(z) \tag{11.69}$$

is constant for all $0 < r < d(x_0, \Omega)$. Without loss of generality we assume $x_0 = 0$. In order to prove (11.69) we recall that since $f \in W^{1,2}_{loc}(\Omega)$ is weakly harmonic, we must have

$$\int_\Omega f\Delta\varphi \, d\lambda(x) = 0, \qquad \text{for all } \varphi \in C_c^\infty(\Omega). \tag{11.70}$$

We want to choose an appropriate test function φ in (11.70). We consider a test function of the form

$$\varphi(x) = \omega(|x|). \tag{11.71}$$

A direct calculation shows that

$$\begin{aligned} \frac{\partial\varphi}{\partial x_i} &= \omega'(|x|)\frac{\partial}{\partial x_i}(x_1^2 + ... + x_n^2)^{\frac{1}{2}} \\ &= \omega'(|x|)\frac{1}{2}(x_1^2 + \cdots + x_n^2)^{-\frac{1}{2}}(2x_i) \\ &= \omega'(|x|)\frac{x_i}{|x|}, \end{aligned}$$

and hence

$$\begin{aligned} \frac{\partial^2\varphi}{\partial x_i^2} &= \frac{x_i}{|x|}\left(\omega''(|x|)\frac{x_i}{|x|}\right) + \omega'(|x|)\left[\frac{|x| - x_i\frac{x_i}{|x|}}{|x|^2}\right] \\ &= \frac{x_i^2}{|x|^2}\omega''(|x|) + w'(|x|)\left[\frac{|x|^2 - x_i^2}{|x|^3}\right], \ i = 1, 2, 3, \ldots. \end{aligned}$$

Therefore,

$$\frac{\partial^2\varphi}{\partial x_i^2} + ... + \frac{\partial^2\varphi}{\partial x_n^2} = \omega''(|x|) + \frac{n}{|x|}\omega'(|x|) - \frac{\omega'(|x|)}{|x|},$$

from which we conclude that

$$\Delta\varphi(x) = \omega''(|x|) + \frac{(n-1)}{|x|}\omega'(|x|).$$

Let $r = |x|$. Let $0 < t < T < d(x_0, \partial\Omega)$. We choose $w(r)$ such that $w \in C_c^\infty(t,T)$, and with this w we compute

$$\begin{aligned}
0 &= \int_\Omega f(x)\Delta\varphi(x)d\lambda(x)\\
&= \int_\Omega f(x)\left[w''(|x|) + \frac{(n-1)}{|x|}w'(|x|)\right]d\lambda(x)\\
&= \int_t^T\int_{\partial B(0,1)} f(r,z)\left[w''(r) + \frac{(n-1)}{r}w'(r)\right]r^{n-1}d\mathcal{H}^{n-1}(z)dr\\
&= \int_t^T\int_{\partial B(0,1)} f(r,z)\left[w''(r)r^{n-1} + (n-1)r^{n-2}w'(r)\right]d\mathcal{H}^{n-1}(z)dr\\
&= \int_t^T\int_{\partial B(0,1)} f(r,z)(r^{n-1}w'(r))'d\mathcal{H}^{n-1}(z)dr\\
&= \int_t^T F(r)(r^{n-1}w'(r))'dr.
\end{aligned}$$

We conclude that

$$\int_t^T F(r)(r^{n-1}w'(r))'dr = 0, \tag{11.72}$$

for every test function of the form (11.71), $w \in C_c^\infty(t,T)$. Given $\psi \in C_c^\infty(t,T)$, $\int_t^T \psi = 0$, we now proceed to construct a particular function $w(r)$ such that $(r^{n-1}w'(r))' = \psi$. Indeed, for each real number r, define

$$y(r) = \int_t^r \psi(s)ds$$

and define η by

$$\eta(r) = \int_t^r \frac{y(s)}{s^{n-1}}ds.$$

Finally, let

$$w(r) = \eta(r) - \eta(T).$$

Note that $w \equiv 0$ on $[0,t]$ and $[T,\infty)$. Since $w'(r) = \frac{y(r)}{r^{n-1}}$, we obtain from (11.72)

$$\begin{aligned}
0 &= \int_t^T F(r)(r^{n-1}\frac{y(r)}{r^{n-1}})'dr\\
&= \int_t^T F(r)y'(r)dr\\
&= \int_t^T F(r)\psi(r)dr.
\end{aligned}$$

We have proved that

$$\int_t^T F(r)\psi(r)dr = 0, \text{ for every } \psi \in C_c^\infty(t,T), \int_t^T \psi = 0. \tag{11.73}$$

We consider the function $F(r)$ as a distribution, say T_F (see Definition 10.3). It is clear that (11.73) implies

$$\int_t^T F(r)\varphi'(r) = 0, \text{ for every } \varphi \in C_c^\infty(t,T),$$

and from Definition 10.7, this is equivalent to

$$T_F'(\varphi) = 0, \text{ for every } \varphi \in C_c^\infty(t,T). \tag{11.74}$$

From (11.74) we conclude that T_F', in the sense of distributions, is 0, and we now appeal to Theorem 10.8 to conclude that the distribution T_F is a constant, say α. Therefore, we have shown that $F(r) = \alpha(x_0)$ for all $t < r < T$. Since t and T are arbitrary, we conclude that $F(r) = \alpha(x_0)$ for all $0 < r < d(x_0, \partial\Omega)$, which is (11.69).

Step 2: In this step we show that

$$\fint_{B(x_0,\delta)} f(y)d\lambda(y) = C(x_0, n), \tag{11.75}$$

for every $x_0 \in \Omega$ and every $0 < \delta < d(x_0, \partial\Omega)$. Without loss of generality we assume $x_0 = 0$. We fix $0 < \delta < d(x_0, \partial\Omega)$. We have, with the help of step 1,

$$\begin{aligned}
\int_{B(0,\delta)} f(x)d\lambda(x) &= \int_0^\delta \int_{\partial B(0,1)} f(r,z)r^{n-1}d\mathcal{H}^{n-1}(z)dr \\
&= \int_0^\delta \alpha(0)r^{n-1}dr \\
&= \alpha(0)\frac{\delta^n}{n}.
\end{aligned}$$

Hence,

$$\frac{1}{\delta^n}\int_{B(x_0,\delta)} f(x)d\lambda(x) = \frac{\alpha(x_0)}{n},$$

and therefore,

$$\frac{1}{\lambda(B(x_0,\delta))}\int_{B(x_0,\delta)} f(x)d\lambda(x) = \alpha(x_0)C(n). \tag{11.76}$$

Step 3: Since almost every $x_0 \in \Omega$ is a Lebesgue point for f, we deduce from (11.76) that

$$f(x_0) = \frac{1}{\lambda(B(x_0,\delta))}\int_{B(x_0,\delta)} f(x)d\lambda(x) = \alpha(x_0)C(n),$$

for λ-a.e. $x_0 \in \Omega$ and every ball $\overline{B}(x_0,\delta) \subset \Omega$.

Step 4: Now f can be redefined on a set of measure zero in such a way as to ensure its continuity in Ω. Indeed, if x_0 is not a Lebesgue point and $\overline{B}(x_0, r) \subset \Omega$, we define, with the aid of Step 2,

$$f(x_0) := \fint_{B(x_0,r)} f(y)d\lambda(y).$$

We leave as an exercise to show that with this definition, f is continuous in Ω. □

11.32. DEFINITION. A function $\varphi\colon B(x_0, r) \to \mathbb{R}$ is called **radial** relative to x_0 if φ is constant on $\partial B(x_0, t)$, $t \leq r$.

11.33. COROLLARY. *Suppose $f \in W^{1,2}(\Omega)$ is weakly harmonic and $\overline{B}(x_0, r) \subset \Omega$. If $\varphi \in C_c^\infty(B(x_0, r))$ is nonnegative and radial relative to x_0 with*

$$\int_{B(x_0,r)} \varphi(x)\, dx = 1,$$

then

$$f(x_0) = \int_{B(x_0,r)} f(x)\varphi(x)\, dx. \tag{11.77}$$

PROOF. For convenience, assume $x_0 = 0$. Since φ is radial, note that each superlevel set $\{\varphi > t\}$ is a ball centered at 0; let $r(t)$ denote its radius. Then with $M \colon = \sup_{B(0,r)} \varphi$, we compute

$$\begin{aligned}\int_{B(0,r)} f(x)\varphi(x)\, dx &= \int_{B(0,r)} (f^+(x) - f^-(x))\varphi(x)\, dx \\ &= \int_{B(0,r)} \varphi(x) f^+(x) d\lambda(x) - \int_{B(0,r)} \varphi(x) f^-(x) d\lambda(x) \\ &= \int_{B(0,r)} \varphi(x) d\mu^+(x) - \int_{B(0,r)} \varphi(x) d\mu^-(x),\end{aligned}$$

where the positive measures μ^+ and μ^- are given by

$$\mu^+(E) = \int_E f^+(x) d\lambda(x) \text{ and } \mu^-(E) = \int_E f^-(x) d\lambda(x).$$

Hence, from Theorem 6.59, we obtain

$$\int_{B(0,r)} f(x)\varphi(x) dx = \int_0^M \mu^+(\{x : \varphi(x) > t\}) d\lambda(t) - \int_0^M \mu^-(\{x : \varphi(x) > t\}) d\lambda(t).$$

Since $\{\varphi > t\} = B(0, r(t))$, we have

$$\begin{aligned}\int_{B(0,r)} f(x)\varphi(x) dx &= \int_0^M \int_{\{\varphi > t\}} d\mu^+(x) d\lambda(t) - \int_0^M \int_{\{\varphi > t\}} d\mu^-(x) d\lambda(t) \\ &= \int_0^M \int_{\{\varphi > t\}} (f^+(x) - f^-(x)) d\lambda(x) d\lambda(t) \\ &= \int_0^M \int_{B(0,r(t))} f(x) d\lambda(x) d\lambda(t).\end{aligned}$$

Appealing now to Theorem 11.31, we conclude that

$$\begin{aligned}\int_{B(0,r)} f(x)\varphi(x)dx &= \int_0^M f(0)\lambda[B(0,r(t))]\, d\lambda(t)\\ &= f(0)\int_0^M \lambda[\{\varphi > t\}]\, dt\\ &= f(0)\int_{B(0,r)} \varphi(x)d\lambda(x)\\ &= f(0).\end{aligned}$$

We have proved (11.77). □

11.34. THEOREM. *A weakly harmonic function $f \in W^{1,2}(\Omega)$ is of class $C^\infty(\Omega)$.*

PROOF. For each domain $\Omega' \subset\subset \Omega$ we will show that $f(x) = f * \varphi_\varepsilon(x)$ for $x \in \Omega'$. Since $f * \varphi_\varepsilon \in C^\infty(\Omega')$, this will suffice to establish our result. As usual, $\varphi_\varepsilon(x){:}\ = \varepsilon^{-n}\varphi(x/\varepsilon)$, $\varphi \in C_c^\infty(B(0,1))$, and $\int_{B(0,1)} \varphi(x)d\lambda(x) = 1$. We also require φ to be radial with respect to 0. Finally, we take ε small enough to ensure that $f * \varphi_\varepsilon$ is defined on Ω'. We have, for each $x \in \Omega'$,

$$f*\varphi_\varepsilon(x) = \int_\Omega \varphi_\varepsilon(x-y)f(y)dy = \int_{B(x,\varepsilon)} \varphi_\varepsilon(x-y)f(y)dy = \int_{B(0,\varepsilon)} \varphi_\varepsilon(y)f(x-y)dy.$$

By Exercise 1, Section 11.7, the function $h(y) := f(x-y)$ is weakly harmonic in $B(0,\varepsilon)$. Therefore, since φ_ε is radial and $\int_{B(0,\varepsilon)} \varphi_\varepsilon(x)d\lambda(x) = 1$, we obtain, with the help of Corollary 11.33,

$$f * \varphi_\varepsilon(x) = \int_{B(0,\varepsilon)} h(y)\varphi_\varepsilon(y)\, dy = h(0) = f(x).$$ □

Theorem 11.30 states that if Ω is a bounded open set and $\psi \in W^{1,2}(\Omega)$ a given function, then there exists a weakly harmonic function $f \in W^{1,2}(\Omega)$ such that $f - \psi \in W_0^{1,2}(\Omega)$. That is, f assumes the same values as ψ on the boundary of Ω in the sense of Sobolev theory. The previous result shows that f is a classically harmonic function in Ω. If ψ is known to be continuous on the boundary of Ω, a natural question is whether f assumes the values ψ continuously on the boundary. The answer to this is well understood and is the subject of other areas in analysis.

Exercise for Section 11.7

1. Suppose $f \in W^{1,2}(\Omega)$ is weakly harmonic and let $x \in \Omega' \subset\subset \Omega$. Choose $r \leq$ dist $(\Omega', \partial\Omega)$. Prove that the function $h(y){:}\ = f(x-y)$ is weakly harmonic in $B(0,r)$.

Bibliography

This list includes books and articles that were cited in the text and some additional references that will be useful for further study.

[1] L. Ambrosio, N. Fusco, D. Pallara, *Oxford Mathematical Monographs*, Functions of Bounded Variation and Free Discontinuity Problems (The Clarendon Press, Oxford University Press, New York, 2000)

[2] St. Banach, A. Tarski, Sur la décomposition des ensembles de points en parties respectivement congruentes. Fund. Math. **6**, 244–277 (1924)

[3] T. Bagby, W.P. Ziemer, Pointwise differentiability and absolute continuity. Trans. Amer. Math. Soc. **194**, 129–148 (1974)

[4] A.S. Besicovitch, A general form of the covering principle and relative differentiation of additive functions. Proc. Camb. Phil. Soc. **41**, 103–110 (1945)

[5] A.S. Besicovitch, A general form of the covering principle and relative differentiation of additive functions II. Proc. Camb. Phil. Soc. **42**, 1–10 (1946)

[6] B. Bojarski, Remarks on some geometric properties of Sobolev mappings, in *Functional Analysis and Related Topics*, ed. by S. Koshi (World Scientific, Singapore, 1991)

[7] J. Bourgain, H. Brezis, On the equation $\operatorname{div} Y = f$ and application to control of phases. J. Amer. Math. Soc. **16**(1), 393–426 (2002)

[8] L. Carleson, On the connection between Hausdorff measures and capacity. Ark. Mat. **3**, 403–406 (1958)

[9] L. Carleson, *Selected Problems on Exceptional Sets* (Van Nostrand Company Inc, Princeton, New Jersey, Toronto, London, Melbourne, 1967)

[10] G.-Q. Chen, H. Frid, Extended divergence-measure fields and the Euler equations for gas dynamics. Comm. Math. Phys. **236**(2), 251–280 (2003)

[11] G.Q. Chen, H. Frid, Divergence-measure fields and hyperbolic conservation laws. Arch. Ration. Mech. Anal. **147**(2), 89–118 (1999)

[12] G.-Q. Chen, M. Torres, Divergence-measure fields, sets of finite perimeter, and conservation laws. Arch. Ration. Mech. Anal. **175**(2), 245–267 (2005)

[13] G.-Q. Chen, M. Torres, W. Ziemer, Gauss-Green theorem for weakly differentiable vector fields, sets of finite perimeter, and balance laws. Commun. Pure Appl. Math. **62**(2), 242–304 (2009)

W.P. Ziemer, *Modern Real Analysis*, Graduate Texts in Mathematics 278,
https://doi.org/10.1007/978-3-319-64629-9

[14] G.-Q. Chen, M. Torres, W.P. Ziemer, Measure-theoretical analysis and non-linear conservation laws. Pure Appl. Math. Quarterly **3**(3), 841–879 (2007)

[15] P.J. Cohen, *Set Theory and the Continuum Hypothesis* (Benjamin, New York, 1966)

[16] P.J. Cohen, The independence of the continuum hypothesis. Proc. Natl. Acad. Sci. U.S.A. **50**, 1143–1148 (1963)

[17] P.J. Cohen, The independence of the continuum hypothesis II. Proc. Natl. Acad. Sci. U.S.A. **51**, 105–110 (1964)

[18] P.J. Cohen, A minimal model for set theory. Bull. Amer. Math. Soc. **69**, 537–540 (1963)

[19] G. Comi, M. Torres, One sided approximations of sets of finite perimeter. Rendiconti Lincei-Matematica e Applicazioni **28**, 181–190 (2017)

[20] T. De Pauw, On SBV dual. Indiana Univ. Math. J. **47**(1), 99–121 (1998)

[21] T. De Pauw, On the exceptional sets of the flux of a bounded vectorfield. J. Math. Pures Appl. **82**, 1191–1217 (2003)

[22] T. De Pauw, W.F. Pfeffer, Distributions for which div $v = f$ has a continuous solution. Comm. Pure Appl. Math. **61**(2), 230–260 (2008)

[23] T. De Pauw, W.F. Pfeffer, The Gauss-Green theorem and removable sets for PDEs in divergence form. Adv. Math. **183**(1), 155–182 (2004)

[24] T. De Pauw, M. Torres, On the distributional divergence of vector fields vanishing at infinity. Proc. R. Soc. Edinb. **141A**, 65–76 (2011)

[25] L.C. Evans, *Graduate Studies in Mathematics*, vol. 19 (Partial Differential Equations (American Mathematical Society, Providence, 1998)

[26] L.C. Evans, R.F. Gariepy, *Measure Theory and Fine Properties of Functions*, Studies in Advanced Mathematics (CRC Press, Boca Raton, 1992)

[27] K. Falconer, *Fractal Geometry* (Wiley, New York, 1990)

[28] H. Federer, *Geometric Measure Theory* (Springer, New York, Heidelberg, 1969)

[29] H. Federer, W. Ziemer, The Lebesgue set of a function whose distributional derivatives are p-th power summable. Indiana Univ. Math. J. **22**, 139–158 (1972)

[30] B. Fuglede, Extremal length and functional completion. Acta Mathematica **98**(130), 171–219 (1957)

[31] E. Giusti, *Minimal Surfaces and Functions of Bounded Variation.* With notes G.H. Williams (ed.) Notes on Pure Mathematics, vol. 10, Department of Pure Mathematics (Australian National University, Canberra, 1977)

[32] K. Gödel, *The Consistency of the Continuum Hypothesis*, vol. 3 (Princeton, Annals of Mathematics Studies (Princeton University Press, 1940)

[33] W. Gustin, Boxing inequalities. J. Math. Mech. **9**, 229–239 (1960)

[34] P. Halmos, *Naive Set Theory* (Van Nostrand, Princeton, 1960)

[35] G.H. Hardy, Weierstrass's nondifferentiable function. Trans. Amer. Math. Soc. **17**(2), 301–325 (1916)

[36] F. Maggi, *Sets of Finite Perimeter and Geometric Variational Problems*, Cambridge Studies in Advanced Mathematics (Cambridge University Press, Cambridge, 2012)

[37] V.G. Maz'ja, *Sobolev Spaces*, Springer Series in Soviet Mathematics (Springer, Berlin, 1985). Translated from the Russian by T.O. Shaposhnikova

[38] N.G. Meyers, W.P. Ziemer, Integral inequalities of Poincare and Wirtinger type for BV functions. Amer. J. Math. **99**, 1345–1360 (1977)

[39] C.B. Morrey Jr., Functions of several variables and absolute continuity II. Duke Math. J. **6**, 187–215 (1940)

[40] A.P. Morse, The behavior of a function on its critical set. Ann. Math. **40**, 62–70 (1939)

[41] A.P. Morse, Perfect blankets. Trans. Amer. Math. Soc. **6**, 418–442 (1947)

[42] W.F. Pfeffer, *Derivation and Integration*, vol. 140 (Cambridge Tracts in Mathematics (Cambridge University Press, Cambridge, 2001)

[43] N.C. Phuc, M. Torres, Characterizations of the existence and removable singularities of divergence-measure vector fields. Indiana Univ. Math. J. **57**(4), 1573–1598 (2008)

[44] N.C. Phuc, M. Torres, Characterizations of signed measures in the dual of BV and related isometric isomorphisms. Annali della Scuola Normale Superiore di Pisa, Volume XVII **5**, 1–33 (2017)

[45] S. Sacks, *Selected Problems on Exceptional Sets* (Van Nostrand Company Inc, Princeton, New Jersey, Toronto - London - Melbourne, 1967)

[46] A.R. Schep, And still one more proof of the Radon-Nikodym theorem. Amer. Math. Mon. **110**(6), 536–538 (2003)

[47] L. Schwartz, *Théorie des Distributions*, 2nd edn. (Hermann, Paris, 1966)

[48] M. Šilhavý, Divergence-measure fields and Cauchy's stress theorem. Rend. Sem. Mat Padova **113**, 15–45 (2005)

[49] E.M. Stein, *Singular Integrals and Differentiability of Functions* (Princeton University Press, Priceton, New Jersey, 1970)

[50] K. Weierstrass, On continuous functions of a real argument which possess a definite derivative for no value of the argument. Kniglich Preussichen Akademie der Wissenschaften **110**(2), 71–74 (1895)

[51] W.P. Ziemer, *Weakly Differentiable Functions*, vol. 120, Graduate Texts in Mathematics (Springer, New York, 1989)

Index

W.P. Ziemer, *Modern Real Analysis*, Graduate Texts in Mathematics 278,
https://doi.org/10.1007/978-3-319-64629-9

T

U

V

W

Printed in the United States
By Bookmasters